RIVER JORDAN

RACHEL HAVRELOCK

River Jordan

THE MYTHOLOGY OF A DIVIDING LINE

To Michael,
In friendship & shared attachment to the antiquated word,
Rachel

The University of Chicago Press • Chicago & London

PUBLICATION OF THIS BOOK HAS BEEN AIDED BY A GRANT FROM THE BEVINGTON FUND.

Rachel Havrelock is associate professor of Jewish studies and English at the University of Illinois at Chicago.

The University of Chicago Press, Chicago 60637
The University of Chicago Press, Ltd., London
© 2011 by The University of Chicago
All rights reserved. Published 2011.
Printed in the United States of America

20 19 18 17 16 15 14 13 12 11   1 2 3 4 5

ISBN-13: 978-0-226-31957-5 (cloth)
ISBN-10: 0-226-31957-1 (cloth)

A shorter version of chapter 1 first appeared as "The Two Maps of Israel's Land," in the *Journal of Biblical Literature* 126:4 (2007): 649–67. Portions of chapter 9 first appeared as "My Home Is Over Jordan: River as Border in Israeli and Palestinian National Mythology," in *National Identities* 9:2 (2007): 105–26, reprinted by permission of the publisher (Taylor & Francis Ltd., http://www.tandf.co.journals), and as "Pioneers and Refugees: Arabs and Jews in the Jordan River Valley," in *Understanding Life in the Borderlands*, ed. I. William Zartman (Athens: University of Georgia Press, 2010), 189–216. © 2010 by the University of Georgia Press. All rights reserved.

Library of Congress Cataloging-in-Publication Data

Havrelock, Rachel S.
    River Jordan : the mythology of a dividing line / Rachel Havrelock.
        p. cm.
    Includes bibliographical references and index.
    ISBN-13: 978-0-226-31957-5 (cloth : alk. paper)
    ISBN-10: 0-226-31957-1 (cloth : alk. paper) 1. Jordan River—Religious aspects. I. Title.
    DS110.J6H38 2011
    220.8'32012—dc23
                                    2011024766

♾ This paper meets the requirements of ANSI/NISO Z39.48-1992 (Permanence of Paper).

*Dedicated to my grandparents of blessed memory
Dr. Meyer Stamell, who always said that water is the only cure
& Rose Stamell (née Rahzel Schwartz) for her stories*

CONTENTS

List of Figures ix
Acknowledgments xi

INTRODUCTION
Maps and Legends 1

CHAPTER ONE
The Two Maps of Israel's Land 17

CHAPTER TWO
Israel and Moab as Nation and Anti-nation 40

CHAPTER THREE
Two Camps: Ancient Israel Between Homeland and Diaspora 64

CHAPTER FOUR
The Book of Joshua and the Ideology of Homeland 85

CHAPTER FIVE
The Other Side 106

CHAPTER SIX
Crossing Over: Prophetic Succession at the Jordan 135

CHAPTER SEVEN
Dipping In: Baptism and the State of the Body 175

CHAPTER EIGHT
Two More Maps of Israel's Land 208

CHAPTER NINE
My Home Is Over Jordan: River as Border in Israeli and Palestinian National Mythology 218

CONCLUSION
The Baptism Business and the Peace Park 275

Suggested Readings 291
Index 303

FIGURES

1. Riverine borders of the Priestly and Deuteronomistic maps   20
2. Canaan and its neighbors to the east   48
3. Gilead and the kingdoms of Israel   73
4. Territory of the twelve tribes according to the book of Joshua   110
5. Judea under the Hasmonean kings   132
6. Administrative districts in the time of Jesus   133
7. Ottoman administrative districts in Syria and Palestine   222
8. Tiberias Camp   224
9. T. E. Lawrence's hand-drawn map of the Middle East following World War I   225
10. Sir Mark Sykes's personal map of the 1916 Sykes-Picot Agreement   226
11. Zionist territorial proposal of 1918–20 and the border from the Franco-British agreement of 1920–23   228
12. Plan of partition of the 1937 Peel Commission Report   236
13. Proposed partition of Palestine, 1943–46   238
14. United Nations 1947 approved regions indicating the Arab state, the Jewish state, and the international area of Jerusalem   240
15. Lower Jordan River Valley   285

ACKNOWLEDGMENTS

Many people have helped me navigate the actual and the imaginary Jordan River. While I acknowledge their help and contributions, I alone take responsibility for any errors or oversights.

My gratitude begins with my teachers. Mishael Caspi initiated me into the world of biblical scholarship as he taught me about Israel, the Arab world, and Islam. As his pupil and apprentice at the University of California, Santa Cruz, I first learned to recognize the multiple forms of biblical interpretation.

Daniel Boyarin, Alan Dundes, Ron Hendel, and Dina Stein both reined me in and set me free. I thank them all for the sage advice, the attention, and the encouragement. Dina Stein animated midrashic texts through the prisms of folklore and literary theory, serving as a guide and mentor. With her loving praise and unsparing criticism, Dina oversaw my growth as a scholar. Ron Hendel schooled me in textual rigor while opening the door to imaginative approaches to the Hebrew Bible. His combination of expertise and openness continues to serve as a treasured resource. Alan Dundes, of blessed memory, trained me in folklore and sent me into the field in search of stories. His wit, intelligence, and encyclopedic knowledge made each meeting an unforgettable experience.

I will always think of myself as coming from the school (or, in rabbinic terms, the "house") of Daniel Boyarin. Theory, politics, and all, Boyarin has established something quite akin to a talmudic academy in Berkeley. Participating in this academy has been a formative experience and I continue to be challenged by his scope and his bravery. Serving as an advisor in every sense of the word, Boyarin has made many things possible for me on a professional as well as conceptual level.

Thanks are also due to the late Louis Owens for teaching me the cultural impact of mixed-bloods and half-breeds, to Ilana Pardes for her remarkable scholarship and her support, and to Robert Alter, Murray Baumgarten, David Biale, Margaret Brose, Ed Greenstein, John Hayes, Rabbi Alan Lew of blessed memory, Ibrahim Muhawi, Meira Polliack, Naomi Seidman, Hayden White, and Rabbi Arnold Jacob Wolf of blessed memory for being such wonderful teachers.

I thank my colleagues at the University of Illinois at Chicago for creating an exciting environment in which to research and to teach. I am particularly indebted to Walter Benn Michaels, Mark Canuel, Sam Fleischacker, Paul Griffiths, Mustapha Kamal, Michael Lieb, Nanno Marinatos, Mary Beth Rose, Alfred Thomas, and Jennifer Tobin. For hiring me and for charging the environment at UIC, I thank Stanley Fish. I would also like to thank the members of the Religion Department at Swarthmore College—in particular Yvonne Chireau, Steven Hopkins, Helen Plotkin, Ellen Ross, and Mark Wallace—for the encouragement during the year I served as a visiting assistant professor while writing my dissertation.

Paolo Asso, Carol Bakhos, Mara Benjamin, Marcy Brink-Danan, Gidon Bromberg, Allen Callahan, Tamara Cohn-Eskenazi, Deborah Dash-Moore, Jorge Egal, Yakir Englander, Maxine Epstein, Michael Feige, Jason Frank, Elizabeth Goldstein, Rhiannon Graybill, Todd Hasak-Lowy, Susannah Heschel, Johari Jabir, Joel Kaminsky, Jumana Kawar, Robert Kawashima, Ari Kelman, Akiba Lerner, Diana Lipton, Alicia Ostriker, Miriam Peskowitz, Abigail Pickus, Na'ama Rokem, Marsha Rozenblit, Susan Shapiro, David Shneer, Eliza Slavet, David Stern, Joan Taylor, Frederick Wherry, Elizabeth Ya'ari, and Yael Zerubavel engaged with my work in oral and written form. Efrat Appel, Corey Capers, Kevin Dwarka, and Seth Sanders read chapters and improved them through their insights. I extend particular thanks to Matti Bunzl, Nathaniel Deutsch, and Laura Levitt for intellectual friendships that ever push me forward.

In the midst of investigating the borders that create a national home, I was fortunate enough to have two families open their homes to me. Since driving me to get a falafel after disembarking from a ship in the Haifa harbor in 1992, Ruth and Yitzhak Rapoport have been my adoptive Israeli parents. Along with the extended Rapoport clan, Ruth and Yitzhak have provided me with a home in Israel where I have always felt like one of the family. I thank them for the generosity, warmth, and willingness to hear me out even when they disagree. I thank the Shehadeh family for opening homes to me in Amman, Bir Zeit, and San Francisco. I fondly remember long Wednesday lunches with Karema Shehadeh in Bir Zeit, during which she helped me progress with my spoken Arabic.

Countless narrators told me Jordan River stories in Israel, the West Bank, Jordan, Chicago, and San Francisco. Some of these stories appear in the present work, while others await their next context.

The San Francisco Bureau of Jewish Education, the Koret Foundation, the US Department of Education Foreign Language and Area Studies Fellowship Program, the Haas-Koshland Award, the American Center for Oriental Research in Amman, Jordan, the UIC Institute for the Humanities, and the American Academy for Jewish Research Special Initiatives Grant all supported either the research or the writing of this book.

Hugh Alexander of the National Archives (UK) made some extraordinary maps available to me, as did Felicity Cobbing of the Palestine Exploration Fund. Thank you to Reuven and Tammy Soffer for drawing the maps from the biblical period. As my undergraduate research assistant, Tim Oravec helped me to bring the project to a close.

I would like to thank Randy Petilos at the University of Chicago Press for leading me through the process of publication, the press's anonymous readers for their helpful insights, and my editor, Alan Thomas, for his discerning eye and great ideas.

My mother, Rhoda Stamell, has read and edited my work since I first took pen to paper. I thank her for her patience and grammatical expertise. As a single mother, an inner-city high school teacher, and a writer, she has always inspired my respect and admiration. Most of all, I thank her for making my education her first priority.

In working with Sharif Ezzat and Yuri Lane, my husband, on the hip-hop play *From Tel Aviv to Ramallah* I was able to say much about borders with fewer words. Yuri and Sharif's artistic insights contributed to this work as well. Yuri has, at times, joined me on Middle East trips and, at others, held things down at home so that I could go on my own. His love, dedication to our family, and delicious vegan meals have made this book possible.

Days of Awe
5771

Introduction

MAPS AND LEGENDS

The mythologist is not even in a Moses-like situation: he cannot see the Promised Land.
—Roland Barthes, *Myth Today*

I first encountered the map of biblical Israel in a fourth-grade classroom of Hillel Day School of Metropolitan Detroit. The map had been present in some form or another ever since I could remember, but in the fourth grade it hung to the left of the blackboard as the central object of study. Without the map, explained our teacher—a woman who had come to us after rising through the ranks of the Israel Defense Forces—we could not understand the biblical books of Joshua, Judges, and Samuel, which concerned the conquest and settlement of the Promised Land. The biblical stories animated the map and the map—a scientific instrument after all—validated biblical miracles as events that had transpired in history.

Geographic symbols and place names ended at the Jordan River, a single line that separated the ordered world to the west from the apparent chaos to the east. Only a spectrum of greens and browns indicating rainfall connected the Promised Land to its nameless outlying areas. Nevertheless, all the places within the borders represented a unified whole. The biblical Israel of the map did not acknowledge tribal divisions or competing nations. This map further assured us of Jewish unity in our own times. Our teacher transposed two narratives on the map and so we learned how the biblical crossing of the

Jordan River paralleled the mass migration of the remnant Jewish people to the State of Israel. "Our history moves in cycles," our teacher told us, "and we are witnesses to a third cycle of miracle in which the Jewish people lives in its homeland." Just as Joshua instructed the People to tell the story of their miraculous Jordan crossing, she instructed us to transmit the story of our times in the ancient-modern tongue of Hebrew.

I would gaze at the map, enchanted by its miracles and longing to visit Jericho, Jerusalem, Hebron, Shiloh, Beth El, and all of the places where prophets had walked. Was it my love of the map or the map's need for devotion that determined its presence in so many subsequent classrooms? When I began my travels through the old-new land, I was shocked to find so many dividing lines. Those internal to the Jewish population—ethnicity, class, level of religious observance—were proof that the Jewish state was a Palestinian nation like all others, but the dividing lines between the Jewish and the Palestine populations—which nothing in my education had prepared me for—made sense to me only through analogy with Eight Mile Road, the Detroit street that separates the mostly black city from the largely white suburbs. Perhaps because my mother had taught me through experience to feel at home in inner-city Detroit or perhaps because I remained an unredeemed lover of the map, I began crossing the dividing lines in and around Israel.

In 1999, I attended the international students program at Bir Zeit University in the Palestinian West Bank. One day, a Palestinian student stuck his head into my colloquial Arabic class and instructed my professor to vacate the room. The professor was annoyed that we would have to relocate our class because of a student strike protesting an increase in tuition. He told us to follow him through the crowds and not to be intimidated by the strike. We walked up a flight of stairs in the engineering building and passed a science lab, a computer center, and the Women's Studies Department before arriving at his office, where I expected to see walls lined with Arabic tomes. When I entered the office, there it was—the map filling an entire wall! Drawn with the very shades of green and brown and ending at the Jordan River, it was *the* map. This map did not depict Israel in any incarnation. This was the map of Palestine, my fourth-grade map in reverse, and there were no signs of Israel. Had a single company produced both maps and turned a profit by issuing identical maps of different places?

In its two manifestations, the map effectively erases an entire population. But the map does not effect the erasure on its own; it works in conjunction with legends—including those told to me in the fourth grade and those told to me in Bir Zeit—to insist upon a indivisible, homogeneous homeland. The map and its legends create a unified Israel and a unified Palestine distinct from one another. They produce the national groups by resisting the fact that Israel and Palestine overlap and by ignoring their respective internal divisions. As part of the same mythic system, the map gives the legends precision, factuality; and the legends imbue the map with collective memory and individual passions.

This book explains how the map of Israel and the map of Palestine assumed coincident forms. The explanation involves texts from the Bible and late antiquity as well as from the modern Middle East. Several alternative maps will become apparent in these texts, implying that the overlapping maps of Israel and Palestine are neither a necessary nor a natural outcome. I am interested first and foremost in the Jordan River, the definitive eastern border according to the map. When does the Jordan operate as a border? Who sees it in terms of a dividing line and what does it divide? What role does the river play in definitions of the nation? I ask these questions of a series of texts and collected stories that I define as national myths.

Maps and myths intersect in numerous ways. Often, maps are embedded in myths—as in the case of the verbal maps in the Hebrew Bible—in order for territorial boundaries to become meaningful in narrative as well as geographic terms. Myths describe the origins and establishment of place and maps show the location of the place. By marking where monumental events transpired, maps direct the emotion stirred up by myth to particular sites. Myths enforce the boundaries outlined in maps by describing division in terms of cultural difference. In myth, the lines on the map signify the difference between us and them. The boundary lists—comprised of words and not pictures—in the Bible work like graphic maps to establish the symbolic limits of the collective. In other words, group identity depends upon an image of a certain line that cannot be crossed. Maps and myths work in different ways to draw this very line.

Although in popular parlance "myth" often means something that is not true, I understand myth as an expression of what people hold to be most true. It is the discourse that defines the group and sets the parameters of inclusion. Myth speaks to who we are and how we got here. As Claude Lévi-Strauss revealed, myth draws social oppositions and portrays the risks and possibilities of mediating these oppositions. Lévi-Strauss's poststructuralist heirs showed myth to be unstable—a system whose oppositions exist in a state of constant mediation—and at the same time protective of its allegedly stable structures. Myth compensates for its fluctuations through the insistence that things have always been as the myth portrays them or that myth depicts a golden age when things were the way they ought to be.

Myth often evokes a distant world—think of the Garden of Eden or the American frontier—in order to enforce hierarchies or to justify the unequal distribution of resources in the present.¹ "Myth can attach itself to any form of social power or social claim. It is used always to account for extraordinary privileges or duties, for great social inequalities, for severe burdens of rank."² Myth persuades and coerces without argument or force through repeated, authoritative accounts of the moments that inevitably led to the way things are. The ubiquity of myth in oral, written, ritual, and visual media magnifies power and saturates experience as knowledge and common sense. Because people rarely believe the truth to be fully realized in society, myth does not conform to standards of reality. Myth contains the symbolic distillation of values and identitarian tenets to which people can have recourse when the complexity and complications of social interactions arise. With its truth claims made through symbol, myth has the ability to justify societal structures as well as destabilize them through images of more authentic, pure, or efficient conceptions.

The analysis of myth allows for a critique of power at the same time that the genre of myth opens power up to transformation. The folklorist Alan Dundes defined myth formally as a story about beginnings that exists in multiple variations. This means that we never quite have a singular myth so much as a complex mythology. Every story (and every map) has many versions. The many versions of a myth can alternately justify, enforce, contest, or restructure power. For my purposes, this means that many sorts of Israels—and pointedly non-Israels—take form in myths about the Jordan River. There is not one Promised Land between the Jordan and the Mediterranean. Instead, let us read homeland, Holy Land, and a string of proper nouns like Canaan, Palastina, Israel, Palestine as variations. By looking at these multiple traditions, we see that homeland is an amalgamated concept that depends upon articulations that themselves differ from speaker to speaker.

I identify these traditions as national myths that not only outline the parameters of the collective, but also define the collective in terms of a nation. Although Benedict Anderson defined the nation as a modern phenomenon, others have attributed national characteristics to the political configurations described in biblical and postbiblical Hebrew documents.³ I build upon this

---

1. Myth speaks to a truth whose authority stems from a source outside of or beyond the present. Distancing techniques allow myth to relate intimately to the present without purporting to reflect the present.

2. Bronislaw Malinowski, *Magic, Science, and Religion* (Garden City: Doubleday, 1954), 84.

3. Benedict Anderson, *Imagined Communities: Reflections on the Origin and Spread of Nationalism* (London: Verso, 1983). Anthony D. Smith cites ancient Israel as a primary example of a national entity. My corrective to Smith is to avoid taking the represented features of the Israelite nation empirically. I understand the durability of Israel's myth(s) to result from an advanced conceptual structure with which physical reality may never match. The narrative is the space of national imagining. See

research as I argue that the three central national criteria identified by Anderson—census, map, and museum—can be identified in texts of the Hebrew Bible.[4] While recognizing the nationalist rhetoric of biblical literature, I want to complicate our sense of this nationalism by highlighting the competing conceptions of the nation Israel that ultimately cohere in the biblical canon. Rather than one nation with one mythology, the literary record of ancient Israel—the Hebrew Bible—contains several national mythologies.

Five biblical national myths describe the borders of the Promised Land. The Priestly myth presents a nation defined by clear territorial and social borders.[5] The Deuteronomistic national myth promotes an expansionist view of territory while emphasizing the importance of a centralized state that revolves around a capital city.[6] The Northern national myth (or myth of the Northern Kingdom of Israel) presents borders as sites of encounter as well as of delimitation.[7] The Northern myth does not imagine Israel as a discrete political unit set apart from other nations. Instead, the territory of Israel results from constant negotiation with neighboring peoples. The fourth national myth, in contrast, promotes an ethno-national agenda in which members of the nation are bound by blood. This myth becomes evident in the books of Ezra and Nehemiah, which narrate the return of exiles to the region of Judea (the former Southern Kingdom of Judah) under the sponsorship of the Persian Empire. In this myth, we see how the experience of exile and dependence on empire leads to ethnic constructions of the nation. I derive the fifth national myth from the book of Ruth and argue that it develops in opposition to Ezra and Nehemiah's more absolutist idea of the nation. As Ruth—a Moabite

---

Smith, "The Problem of National Identity: Ancient, Medieval, and Modern?" *Myths and Memories of the Nation* (Oxford: Oxford University Press, 1999), 106–8. In support of viewing biblical Israel as a nation see Steven Grosby, *Biblical Ideas of Nationality: Ancient and Modern* (Winona Lake, IN: Eisenbrauns, 2002); Ilana Pardes, *The Biography of Ancient Israel: National Narratives in the Bible* (Berkeley: University of California Press, 2000); in relation to the Second Temple period, see David Goodblatt, *Elements of Ancient Jewish Nationalism* (New York: Cambridge University Press, 2006).

4. For evidence of the census: Numbers 1–4, 26; Havrelock, "B'midbar: Numbers 1:1–4:20," in *The Torah: A Women's Commentary*, ed. Tamara Cohn Eskenazi and Andrea L. Weiss (New York: Union of Reform Judaism Press, 2008), 789–807, and Havrelock, "Outside the Lines: The Status of Women in Priestly Nationalism," in *Embroidered Garments: Priests and Gender in Biblical Israel*, ed. Deborah W. Rooke (Sheffield: Sheffield Phoenix Press, 2009), 89–101. For evidence of the map, see chapter one; for evidence of the museum, see chapter 5.

5. What I call the Priestly myth contains traditions that biblical scholars identify as belonging to the Priestly (P) source as well as the Holiness (H) Code.

6. I derive this national myth from traditions in the books of Deuteronomy, Joshua, Judges, Samuel, and Kings. Biblical scholars tend to refer to this body of biblical literature as the Deuteronomistic History.

7. These traditions for the most part coincide with what biblical scholars called the Elohist (E) source.

woman from the other side of the Jordan—crosses an ethnic as well as a territorial boundary, a nation that incorporates others and includes women finds its voice.

The five national myths are all found in the Hebrew Bible. I arrive at the number five through a combination of the methods of folklore and biblical source criticism. Folklore methods, and methods of biblical folklore in particular,[8] help me to classify variant stories about the nation of Israel as myths that articulate changing ideas of the collective. Source criticism, which argues that the Bible is comprised of distinct literary sources written by different authors, allows me to associate the five myths with particular political parties or scribal ideologies. Source-critical studies tend to ask when parts of the Bible were written and to advance dates for various texts of the Bible. I do not attempt to provide dates for the five national myths. Instead, I define the sources as mythic variants and inquire into their different conceptions of Israel. Without insisting on when they wrote, we can say that an elite, hereditary class of priests advanced the Priestly national myth; statist, pro-monarchical authors wrote the Deuteronomistic national myth; proponents of a syncretic state disseminated the Northern national myth; returnees from Babylonian exile promulgated the Ezra-Nehemiah myth; and, I propose, a scribal group advocating female inclusion in the annals of the nation wrote the book of Ruth.

Why would the editors of the biblical canon (the Tanakh) include contradictory positions and support different versions of the past, instead of advancing a singular, unambiguous narrative to promote the interests of the editors or their patrons? Scholars have argued that the inclusion of variant traditions may be part of a recurring salvage project: a remnant of Israel absorbs the traditions of their more beleaguered counterparts so that no witness to Israel's history is lost. In this scenario, Northern traditions would be incorporated after the Northern Kingdom falls to Assyria and some of its refugees reach Judah and Jerusalem; the chronicle of deviant and upright kings in the

---

8. On patterns and variants in biblical texts, see Robert C. Culley, *Studies in the Structure of Hebrew Narrative* (Philadelphia: Fortress Press, 1976); Alan Dundes, *Holy Writ as Oral Lit* (Lanham: Rowman & Littlefeld Publishers, 1999); Dundes, "The Hero Pattern and the Life of Jesus," in *Interpreting Folklore*, ed. Alan Dundes (Bloomington: Indiana University Press, 1980); James Frazer, *Folklore in the Old Testament* (London: Macmillan, 1918); Hermann Gunkel, *The Legends of Genesis*, trans. W. H. Carruth (Chicago: Open Court Publishing,1901), which is the English translation of Gunkel's *Elias, Jahve und Baal* (Tübingen: J. C. B. Mohr, 1906); Ronald S. Hendel, *The Epic of the Patriarch: The Jacob Cycle and the Narrative Traditions of Canaan and Israel*, HSM 42 (Atlanta: Scholars Press, 1987); Susan Niditch, *Folklore and the Hebrew Bible* (Minneapolis: Fortress Press, 1993); Niditch, *Oral World and Written Word: Ancient Israelite Literature* (Lousiville, KY: Westminster John Knox Press, 1996).

Deuteronomistic history would be preserved as a moral as well as a political example; and Priestly editors would order local traditions according to a genealogical schema after the places sanctified by the traditions were lost to war or invasion. Such scenarios are difficult to imagine because an imperiled group does not tend to preserve the interests of others. It is certainly the case that occupied and exiled people guard their past vigilantly in order to preserve group identity and cohesion. However, the preservation of variant and competing traditions alongside a dominant narrative runs counter to an identity narrative that resists outside influence and establishes cultural barriers when geographic boundaries are imperiled or effaced.

The scenario described above relies on a notion of a consistently unified Israel, one that does not emerge in the Hebrew Bible. Instead, following Martin Noth, the German Bible scholar who saw canon formation and national formation as parallel processes, I argue that the variant depictions of the national past reflect the ongoing process of national formation.[9] According to Noth, the variant traditions were not incorporated into Israel's national history as segments of Israel fell to empires, but became part of history as different groups affiliated with Israel. This conception suggests that we should not see ancient Israel as a stable group whose achievements or ideology are simply recorded by the biblical text. Instead, when groups became incorporated into Israel (or into the tribal league as Noth imagines it) through defeat, alliance, or cultural blending, they brought their regional traditions to bear on the national narrative. As this cultural blending occurred, the recorded and recited past changed as well. When new peoples were absorbed into Israel, their stories were woven into the national history. At the same time, foundational stories like the Exodus and the crossing of the Jordan River were incorporated into the histories of the newcomers. Without this dual process of incorporating new traditions and conferring extant ones, the past and consequently the present would not be shared.

With ever-changing constituent parts, the chronicle as well as the nation of Israel is a work in progress. The continual morphing of ancient Israel through alliance, absorption, and the defection or defeat of its component groups involves recurrent acts of boundary drawing. Groups of people move in and out of insider and outsider positions. And as the group undergoes shifts, so the nation and its territory change. Who were the people who made up Israel? Each time the question was asked, the response created new borders, as well as new relationships to Israel's most significant border, the River Jordan.

All of the myths include some kind of mapping of national territory or designation of boundaries. My analysis focuses on how each myth describes

---

9. Martin Noth, *Überlieferungsgeschichte des Pentateuch* (Stuttgart: Kohlhammer, 1948).

the Jordan River and on how different expressions of the nation Israel result from these articulations of a border. Chapters one through six concern the biblical myths themselves; chapters seven through ten focus on the use of these myths by interpreters of the Bible. The first chapter presents the Priestly and the Deuteronomistic maps of the Promised Land. In the Priestly map, the Jordan serves as an eastern border that distinguishes territory conferred by God from the places where Israel wandered and transgressed. The Priestly ideal of a distinct and purified Israel relies upon the Jordan as a foundational geographic limit. At the same time, the Jordan operates as one border in a complex system of borders.

Priestly thought imagines the world in terms of interlocking binary oppositions. For example, the Priestly creation myth, narrated in Genesis 1:1–2:4, describes God creating the world by separating chaotic elements into the binary categories of darkness and light, water and air, bodies of water and earth. The boundary between these categories is vital because should, say, the waters of the sea break their bounds and flood the earth, the result is a return to a primeval chaos. Boundaries between the elements protect creation, which is further divided into the classifications of sea and air creatures, animal and human, human and the (uncreated) divine. The Priestly creation myth is key to understanding Priestly law and narrative, which works under the premise that just as God created an ordered universe through dualities, so Israel must recognize and maintain social dualities. In addition to the boundary between human and divine, Priestly texts instruct Israel to uphold distinctions between Israel and other nations, a Priestly class and the rest of the People, the holy and the profane, the pure and the impure. The creation myth serves as a reminder that should Israel transgress social and ritual boundaries, God may suspend the boundaries of creation and return the world to chaos.

In Priestly literature, delimitation defines the holy. The holy is set apart from the ordinary through boundaries policed by the priests. The Jordan River, which bifurcates territory into two recognizable banks, becomes an image that helps the priests to advance their idea of a distinct, circumscribed homeland. Because in Priestly ritual water works as a transformative agent that facilitates the change from impurity to purity, the Jordan River further enhances the image of a sanctified Israel. By delimiting territory, the Jordan makes Israel holy and by representing water, it suggests that Israel's territory is pure. By looking at the Jordan as an important border in a highly structured binary system, we can see to what degree Priestly writers favor a limited homeland ordered according to divine categories rather than a vast kingdom that absorbs many nations and peoples. From the Priestly perspective, it is more important to safeguard the system than to enlarge the national territory. In fact, expansion—which by definition overruns borders—threatens the system.

In contrast to the limited, symbolic conception of territory advanced by Priestly writers, the Deuteronomistic writers promote an ironically inclusive map of Israel's land that stretches from the Mediterranean Sea to the Euphrates River. As they advance the Euphrates map, the Deuteronomists also portray the Jordan as a kind of normative limit of the legitimate homeland and span of the law. Exactly as they guide political leaders to set their sights on expanding their territorial holdings right up to the edge of Mesopotamia, the Deuteronomistic writers also limit national membership, primarily to those west of the Jordan.[10] Concurrently, Deuteronomy promotes a temple capital as the fixed national center. Although this center is most properly established in Jerusalem and is not transferable in the manner of the Priestly sanctuary, the Deuteronomists convey less anxiety than the priests about the need for definitive borders. As long as the nation remembers Jerusalem as it fights together, then the contingency of borders need not be viewed as a threat. In presenting a national system based on the "horizontal integration of individuals who determine their sovereignty," the Deuteronomists seek to efface tribal and regional affiliations.[11] This means that the assertion of tribal and regional identities by characters in Deuteronomistic literature usually signals the onset of disruption and fragmentation. Perhaps because centralization is the name of the game in the Deuteronomistic national myth, there is less investment in fixed boundaries.

The Priestly and the Deuteronomistic maps tell us important things about how the two schools think of the nation. They also point toward how the respective writers viewed Israel's place in regional politics. The two maps of the land—the Jordan map and the Euphrates map—cite a river as the eastern border although their rivers are different. The similarity depends upon shared ideas of the cosmos and the difference arises from the fact that the two maps mimic different imperial taxonomies. The Priestly map imagines Israel as replacing an Egyptian-ruled land of Canaan, and the Deuteronomistic map presents a mighty Israel that rivals Babylonia. Insofar as empires advance the technology of cartography, national maps simply place the nation in an imperial scheme. Despite the fact that nativist myths give nationalist maps their legitimacy, such maps actually depend upon the paradigms of empire.

---

10. As we will see, the Israelite tribes east of the Jordan pose a distinct problem to Deuteronomistic conceptions of the nation.

11. "There is good reason to think that already in the seventh century BCE, under the pressure exerted by the Assyrian empire, Judah's Deuteronomic theologians reinterpreted earlier legal traditions in a manner that established the priority of national 'brotherhood' above the obligations of clans and tribes." Mark G. Brett, "National Identity as Commentary and as Metacommentary," in *Historiography and Identity (Re)formulation in Second Temple Historiographical Literature*, ed. Louis Jonker (London: T & T Clark International 2010), 39.

The second chapter shows how the national myth in the book of Ruth counters the myth in the books of Ezra and Nehemiah. Neither myth presents a map in the manner of the Priestly and Deuteronomistic myths, but both imagine the nation by describing its boundaries. On the one hand, the boundaries in question are territorial and include the Jordan River, but on the other hand, these physical boundaries reify gendered and ethnic categories. This means that the Jordan River is important because it separates Israel from Ammon and Moab, nations with a reputation for deviance and perversion in biblical literature. While Yahwistic, Priestly, and Deuteronomistic sources demonize Ammon and Moab, the books of Ezra and Nehemiah advocate the expulsion of Moabite, Ammonite, and all foreign women from Israel's midst. Ezra and Nehemiah lead a group of exiles back to a reduced province of Judea, where Judeans and returnees from Babylon live amongst other people without any clear lines of separation. In response, Ezra and Nehemiah propose the idea of a "holy seed"—an unadulterated Jewish lineage purified through the removal of foreign women and their mixed-blood children—as a form of compensation for the lack of territorial limits.

Applying anthropologist Frederik Barth's notion "that ethnicity is a matter of social organization," we see how Ezra and Nehemiah's redefinition of Israel works to consolidate their power and to root out elements resistant to this power.[12] Along these lines, the chapter construes "Moabite" not only as an ethno-national designation, but also as a term applied to those who resist dominant power structures, particularly the gendered power dynamics of nation. The book of Ruth not only subverts Ezra and Nehemiah's program of expulsion, but also reverses the theme of deviant women east of the Jordan that runs through biblical texts. By showing the benefits to Judah that result from Ruth the Moabite's crossing of the Jordan, the book of Ruth contests exclusions from the nation based on norms of ethnicity and gender.

Where the book of Ruth advocates for greater inclusion, the Northern national myth presents Israel as a nation characterized by fluid borders. The attendant mapping of national territory depicts the Northern Kingdom as spanning both banks of the Jordan River with dynamic frontiers to the east with Aram (Syria), to the north with Sidon, and to the south with the Southern Kingdom of Judah. Contact with these other groups constantly transforms Israel. Chapter three considers traditions about the Northern Kingdom's founding father, Jacob. In an encounter with the Divine just east of the Jordan, Jacob wrestles with God and is renamed Israel as a result. The momentous transformation occurs at a border and implies that the nation Israel, like its

---

12. Frederik Barth, introduction to *Ethnic Groups and Boundaries: The Social Organization of Culture Difference*, ed. Frederik Barth (Long Grove: Waveland Press, 1969), 13.

founding father, encounters God on the road between homeland and exile and acquires its character by crossing borders. Before establishing a home, Jacob negotiates borders of sorts with his Aramean father-in-law, Laban, and his Edomite twin, Esau. In contrast to the border between Israel and Moab/Ammon imagined in other national myths, those established by Jacob remain open rather than absolute.

The Jacob stories constitute the founding myths of the Northern or Israelite Kingdom, a wealthy and syncretistic monarchy with subjects on both sides of the Jordan River. Because forces of the Assyrian Empire destroyed the Northern Kingdom, its national myth becomes subordinate to traditions about the Southern Kingdom of Judah and its sanctuary in Jerusalem. The Northern tradition with its fluid boundaries between Israel and its neighbors represents a national strategy different from Southern traditions of sharp distinctions and ongoing wars.

Ongoing war is precisely the subject of the Deuteronomistic national myth as it appears in the book of Joshua. In the fourth chapter, I treat the book of Joshua, in which crossing the Jordan ends Israel's exile and inaugurates its conquest of the land as the most violent version of the biblical national myth. However, by building on the premise of myth theory that all myths exist in variation, I show that the book of Joshua contains competing versions of the land and its constituent population. Indeed, the myth of conquest depends upon the story of a unified nation reborn by crossing the Jordan River. The staunch general unifies the tribes through the shared national experience of crossing the Jordan, defeats all the peoples in Israel's path, and then settles Israel in the emptied regions. The book casts an alleged extermination and ban on the property of the nations populating Canaan as the ultimate fulfillment of God's plan and the redemption of a fragmented, rebellious people. But rather than a record or reflection of historical events, the book of Joshua is a founding myth that establishes the parameters of inclusion and exclusion. It writes allies into the national beginning at the Jordan River and accounts for non-Israelite neighbors as those who withstood an ancient conquest. National unity dissipates in the patchwork nature of the national myth.

In the biblical texts considered in chapter five, strongly antinationalist positions are associated with the east bank of the Jordan River. This region labeled "the other side" displays paradoxical characteristics because legitimate members of Israel reside on territory considered illegitimate. Building on David Jobling's conclusion that the problem of foreigners living west of the Jordan may be correlative with Israelites living east of the Jordan, the chapter uses his structuralist analysis in order to read the biblical construction of opposition between the two banks in terms of the problematic of diaspora. As different biblical and postbiblical sources struggle with the idea of partial affiliation with the nation, the tradition of Jewish diaspora takes form. Chapters four and five, which focus on specific texts in the book of Joshua, recommend that the national myth and countermyths that run through the Deuteronomistic history (books of Joshua, Judges, Samuel, and Kings) be understood not only as parallel, but also as dialogic. While the countermyths position themselves in opposition to presiding norms, the national myth itself assumes a didactic, authoritative tone in order to compensate for the absence of a desired unity. The book of Joshua's extreme assertions that Israel exterminated peoples and marched in line behind an exemplary general intend to obscure the disparate beginnings and affiliations that fall under the term "Israel."

Chapter six considers the Jordan River as a ritual threshold where divine spirit is transmitted. Testing the limits of Robert Alter's theory of the biblical type-scene, a recurrent plot that establishes correspondences among figures while its details differentiate character, I compare parallel scenes in the Old and New Testaments. Joshua, Elisha, and Jesus all gain authority, outside of hereditary and monarchical structures, on the banks of the Jordan. Ernst Kantorowicz has noted medieval versions of this myth in which a king's spirit is said to gain immortality at the same time that it authorizes his successor. The biblical stories calibrate the boundary between heaven and earth in order to present a metaphoric space for a collective defined by shared ritual. The immortality of Moses and Elijah signals the historical immortality of the People of Israel. The promise of resurrection conferred by baptism, in contrast, transpires in the posthistorical space of the Kingdom of Heaven. These orientations ultimately differentiate Jewish and Christian redemption as either a return or a rebirth.

• • •

Biblical national myths serve as an instrumental past for a wide range of groups seeking legitimacy through recourse to the Hebrew Bible. In the second half of *River Jordan*, I illustrate which of the biblical national myths have had political currency and which have been neglected. I begin with the binary divisions and investments in purity of the Priestly national myth that provide the conceptual structure for early Christian and rabbinic imaginings of the collective. The proposal that early ideas of a Christian collective depend upon Priestly national conceptions will no doubt surprise those accustomed to the claim that Christianity, by definition, rejected Priestly law.[13] In opposition to such claims, I suggest that early Christian ideas about sexual purity and the

---

13. That rabbinic literature makes use of Priestly categories will seem somewhat less surprising despite the avowed rabbinic disdain for the Priestly caste of Sadducees.

avoidance of sin extend and transform Priestly categories. Furthermore, the image of a community brought together through the initiatory ritual of baptism relies on the Priestly sense that ritual produces the categories of self and other and that water affects a transition between distinct states of being.

Priestly thought becomes a strategy for Jewish communal definition in the diaspora. Like the early Christian thinkers, the Rabbis adapted the Priestly structure of purity maintained through correlative distinctions. This structure that situates holiness at shifting centers proves particularly useful for the Rabbis because, as Jonathan Z. Smith noted, Priestly boundaries do not require fixed geography, but can be realized in multiple spaces and times. In this way, the Jordan persists as a symbol of communal distinction even when the river itself no longer has relevance as a physical boundary of a sovereign state.

Chapter seven examines the Jordan as a symbolic border between Christians and Jews. In the Christian case, baptism enacts an individual rebirth; in the Jewish case, continual purification maintains communal sanctity. As the different biblical myths of national origin concern themselves with the Jordan as a border, so the myths that accompany the rituals of baptism and purity articulate the limits of religious community through the trope of the Jordan River. In other words, Christians and Jews distinguish themselves through parallel interpretations of the same biblical national myth. Chapters seven and eight show the persistence and the transformation of the Jordan myth. In Christianity, the Jordan is the river crossed to enter heaven. In rabbinic Judaism, the Jordan no longer signifies Jewish national territory or political sovereignty. It is upheld as a conceptual limit within a legal framework as well as a symbol of eschatological redemption.

Where the Priestly national myth informs the ritual and communitarian structures of early Christianity and Judaism, the Deuteronomistic national myth achieves its longevity as an applied political program. When those promoting a political agenda look to the Bible for legitimization, the precedent in question usually originates in the books of Deuteronomy, Joshua, Judges, Samuel, or Kings. A prominent example can be found in Christian colonial appeals to Deuteronomy's command to expel indigenous populations. At the risk of oversimplifying, I suggest that the Priestly myth more easily lends itself to charters of religious authority, while the Deuteronomistic myth tends to be mobilized politically.[14]

The use of the Deuteronomistic national myth becomes particularly rel-

---

14. The distinction between the religious and the political is as difficult to sustain in the biblical period as it is in contemporary times. The fact that the distinction so easily collapses does not trouble my argument, but rather seems to support an approach that considers biblical national myths in political terms.

evant to a discussion of the Jordan River inasmuch as early Zionist thinkers directly take up the myth as a political program. Although variation, dispute, and alternate proposals characterize early Zionist thought, a dominant narrative ultimately emerges involving Jewish redemption through labor, territorial acquisition, and military heroism. There are many figures involved in forging and advancing this narrative, but David Ben-Gurion is central to the creation and dissemination of a Jewish national narrative. Ben-Gurion and his colleagues based their new Jewish national myth on a revivification of the conquest, settlement, territorial distribution, and national brotherhood as described in the book of Joshua. Biblical maps authorized the Jewish claim to Palestine and biblical myth served as the charter for an independent Jewish state. For example, Ben-Gurion hosted a weekly study group on the book of Joshua at the Prime Minister's residence attended by biblical scholars, politicians, and military officials. In light of the centrality of biblical texts to early Zionist thought, it is not surprising that the Jordan came to be seen as the necessary eastern border of the Jewish State.[15]

However, Zionist and Palestinian national movements standardized their territorial aspirations only after the British Empire fixed the borders of Mandate Palestine. Islam recognizes the Jordan as a place associated with biblical prophets, and Muslim armies fought against crusaders along the Jordan, but in Islam it does not form a substantial geographic or conceptual boundary. The Palestinian claim to the Jordan originates largely from national aspirations. This is also the nature of the hegemonic Zionist claim, which relies on secular, political interpretations of the biblical national myth. Neither claim should be considered indigenous because the Jordan was not an operative border for Arab and Jewish nationalists until the British, who were awarded the region of the Holy Land when the European powers carved up the Ottoman Empire, drew a line at the Jordan River. Jewish and Arab national movements absorbed the borders of the British Mandate in Palestine, and then turned their efforts to ending the mandate and outmaneuvering one another.

The final chapter presents Palestinian and Israeli national myths that claim the Jordan as a necessary border of an indivisible homeland. In the same way that Jewish adaptations of the Priestly national myth developed in dialogue with their Christian counterparts, the Zionist national myth interacts with the contesting and parallel Palestinian national myth. Suggesting that neither one of these national myths can be fully understood in isolation does not in-

---

15. Ze'ev Jabotinsky's break from the Labor Zionists and resignation from the executive council of the World Zionist Organization was motivated largely by opposition to the British setting a boundary at the Jordan River in 1923. Jabotinsky advocated a Jewish state on both banks of the river. The national myth he advocated can be seen as counter to that of Labor Zionism from the beginning.

dicate an intention to equate them. How the two traditions mobilized aspects of the Deuteronomistic myth reveals their differences. In the Israeli context, the founding wars with the Arab states in 1948 and 1967 were mythologized as Joshua-like battles to reclaim Israel's Promised Land. For the Palestinians these same wars mark the *Nuzuh*, the first and the second "Exodus" from the homeland, events which begin the tragic period of exile and occupation. While acknowledging the asymmetry of the Israeli and the Palestinian national experiences, the final chapter shows the similar types of stories in which Palestinians and Israelis claim the Jordan as a necessary boundary and portray the events that transpire there as constitutive of the nation.

If the Priestly and the Deuteronomistic national myths have persevered in religious practice and in modern nationalist rhetoric, then what of the Northern national myth or the myth expressed in the book of Ruth? Neither of these myths has been taken up as an authoritative model in quite the same way. But they can provide potential alternative ways of understanding ancient Israel and imagining a modern state between the Jordan and the Mediterranean Sea. In the context of contemporary Israel, exclusivist and right-wing groups have used the Priestly and the Deuteronomistic national myths to justify their actions. In addition, contemporary Christian adaptations of biblical myths as well as Palestinian versions of the Deuteronomistic myth have assumed exclusivist and reactionary forms. The solution, I suggest, may lie in the biblical national myths not currently circulating in public discourse.[16]

There is a violent conflict and a militarized occupation in the space between the Jordan River and the Mediterranean Sea. But for all of the force and humiliations, the conflict and occupation are techniques of denying what is already a binational or transnational state. So much effort is poured into making Israel and making Palestine, and yet many states find expression in the variations among these myths and the differences between them. Here is the paradox: I write quite literally about the limits of ethno-nationalism while seeing it as a site and source of variation. Some myths construct the borders of homeland and blood, others tear them down, and others never recognize the borders to begin with. I recommend no universalist correction to ethno-national myths and wouldn't dispense with any of them in order to widen the market, save souls or bring the revolution. I would, however, change the frame and not construe the myths as oppositional so much as variant.

Current discussions of the conflict present the following options as op-

---

16. Here I enter into the realm of myth making. As identified by Bruce Lincoln, "The enunciation of any mythic variant opens up an arena of struggle and maneuver that can be pursued by those who produce other variants of the myth and other interpretations of the variant." *Theorizing Myth: Narrative, Ideology, and Scholarship* (Chicago: University of Chicago Press, 1999), 151.

positional: will Israel become a state of all its citizens and fully enfranchise Palestinians on both sides of the Green Line (and even east of the Jordan River), or will it maintain its specifically Jewish character? I suggest that ancient Jewish articulations of nationalism promoted, either explicitly or by default, a state for all of its citizens—Jewish and Gentile alike. This is to say that the enfranchisement of the Other is perfectly consistent with Jewish political traditions. While it is indeed the case that most of the conceptual boundaries that have defined the Jewish religion enact Jewish separation and distinction, the longstanding territorial boundary of the Jordan River—in conceptual and material terms—more often than not encompassed or connected outsiders to Israel. In my own times many acts of aggression, settlement, or excavation that displace current residents find justification through ancient claims. These events necessitate close analysis of biblical texts, which promise that God will dispossess other nations and exhort Israel to reject both treaties and mercy to the Other (Deut 7:1–2). Yet the texts also speak of a Jerusalem shared in perpetuity by Judah and its Jebusite neighbors (Josh 15:63) as well as the legal rights of resident outsiders who live among Israel.[17] So if the warrant for the Jewish state (and, as some contend, for the nation-state in general) derives from the Bible, then we had best understand all of the political possibilities that the Bible extends.

---

17. I take the term "resident outsiders" from Saul Olyan, *Rites and Rank: Hierarchy in Biblical Representations of Cult* (Princeton: Princeton University Press, 2000).

Chapter One

THE TWO MAPS OF ISRAEL'S LAND

They say that Ocean runs around the whole earth
—Herodotus 4.8.2

Demarcation and naming divide the world into distinct places with which people can identify. The map and the name of a place become emblems that cover up the disunity, lack of clear borders, and proliferation of titles ascribed to a location. Acts such as drawing borders, naming natural features, building memorial structures, and telling stories of pioneering ancestors play central roles in colonization and settlement.[1] This chapter investigates the process through which the map of a nation comes into being. The relevant examples derive from the Hebrew Bible and the context of antiquity, yet similar processes also determine the nature of maps from subsequent eras. Biblical maps display how the emblematic representation of the nation relies on intersecting mythic and political standards. The question

---

1. The biblical book of Joshua, for example, describes such actions. The extensive geographical description in Joshua 15–21 is replete with acts of demarcation, identifying and naming. Additional examples of naming as a mode of staking claim can be found in Josh 5:9 and 14:12–15. Memorial structures are erected in Josh 4:4–8, 4:19–24, 8:30–32, and 24:26–27. In addition to the chronicle of pilgrim ancestors represented by the book of Joshua, the passage in Josh 24:1–13 tells of this generation's intrepid forerunners.

of why there are two different maps of Israel's land stands at the center of the analysis. One set of maps spans from the Mediterranean Sea in the west to the Jordan River in the east and a second set reaches from the Sea to the River Euphrates. A conceptual stability results from the fact that the land in both cases spans from river to sea, while conflicting notions of the state arise from their discrepancies. I argue that the seeming paradox of conflicting versions of national territory illustrates how maps reconcile the idea of the nation with regnant mythic conceptions as well as how the nation borrows the means of self-presentation from empire.

The maps to which I refer are narratives that evoke place by consecutive enumeration of limits rather than by graphic symbols.[2] We know of pictorial maps from the Ancient Near East such as the Babylonian *Mappa Mundi* and the Egyptian map of Turin. In contrast the maps of Israel's land are boundary lists, mediated in language. Although they first read like an inventory—a geographical corollary of the genealogy—the maps are rich in literary nuance and historical suggestion. From maps we learn how those in power such as monarchs or priests circumscribe space in order that institutions like the court or the priesthood be perceived as the center of state and cosmos alike. At the same time, the grandiose span of maps often signals a tremulous hold on power and territory alike and incongruous depictions suggest fronts of resistance.

The structure of this chapter follows J. B. Harley's suggested analysis of both "the cartographers' rules" for how maps should look and a map's "'signifying system' through which 'a social order is communicated, reproduced, experienced, and explored.'"[3] I discuss these issues in two subsections: the first deals with the "role of measured maps in the making of myth" and the second with the imperial standards by which smaller nations measure themselves.[4]

Mythic Geography

What to call the place promised to Israel presents its own challenge: is it Canaan, the land of Amorites, the kingdom of Israel, or more generally the land bequeathed to the ancestors? In order to leave the concept sufficiently open, I call it "the land" throughout—the definite article anchors the concept, while the absence of a proper noun allows for its shifting nature. Since the land, its

---

2. Such boundary lists seem to be among the oldest cartographic relics; Sumerian documents of this nature from 2500–2200 BCE have been discovered. A. R. Millard, "Cartography in the Ancient Near East," in *The History of Cartography*, vol. 1, *Cartography in Prehistoric, Ancient and Medieval Europe and the Mediterranean*, ed. J. B. Harley and David Woodward (Chicago: University of Chicago Press, 1987), 107.

3. J. B. Harley (quoting Foucault), in *The New Nature of Maps: Essays in the History of Cartography*, ed. Paul Laxton (Baltimore: Johns Hopkins University Press, 2001), 45.

4. Harley, *The New Nature of Maps: Essays in the History of Cartography*, 77.

acquisition, and its contingencies constitute the central thrust of the Hebrew Bible, the question of where exactly this land lies requires an answer at several junctures. Just as the collation of texts by different authors from diffuse periods leaves its mark as textual strata whose meanings are still being mined by scholars, so it has left us many lands.

The territorial referent key to understanding ancient Israel's place is unstable, made up of shifting borders and fluctuating dimensions. Despite the myths to the contrary, this is the nature of national as well as holy ground—subject to war and migration, historically alternating, disrupted by diaspora and mixed populations—that constantly undergoes change since borders exist in a state of flux. To the degree that national identities in general and those of biblical Israel in particular depend upon territory and/or the representation of territory, a land with shifting coordinates signals an identity under constant production. I suggest that the variation of the maps is not the result of imprecision or confusion, but rather the condition of different possibilities of identity.

Jordan Maps

The Jordan maps exist in only two versions, but enjoy thematic dominance because they conform to the idea of the land produced in exodus narratives where the experiences of wandering and homecoming correlate with the east and the west banks of the Jordan. Throughout these narratives and their accompanying laws, crossing the Jordan becomes synonymous with national reintegration. The books of Numbers, Deuteronomy, and Joshua all stage the homeland west of the Jordan and employ the river as a legal, temporal, and territorial boundary. Numbers 34 contains the most prominent map in which the Jordan serves as the eastern boundary.

> God spoke to Moses saying: "Instruct the Israelites by saying to them: When you enter the land of Canaan, this is the land that will constitute your property, the land of Canaan as defined by its borders ... Your western border will be the Great Sea; this border will be your western border ... Mark your eastern border from Hazar-enan to Shepham. The eastern border will go down from Shepham to Riblah on the east side of Ain, from there the boundary will continue down to skirt the eastern edge of the Kinneret Sea [the Sea of Galilee]. Then the border will descend along the Jordan until it reaches the Dead Sea; this will be your land as defined by its borders." (Num 34:1–2, 6, 10–12)

The Mediterranean serves as the western boundary and the Jordan as the clearest eastern boundary, although the inclusion of northern lands considerably

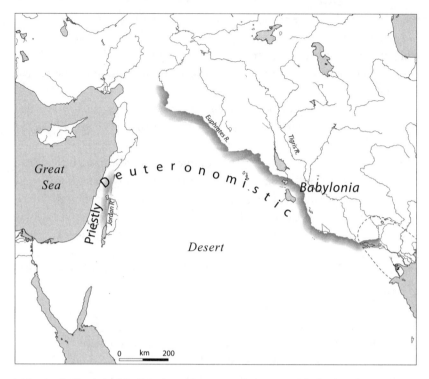

1. The riverine borders of the Priestly and Deuteronomistic maps of the land. Soffer Mapping.

east of the Jordan means that the Jordan operates as the eastern border only from the Sea of Galilee to the Dead Sea (figure 1). This stretch of the river most prominently delineates the land and serves as the setting for the Bible's Jordan crossing stories (Gen 28:10–22, 32; Josh 1–4; Judg 12:1–6; 2 Sam 17:22, 19:16–41; 2 Kings 2:2). The detail of the map transforms the land from a domain of nurture, "the land flowing with milk and honey,"[5] into a domain of ownership, "the land that will constitute your property" (Num 34:2). The coordinates determine Canaan—with or without the presence of Israel—and allow it to be grasped in conceptual as well as military terms.

The other map in which the Jordan forms the eastern frontier occurs in the concluding vision of the book of Ezekiel.[6] This exilic book assures the persis-

---

5. This reference abounds in biblical narrative, see, for example, Exod 3:8, 17, 33:3; Num 13:27, 14:8, 16:13, 14; Deut 6:3, 11:9, 26:9, 15, 27:3, 31:20; Josh 5:6; Lev 20:24.

6. Bodies of water set three of the four boundaries of Ezekiel's land, including the Mediterranean to the west and the waters of Meribah along with the river of Egypt to the south. While Ezekiel 47:19 speaks only of the "river" to the south, the parallel in Numbers 34:5 suggests that the river of Egypt is likely Wadi el-Arish.

tence of homeland by mapping it in scrupulous detail and portraying its borders as able to encompass overlapping claims.[7] Self-consciously utopian, the map homologizes the land, the Temple, and paradise as interchangeable topoi of symmetry and abundance. The map moves from north to east to south to west delimiting "the land that the twelve tribes can claim as an inheritance" (47:13) and then allots territory with exacting equality to twelve non-priestly tribes. Even the nettlesome "strangers in your midst," who prove problematic in other biblical texts and other sections of Ezekiel, are granted citizenship and ceded territory in the virtual land (Ezek 47:22–23).[8] The tribes of Israel are inscribed in "thirteen longitudinal strips" around a central portion reserved for Yahweh, the Zadokite priests, the Levites, and the archetypal monarch called *nāsî* (48:1–29).[9] The map stations all tribes west of the Jordan with Dan (48:1), Asher (48:2), Naphtali (48:3), the two Joseph tribes of Manasseh (48:5) and Ephraim (48:6), Reuben (48:7), and Judah (48:8) north of the sacred sphere, and Benjamin (48:23), Simeon (48:24), Issachar (48:25), Zebulun (48:26), and Gad (48:28) to the south.[10] Although several tribes had been "lost" by the time of Ezekiel's composition, all have a place in the ideal national configuration. The mention of specific mountains and waterways suggests that the map is not only a utopian social vision, but also an assertion of a territorial homeland. However idealized, the tribal reunion and national restoration cannot be realized just anywhere—they require the land between the Jordan and the Mediterranean Sea. In the absence of this land, the memory of its boundaries preserves a sense of social coherence and collective destiny.[11]

Jonathan Z. Smith understands Ezekiel's maps as pragmatically survivalist. Their geographic and architectural images set up systems of distinction that do not depend on the places evoked. Instead, the distinctions can be overlaid on the calendar, on notions of kinship and identity, and onto ritual practice. The representation of sacred geography then operates to marry memory to transposable distinctions, not to communicate that the absence of place entails the demise of identity. Of the four maps that Smith identifies in Ezekiel

---

7. That Ezekiel's final vision presents an alternative to exile is suggested by the way in which rivers function as a framing device. The book opens by the Babylonian River Chebar (Ezek 1:1) and concludes by the homeland's River Jordan (Ezek 47:8, 18).

8. "This radical reform envisioned by Ezek 47:22–23 represents the granting of (fictive) kinship to the resident outsider; he becomes at last a brother rather than an other." Saul Olyan, *Rites and Rank: Hierarchy in Biblical Representations of Cult* (Princeton: Princeton University Press, 2000), 73.

9. Kalinda Rose Stevenson, *The Vision of Transformation: The Territorial Rhetoric of Ezekiel 40–48* (Atlanta: Scholars Press, 1996), 46.

10. Hierarchy finds its place even amidst equality: "The sons of the concubines Zilpah and Bilhah are farthest away from the Portion." Stevenson, *The Vision*, 86.

11. See D. L. Smith, *The Religion of the Landless: The Social Context of the Babylonian Exile* (Bloomington, IN: Meyer-Stone Books, 1989).

40–48, three (40:1–44:3, 45.1–8, and 47:13–48:35) outline "a hierarchy of power built on the dichotomy sacred/profane," and one (44:4–31) "is a hierarchy of status built on the dichotomy pure/impure."[12] Stressing the transferability of the "complex and rigorous systems of power and status," Smith intimates that their potentially limitless replication arises from their mythic character.[13] The dichotomies, not the places, are upheld as eternal and necessary. Ezekiel's maps and their systemic boundaries are mythic not only in their apocalyptic promise of a future Eden and in their potential for reproduction, but also in the structural sense of homologous oppositions evident in other biblical myths and other mythic systems.

The Jordan and Creation

Mythic allusions launch Ezekiel's narrative of transport. God lets him down on "a very high mountain" whose panoramic views recall Moses's final vision (Num 27:12; Deut 32:49, 34:1–4) and whose centrality emphasizes both the Temple's sacredness and its similarity to the garden of God (Ezek 28:14).[14] The carved cherubs and accompanying date palms that line the Temple interiors (41:18–20, 25) "re-create Eden's ambiance" (Gen 3:24; Ezek 28:14) and the presence of God moves in from the east, the primal direction, to illuminate the world and resonate like the crashing of water (43:1).[15] The new Eden is arable, with abundant trees (Ezek 47:7; Gen 2:9), swarming creatures (Ezek 47:9; Gen 1:20), and potential immortality offered by leaves that heal instead of withering and fruits that never rot (Ezek 47:12). The replenishing fruit trees beside sanctified waters promise an imminent and inclusive paradise.

Water is the dominant feature in the paradisical vision. As a river rises from Eden and branches out into four courses (Gen 2:10), so a single stream bubbles from beneath the Temple and swells into an uncrossable river (Ezek 47:5).[16] The surging waters of Jerusalem symbolize a future of surpassing

---

12. Jonathan Z. Smith, *To Take Place: Toward Theory in Ritual* (Chicago: University of Chicago Press, 1987), 56. Smith employs Louis Dumont's distinction between systems of status and systems of power and argues convincingly that both systems are evident in Ezekiel's final vision. I also see how the distinction could usefully characterize Priestly versions of land as systems of status and Deuteronomistic ones as systems of power.

13. Ibid., 73.

14. For the similarities between Ezekiel's mountain and Sinai, see Jon Douglas Levenson, *Theology of the Program of Restoration of Ezekiel 40–48*, HSM 10 (Missoula, MT: Scholars Press, 1976), 43–44. On its Edenic aspects, see Stephen L. Cook, "Cosmos, *Kabod*, and Cherub," in *Ezekiel's Hierarchical World: Wrestling with a Tiered Reality*, ed. Stephen L. Cook and Corrine L. Patton (Atlanta: Society of Biblical Literature, 2004), 185.

15. Cook, "Cosmos, *Kabod*, and Cherub," 185.

16. Picking up on the parallel between this "fructifying river" and the rivers of Genesis 2, Susan Niditch observes that the ideology of hierarchy expressed here is more in line with the boundaried

Babylon,[17] a collective purification, and a national revivification catalyzed by a restored Temple (Isa 33:21; Joel 4:18; Zech 14:8). The river that emerges from the Temple, like the Jordan in Exodus narratives, divides terrain and epochs alike. The redemptive river that heals staid waters and revives fish and fruit trees (Ezek 47:9–12) morphs into the Jordan as it flows in the eastern region through the Arava and Dead Sea (47:8). As the unnamed, eschatologic river assimilates to the Jordan River, the Jordan accrues apocalyptic associations. More to our purposes, however, the merging of the two rivers shows the codependence of geography and myth. Ezekiel 47 juxtaposes two visions with a coursing river: the burgeoning paradise of the restored Temple and the division of tribal territories. The river of the paradisical vision follows the southern leg of the Jordan, and the Jordan of the territorial vision delimits the scope of the land (47:18). The twin rivers with a parallel course merge into a symbolic whole that endows the Jordan with eternal legitimacy as the eastern border of Israel's land. Thus, Ezekiel's serial visions lay bare a complex process always at work with borders in which authoritative accounts of origin compensate for their arbitrary nature.

As the Judean Desert and Jordan River Valley transform into the new Eden (Ezek 47), paradisical themes from Genesis 1–2 and Ezekiel 28 coalesce. The political tenor of Ezekiel's map has most in common with the myth of Genesis 1 and with Priestly programs in general.[18] Ezekiel's Priestly status and the book's connection with the Holiness Code have long been recognized,[19]

---

cosmos of Genesis 1. She avoids source-critical somersaults with the brilliant proposal that Ezekiel 37–48 parallels Genesis 1–11, "the main corpus of cosmogonic material in the OT." Susan Niditch, "Ezekiel 40–48 in a Visionary Context," *CBQ* 48 (1986): 217, 216.

17. At the same time, this Jerusalem through which a river runs resembles the Babylon described in the Akkadian Topography of Babylon. That Babylon seems to reflect the centralized power and wide-reaching empire of Nebuchadnezzar II. See Wayne Horowitz, *Mesopotamian Cosmic Geography* (Winona Lake, IN: Eisenbrauns, 1998), 27.

18. "Indeed, the cosmogonic process of creating and rightly ordering the new world of Ezek 40–48, in which Ezekiel participates, is a task that resonates with priestly overtones." Iain M. Duguid, "Putting Priests in Their Place," in *Ezekiel's Hierarchical World: Wrestling with a Tiered Reality*, ed. Stephen L. Cook and Corrine L. Patton (Atlanta: Society of Biblical Literature, 2004): 56.

19. "That there is a particularly close relationship between Ezekiel and the Holiness Code is undisputed." Andrew Mein, *Ezekiel and the Ethics of Exile* (Oxford: Oxford University Press, 1991), 107. Jacob Milgrom notes that Ezekiel differs from P in its greater restriction of access to holy spheres. Milgrom, *Leviticus 1–16: A New Translation with Introduction and Commentary* (New York: Doubleday, 1991), 452. There are "point by point similarities between the instructions to the priests in Ezekiel 44:15–31 and the instructions to the priests in the Holiness Code (especially Lev 21:1–22:9), which make it clear at least that a common tradition underlies these two texts." Steven Shawn Tuell, *The Law of the Temple in Ezekiel 40–48* (Atlanta: Scholars Press, 1992), 139. For P's predating of Ezekiel, see Avi Hurvitz, *A Linguistic Study of the Relationship between the Priestly Source and the Book of Ezekiel: A New Approach to an Old Problem* (Paris: J. Gabalda, 1982).

while less noted is the interchangeability of ritual and spatial boundaries.[20] The Priestly writers of the maps of Numbers 34 (Priestly source) and Ezekiel 47–48 (Holiness Code or a related Ezekiel source) desire that the Jordan be the border.[21] Putting aside the questions of if, when, and how the Jordan functions as a border, it can be said with certainty that the Priestly school in its various avatars would very much like this to be the case. The motivations include the fact that as a topographical feature, a river naturalizes the sort of religious and ethnic divisions that the Priestly class puts in place and that the Jordan, associated with Israel's beginnings, authorizes the very premise of necessary borders. "A certain circularity obtains here: cosmogonies reinforced existing power structures by presenting them as derived from the divine order asserted by the cosmogony."[22] Traditions about mixing and contamination trouble biblical representations of the east side of the Jordan, whereas notions of a river-bounded land clarify the scope of purity. Although the Jordan as a border narrows the land's midsection, when it is upheld, Priestly systems of differentiation including Israel's distinction among nations and the priests' distinction among Israel correlate with creation and appear unassailable.

Euphrates Maps

The second, more ubiquitous set of maps fixes the land's eastern boundary at the Euphrates and appears in Genesis and Exodus and throughout the Deuteronomistic History.[23] To Abraham, God defines the land intended for his descendants as spanning "from the River of Egypt to the great river, the Eu-

---

20. The architectural mappings in Ezekiel enforce the primacy of the Zadokite priests in Jerusalem, their proximity to the Divine, and their political centrality (Ezek 40:46, 43:19). Part of this positioning means that rigorous gradations of purity must be upheld. At the base of the gradations are three sets of distinctions: between Israel and Others, Levites and Israelites/Judeans, and Zadokites and Levites. While Ezekiel's map of the land accommodates Others as citizens (Ezek 47:22–23), "aliens, uncircumcised of heart and uncircumcised of flesh" are barred from the sanctuary complex and Temple (Ezek 44:7, 9). Implicit in the formulation is a strategy for accepting equality with strangers on the civic level while maintaining Israel's distinction on an ethnic and religious basis.

21. Steven Tuell highlights the discrepancies between the maps in Numbers and Ezekiel: "The two accounts are almost direct opposites, beginning at opposite points, moving in opposite directions, each strong where the other is weak and weak where the other is strong." Tuell, *The Law of the Temple in Ezekiel 40–48*, 155.

22. Julie Galambush, "God's Land and Mine: Creation as Property in the Book of Ezekiel," *Ezekiel's Hierarchical World: Wrestling with a Tiered Reality*, eds. Stephen L. Cook and Corrine L. Patton (Atlanta: Society of Biblical Literature, 2004), 91.

23. "Deuteronomy and Josh. 1:3–4 claim that Yahweh promised Israel the land from Canaan to the Euphrates . . . Deuteronomy is followed by Chronicles, which in the 5th century claims that David reached the Euphrates and that Solomon maintained a hold there." Baruch Halpern, *David's Secret Demons* (Grand Rapids, MI: Eerdman's Publishing Company, 2001), 248.

phrates" (Gen 15:18).[24] In the book of Exodus, God promises His people land "from the Sea of Reeds to the Sea of Philistia and from the wilderness to the river" (Exod 23:31).[25] Solomon's kingdom, the text maintains, extends "over all the kingdoms from the Euphrates to the land of the Philistines and the boundary of Egypt" (1 Kings 5:1). The Euphrates also figures as the eastern boundary in Moses's recapitulation of God's promise made at Sinai (Deut 1:7) and in God's delineation of the boundaries "from the wilderness to the mountains of Lebanon and from the river, the Euphrates, to the Western Sea" (Deut 11:24). As it is highly improbable that any configuration of ancient Israel encompassed the northeastern expanse outlined in Numbers 34 (a Jordan map), it is even more implausible that Israel at any stage included land to the east of Transjordan, let alone to the Euphrates.

The Euphrates maps offer a glimpse of an unfulfilled vision of military strength and imperial influence. This vision runs alongside the insistence that the Jordan distinguishes the land from foreign lands. No matter the informing map, the east bank is always the other side of the Jordan and the Israelites cross over to enact a return. Whereas the Jordan map presents an image of the land that corresponds to social dichotomies, the Euphrates map generates tension between these dichotomies and the boundary of the land. Such tension becomes particularly apparent in chapters that include a map spanning from the Mediterranean to the Euphrates (Deut 11:24; Josh 1:4), yet proclaim that possession of the land begins only after the national westward crossing of the Jordan (Deut 11:31: Josh 1:2).[26] This tension renders the space between the Jordan and the Euphrates, particularly the east bank, ambiguously both Israel and Other. Transjordan, included in one vision of the land and excluded in another, becomes suspended in the pull of conflicting ideologies.

Why does the Euphrates persist as a represented border? This site so central to exilic history (Ps 137:1) is remembered as the place of Israel's origin (Gen 11:31). Abraham's Euphrates crossing, which in many ways prefigures

---

24. "The river of Egypt is the Pelusaic branch of the Nile and corresponds to the role of Pelusium as the Egyptian border station facing the fifth satrapy in Herodotus' description." Anson F. Rainey and R. Steven Notley, *The Sacred Bridge: Carta's Atlas of the Biblical World* (Jerusalem: Carta, 2006), 280.

25. "The river" indicates the Euphrates here as in Gen 31:21; Josh 1:4, 24:2, 3, 14, 15; 1 Kings 5:4; 1 Chr 5:9. On only one occasion in the Hebrew Bible (Gen 36:37) does "the river" not signify the Euphrates. "The river" also indicates the Euphrates in Mesopotamian documents. See Nili Wazana, *All the Boundaries of the Land: The Promised Land in Biblical Thought in Light of the Ancient Near East* (Jerusalem: Bialik Publishing House, 2007), 105–6 (Hebrew).

26. Although the book of Joshua opens with a Euphrates map, by book's end Joshua bequeaths territory "from the Jordan to Mediterranean Sea in the west" (Josh 23:4) to the tribes of Israel.

the national crossing of the Jordan, inaugurates Israelite history.[27] Because the land east of the Euphrates carries associations with a retrograde era of idol worship (Josh 24:15), it cannot be included in any definition of the Promised Land. The Euphrates then acts as a temporal border in its own right that separates shameful "beginnings" from Israel's patriarchal era. The Euphrates maps, particularly those in which the "River of Egypt" or "Sea of Reeds" constitutes the southern border (Gen 15:18; Exod 23:31), include the lands through which Israel's heroes wandered as part of its territory. The areas west of the Euphrates are legitimized through contrast with those to the east.

The River as Cosmic Boundary

In both the Jordan and the Euphrates maps a river serves as the eastern boundary. The correspondence suggests that the maps share a framework in which the Mediterranean marks the western boundary and one of two rivers the eastern boundary.[28] The land is surrounded on at least two and often three sides by bodies of water.[29] In both sets of maps, the portrait of land bounded by water resonates with cosmological descriptions in which the world spans "from sea to sea" or "from the river to the ends of the earth" (Zech 9:10; Ps 72:8). The image of the cosmos behind these descriptions is a three-tiered universe in which the sea encircles the disc of the earth and the heavens rest both beyond and above the earth and sea.[30] This type of cosmic map is well documented in the ancient Near East and Mediterranean. In the Babylonian *Mappa Mundi* dated around 600 B.C.E. and displayed at the British Museum (BM 92687), the earth is surrounded by a river from which otherworldly re-

---

27. The Euphrates also plays a central role in the Babylonian world map. "Despite the absence of a name, it is clear that the parallel lines running to and from Babylon represent the river Euphrates." A. R. Millard, "Cartography in the Ancient Near East," 111–12.

28. The Mediterranean Sea operates less prominently in the biblical imaginary. Where rivers like the Jordan and the Euphrates represent limits, the sea indicates a lack of differentiation akin to the chaos that precedes creation. As evident in biblical allusions and exegetical expansions, the Mediterranean figures as a place beyond boundaries where the edges of creation fray. See John Day, *God's Conflict with the Dragon and the Sea: Echoes of a Canaanite Myth in the Old Testament* (Cambridge: Cambridge University Press, 1985), and A. J. Wensinck, *The Ocean in the Literature of the Western Semites* (Amsterdam: Johannes Müller, 1918).

29. The map in Exodus 23:31 identifies bodies of water as three of the borders: "I have placed your borders from the Sea of Reeds to the Sea of the Philistines and from the wilderness to the River." According to Zechariah Kallai, the Egyptian wadi identified as southern border in the map in Numbers 34 "refers to the southernmost branch of the Nile." Kallai, "The Borders of the Land of Canaan and the Land of Israel in the Bible: Territorial Models in Biblical Historiography," *Eretz Israel* 12 (1985): 21–37 (Hebrew).

30. See Isa 40:22, 28; Job 22:7, 26:10; Prov 8:27. See Wensinck, *The Ocean in the Literature of the Western Semites*, 23, and P. S. Alexander, "Geography and the Bible," *The Anchor Bible Dictionary*, vol. 2, ed. David Noel Freedman (New York: Doubleday, 1992), 977–87.

gions stem.³¹ As examples of ancient cartography, the Phoenician bowl found at Praneste and Egyptian papyri depict the encircling ocean or river as a serpent that surrounds the world and swallows its own tail.³² The "earliest literary reference for cartography in early Greece . . . is the description of the shield of Achilles in the *Iliad* of Homer, thought by modern scholars to have been written in the eighth century B.C."³³ Achilles' shield shows "the Ocean River's mighty power girdling round the outmost rim of the welded indestructible shield" (*Iliad* 18.606–8).³⁴ The combination of biblical allusions and parallels in other ancient Mediterranean cultures supports the idea that a mythic view of the world as encompassed by a world ocean/river is the common framework for the two biblical maps.

The designation of seas and rivers as boundaries conveys a sense that the order of the land reflects the structure of the cosmos. The parallel asserts that the land, implicitly associated with the state and the cult, is natural, divine, and as inevitable as creation. The descriptions by Nebuchadnezzar, the Babylonian king, of his imperial influence similarly speak of a span "from the Upper Sea to the Lower Sea."³⁵ Geography, here a subfield of cosmology,

---

31. The map "is universally admitted to be a copy made after 600 B.C." Robert North S.J., *A History of Biblical Map Making* (Wiesbaden: Dr. Ludwig Reichert Verlag, 1979), 13. "*The Babylonian Map of the World* and *The Bilingual Creation of the World by Marduk* demonstrate that Babylonians, at least, believed that a cosmic ocean encircled the continental portion of the earth's surface. The most familiar parts of this ocean were the Upper Sea (Mediterranean) and the Lower Sea (Persian Gulf, Indian Ocean)." Wayne Horowitz, *Mesopotamian Cosmic Geography* (Winona Lake, IN: Eisenbrauns, 1998), 321. For the otherworldly regions, see Millard, "Cartography in the Ancient Near East," 111, and Nanno Marinatos, "The Cosmic Journey of Odysseus," *Numen* 48, no. 4 (2002): 9.

32. See Marinatos, "Cosmic Journey," 4, 11.

33. Millard, "Cartography," 131. See also P. R. Hardie, "Imago Mundi: Cosmological and Ideological Aspects of the Shield of Achilles," *Journal of Hellenic Studies* 105 (1985): 11–31.

34. Homer, *The Iliad*, trans. Robert Fagles (New York: Penguin Books, 1990), 487. The Shield of Heracles in Hesiod's *Shield* 314–15 portrays a similar encircling ocean. See James S. Romm, *The Edges of the Earth in Ancient Thought: Geography, Exploration, and Fiction* (Princeton: Princeton University Press, 1992), 13–14. "The idea of an encircling Ocean was a very old one, perhaps inherited from early Babylonian maps and reinforced by Greek mythology as interpreted by Homer." O. A. W. Dilke, *Greek and Roman Maps* (Baltimore: Johns Hopkins University Press, 1985), 24. Prudence Jones adds a metatextual layer of analysis: "Okeanos forms the outermost rim of the shield and also marks the ultimate boundary of the world. In addition, the description of Okeanos concludes the poetic description of the shield, thus functioning as a boundary within the poem." She further observes that Virgil replaces Okeanos with the rivers of known empires, including the Euphrates, on the shield of Aeneas. With this, "Vergil shows how the Romans have moved beyond imagining the edges of the earth and now touch them with their empire." Jones, *Reading Rivers in Roman Literature and Culture* (Lanham: Lexington Books, 2005), 72, 74.

35. David Stephen Vanderhooft, *The Neo-Babylonian Empire and Babylon in the Later Prophets* (Atlanta: Scholars Press, 1999), 36. Vanderhooft further picks up on some scribal dissent: "the rhetoric of universal hegemony in Nebuchadnezzar's texts is meant to point to his newly expanded imperial power, but the scribes appear to have recognized a disjunction between the rhetoric and reality," 40.

sanctions states through symbols of primordial beginnings.[36] As the boundary between the earth and the sea separates the order of creation from the primordial chaos associated with water, so the land is separated from the threat of the foreign by boundaries of water. The image of the Euphrates, like that of the parallel rivers of Ezekiel 47, further links the land with the Garden of Eden, the river's source (Gen 2:14).[37]

Since the social configuration of Israel claims divine order as its root, the correspondence between the land and creation serves as a necessary precondition for the territorialization of the divine promise. Therefore, even when the borders of the land are construed differently, the east-west axis must span from sea to river in order that the land appear as a microcosm of the cosmos itself. These borders offer geographic proof of the enveloping character of God and state alike. The flexibility concerning which river forms the eastern boundary results from the fact that the mythic morphology prevails over cartographic specifics. The two sets of maps can coexist because their configurations of the land do not conflict, both corresponding to the authorizing cosmological system.

Imperial Geography

While ordering space and orienting conviction, maps also mark locales with various forms of economic and military power. The cartographic impulse, it seems, arose from a dual motivation to demarcate ownership and to survey lands for conquest. Mapping was tied up with kingship, which perpetuated itself through colonization, raids, and temporary alliances with future opponents. "Maps were used to legitimize the reality of conquest and empire. They helped create myths which would assist in the maintenance of the territorial status quo."[38] Ancient monarchs' sense of destiny, which elevated gods, drove

---

36. In Hittite royal ideology, the empire extended from "sea to sea," from the Black Sea in the north to the Mediterranean in the south. Wazana, *All the Boundaries of the Land*, 21. Similarly in the *Aeneid*, the Roman Empire is prophesied to span the shores of the world ocean (1.287, 7.101). The America that extends "from sea to shining sea" in the patriotic song "America the Beautiful" seems to operate under a similar cosmological/imperial premise.

37. In the Babylonian world map, the Euphrates "within the inner circle is portrayed as a band nearly vertical and almost as broad as the ocean." North, *A History of Biblical Map Making*, 20. The equivalence between the Euphrates and the ocean thus appears as a trope in ancient Near Eastern geography. Josephus sees the rivers of Eden and the world ocean as constituting one waterway. In his map, the four rivers of Eden have their source in the "the one river which encircles the whole earth," and branches from the Garden of Eden (*Antiquities* 1.37–39). Philip S. Alexander, "Geography and the Bible: Early Jewish Geography," *The Anchor Bible Dictionary*, vol. 2, ed. David Noel Freedman (New York: Doubleday, 1992), 979.

38. Harley, *The New Nature of Maps*, 57.

urban architecture, and necessitated memorials as well as court literature, also found expression in the measurement of their spatial sovereignty. Territories were thus indexed as part of the royal core, the conquered or unconquerable, or part of an amorphous and unknown beyond. By creating a spectrum of proximity, maps emplaced home between enmity and alliance and brought variegated relationships into a unified spatial system. Perhaps their hyperbolic dimensions and approximations are not a result of inchoate cartographic technologies, but rather are born of the necessity that a range of incongruous relationships fit into a larger scheme.

Although the maps of ancient Israel emulated those of local empires, they attest to a national, rather than an imperial, self-conception. The span of the promised land tends not to be associated with any one leader. Although the very notion of promise evokes ancestral recipients and the Euphrates as a border conjures up the golden age of Solomon, the maps purport to represent future attainments rather than present accomplishments. In the book of Numbers, the map stipulates the place that the People of Israel will reach at the conclusion of their wanderings, and in Ezekiel it functions as an eschatological palimpsest. In their narrative contexts, the Euphrates maps predict the future in some cases (Gen 15:18; Ex 23:31; Deut 1:7, 11:24; Josh 1:3–4) and declare the accomplishment or potential of a Davidic monarch in others (1 Kings 5:1). For the most part, however, the maps do not describe an Israel as it is now, but point to a glorious state to come. Rather than exalting kings, the Jordan maps seem to sideline them in order to promote Priestly ideologies. The Euphrates maps enunciate more support for monarchs, but only for the kind of whom the Deuteronomists approve. Biblical maps then concern the idea of the nation much more than they concern the manifestation of the nation under any one ruler.[39] As we saw in the previous section, they promote a certain mythic worldview in line with a larger Mediterranean/Near Eastern pattern and, as we will see here, they measure Israel's importance in imperial terms.

Moshe Weinfeld accounts for the two sets of maps as the products of divergent views held by different schools of biblical scribes. The Priestly school, with geographic roots in Shiloh, draws the maps in which the Jordan is a boundary, and the Deuteronomistic school, comfortable with the idea of ter-

---

39. Ed Noort distinguishes between Priestly and Deuteronomistic versions of the land on the basis that Priestly texts present the land as the place where God is present and Deuteronomistic texts see the land as "a function of the law," where the Torah can "be lived and where without Torah there would be no land." "'Land' in the Deuteronomistic Tradition," in *Synchronic or Diachronic? A Debate on Method in Old Testament Exegesis*, ed. Johannes C. de Moor (Leiden: E. J. Brill, 1995), 137.

ritorial expansion, extends the border to the Euphrates.⁴⁰ In Weinfeld's opinion, it is the disputed status of Transjordan that leads to the cartographic discrepancy.⁴¹ In other words, the Priestly school does not recognize the east bank as legitimate Israelite territory, while the Deuteronomistic school both recognizes and includes the east bank in its conception of the land. As far as the Shilonite priests are concerned, the Jordan separates pure and impure territory and thus corresponds to other spatial and symbolic borders.⁴² The Deuteronomistic writers, in contrast, are comfortable with a Transjordanian land claim and even associate it with their near-paradigmatic kings David and Solomon (1 Kings 4–5:1).⁴³ Weinfeld does not explain why the Deuteronomists reach all the way to the Euphrates only to absorb the east bank, but I have accounted for why the eastern boundary must be a river. Building on his thesis, I want to push it a bit further and propose that the Jordan maps conceive of ancient Israel in Egyptian imperial terms and that the Euphra-

---

40. In his study on the geography of the Davidic state, Baruch Halpern accounts for the Euphrates map as a result of the intentional vagueness concerning the river at which David established a stela (2 Sam 8:3). The omission of the river's proper name (filled in by the qere as the Euphrates), according to Halpern, aims to give the impression that David's empire reached the Euphrates when in fact it spread only as far as the Jordan or just beyond it. This missing name, in his estimation, is the seed that grows into the Euphrates maps. It seems to me that too much of this explanation rides on a missing term. In addition, the Jordan is never referred to in the text of the Hebrew Bible as "the River Jordan," but only as "Jordan" ירדן or "the Jordan" הירדן. Thus if, as Halpern believes, the stela was set up at the Jordan, then the omission of the proper name is an instance not of ambiguity but of outright deception, since the word "river" before the name of a river always indicates a river other than the Jordan. He makes a similar argument about the lack of specification of the river in 2 Sam 10:16. Again, the river here named cannot be confused with the Jordan since it is called "the river" הנהר, another designation that never refers to the Jordan, but more often than not to the Euphrates. Despite this flaw in his argument, Halpern's discussion of how national memory perpetually recharts its geographic past is compelling. Baruch Halpern, *David's Secret Demons: Messiah, Murderer, Traitor, King* (Grand Rapids, MI: Eerdmans Publishing Company, 2001): 164–259.

41. Nili Wazana stresses the degree to which the Deuteronomistic maps are literary creations that lack definitive or authoritative demarcation lines. She inquires whether the Euphrates with its impressive span should be understood as the northern or the eastern border. Should it be understood as the northern border, then Transjordanian lands would not be included in the Promised Land. She stresses that such geographic generalizations do not operate under strict directional principles. At the same time that I continue to speak of the Euphrates as an eastern border in the Deuteronomistic maps, Wazana's point furthers my sense that the maps follow mythic rather than topographic logic. Wazana, *All the Boundaries of the Land*, 107.

42. "The land of Canaan according to its borders' assigned to the Israelites does not include the eastern side of the Jordan." Moshe Weinfeld, *The Promise of the Land: The Inheritance of the Land of Canaan by the Israelites* (Berkeley: University of California Press, 1993), 54.

43. The connection between David and Transjordan recurs; he rules the terrain (2 Sam 24:5–6), escapes there during Absalom's rebellion (2 Sam 17:16–22, 19:16–20:2), and requests sanctuary for his parents from the king of Moab (1 Sam 22:3–4), perhaps, as alleged in the book of Ruth, because of his Moabite ancestress (Ruth 4:22).

tes maps configure Israel as a counterpart to a Mesopotamian Empire, most likely Babylon.

The maps are different because they measure ancient Israel against particular imperial forces. The Jordan maps adopt an ancient Egyptian model of Canaan, but replace pharaonic rule with Israelite hegemony. The Euphrates maps imagine an Israel mirroring Babylonia with vast stretches of terrain defined by a mighty river that originated with creation (Gen 2:14). The lexicon of empire then helps Israel, caught more often than not in the pull of its tides, to constitute and perpetuate a national identity.[44] Where the Jordan maps inscribe Israel's emergence and differentiation from Egypt in represented space, the Euphrates maps coalesce various sorts of Babylonian memories like Abram's departure and Israel's exile.

Israel in Terms of Egypt

The Jordan map of "the land of Canaan and its borders" in Numbers 34 is, according to scholars such as Benjamin Mazar and Weinfeld, "simply the designation then customary for the Egyptian province in Syria and Eretz-Israel," which underwent a series of changes but was "more or less stabilized by the treaty signed between Ramesses II and the Hittite king in ca. 1270 B.C.E."[45] Biblical writers borrowed the Egyptian concept of Canaan and made it their own. This observation, according to Weinfeld, conveys literary as well as historical meaning.

> The land of Canaan as given to Israel encompasses the same boundaries as the province of Canaan that had been delineated beforehand under the rule of Egypt. Just as God took the Israelites out of Egypt, so he took away the land of Canaan from the hand of Egypt and gave it to Israel. Therefore, "the land of Canaan with its boundaries" in Num. 34 corresponds to the land of Canaan as it was in the days of the Egyptian empire.[46]

By assuming the Egyptian map, the Priestly writers stake a claim in which the land belongs to Israel as reparation for the suffering of slavery. Because the corruption of the Egyptians caused them to lose the land, Israel is assured

---

44. For the intensity and realpolitik of the pull, see Anthony Spalinger, "Egypt and Babylonia: A Survey (c. 620 B.C.–550 B.C)," *Studien zur Ägyptischen Kultur* 5 (1977): 221–44.
45. Benjamin Mazar, *The Early Biblical Period: Historical Studies*, ed. Shmuel Ahituv and Baruch Levine (Jerusalem: Israel Exploration Society, 1986), 115. For a qualification of the theory, see Wazana, *All the Boundaries of the Land*, 154.
46. Weinfeld, *The Promise of the Land*, 64.

as it inherits the territory that, should they corrupt it, Israel too will forfeit the land.

The land due to Israel does not exceed the Egyptian holdings in Canaan nor does the claim diminish according to the outcomes of war and annexation.[47] Political fortunes may rise and fall, the map seems to suggest, but the land remains static. The Egyptian purview is significant since the map is oriented around the relationship between Israel and Egypt.[48] It follows a kind of narrative logic that, in a story about Israel leaving Egypt for the land of Canaan, Canaan would conform to Egyptian standards. The Priestly writers, absorbing an Egyptian Canaan, initially exclude Transjordan in order to put Israel entirely in Canaan's place.[49] When placed in an Israelite context, the Jordan as the eastern border facilitates the central premise of a holy land distinct from other lands. The geographic border further serves Priestly notions of Israel's difference and the need for ritual separation from other peoples. The map as described in Ezekiel shows how the images of a bounded Israel and a holy land become elements of utopian imaginings.

At most narrative junctures Egypt maintains the status of a paradigmatic Other. Why then would Israel derive its map from Egypt of all places? Here I do not wish to debate whether some portion of an Israelite or later-to-become Israelite population or priesthood actually lived or served Egypt or whether memories of an anti-Egyptian uprising in the mountains of Canaan can be recovered from the biblical texts. Instead, I want to make an alternative proposal that biblical writers, influenced and impressed by imperial Egypt, borrowed its map as a means of asserting the greatness of Israel. I am not stressing that empires set the parameters of political discourse—although this is certainly true—, but rather arguing that those who conceived and constructed ancient Israel did so with elements derived from empire. Yet Israel, for the most part, neither sought to become an empire nor saw itself as such. Disdain for Egypt and Babylon, the Assyrians and the Hittites runs through narrative and proph-

---

47. Moshe Weinfeld, "The Extent of the Promised Land—the Status of Transjordan," in *Das Land Israel in biblischer Zeit* (Göttingen: Vandenhoeck & Ruprecht, 1983), 65. Also, "The priestly tradition consistently adheres to the map of Canaan as it existed up to the thirteenth century and does not admit to its slightest alteration in the light of subsequent events." Milgrom, *The JPS Torah Commentary to Numbers*, 502.

48. The ambivalent status of Transjordan may also be an inheritance from Egypt. According to Zechariah Kallai, at least two Transjordanian cities (Pahal and Zaphon) and potentially Transjordan as a whole are claimed in some Egyptian sources and not claimed in others. See Kallai, "The Borders of the Land of Canaan and the Land of Israel in the Bible: Territorial Models in Biblical Historiography," *Eretz Israel* 12 (1985): 28 (Hebrew).

49. "Since Egyptian records never mention the Gilead or southern Transjordan—archaeology informs us that they were unsettled until the thirteenth century—it is clear that the Jordan was the eastern border of Egyptian Canaan." Milgrom, *The JPS Torah Commentary to Numbers*, 501.

ecy alike. This clever turn is a wonderful example of adaptation—cooptation even—in which biblical writers access the imperial lexicon in order to portray a nonimperial but nonetheless momentous and mighty nation of Israel. What Israel lacks in territory, it makes up in narrative.[50]

My argument here operates on different levels. As I have shown, the Egyptian map of Canaan became the map of Israel's land both because Israel's state institutions were influenced by those of Egypt and because a tightly circumscribed land embodied Priestly ideologies of Israel's ethnic and religious distinction.[51] Perpetuated by Priestly schools, the map persisted and placed Israel in an imperial context from which it could maneuver among empires. The survival of the kingdoms of Israel and Judah depended on the ability to correctly appraise the ascendant empire and to position diplomacy, the military, and the economy accordingly. Although representative of a pre-Israelite period, Egypt's map of Canaan helped launch the territorial idea of Israel and remained a relevant point of reference. The question of Egypt's strength combined with speculation about whether it could match that of Babylonia was a particularly vital issue in the latter days of the Judean monarchy.[52] Such speculation in tandem with the ongoing reality of vacillating alliances led to the coexistence and preservation of the two maps.

The bidirectional self-figuration makes sense because "ancient Israel historically developed, came to an end, and was reconstituted within the bipolar system of political contestation in the Fertile Crescent between Egypt, on the one hand, and various Mesopotamian and Syrian states, on the other."[53] Facing Egypt at certain times and Mesopotamia at others arose from the need to manage the potential influence and military menace from either direction.

---

50. The way in which the maps emulate empire while representing an Israel that is "strategically an insurgent counter-appeal" indicates an instance of mimicry. Homi Bhabha, "Of Mimicry and Man: The Ambivalence of Colonial Discourse," *The Location of Culture* (London: Routledge, 1994), 85–92.

51. For the argument of Egyptian influence on Israel, see R. J. Williams, "'A People Come Out of Egypt': An Egyptologist Looks at the Old Testament," in *Congress Volume: Edinburgh 1974*, VTSup 28 (Leiden: Brill, 1975), 231–52. For the argument against such strong influence, see Donald B. Redford, "Specter or Reality? The Question of Egyptian Influence on Israel of the Monarchy," *Egypt, Canaan, and Israel in Ancient Times* (Princeton: Princeton University Press, 1992), 365–94. Redford does, however, concede to parallel geographic divisions "between Solomon's twelve districts, designated one per month to supply the court with food, and the Egyptian practice of dividing the tax base into twelve parts to meet an ongoing budgetary requirement on a calendrical basis," 372.

52. This question is also at stake in the "riddle and parable" of Ezekiel 17.

53. F. V. Greifenhagen, *Egypt on the Pentateuch's Ideological Map: Constructing Biblical Israel's Identity* (Sheffield: Sheffield Academic Press, 2002), 3, rephrasing A. Malamat, "The Kingdom of Judah between Egypt and Babylon: A Small State within a Great Power Confrontation," in *Text and Context: Old Testament and Semitic Studies for F. C. Fensham*, ed. W. Claassen (Sheffield: Sheffield Academic Press, 1988), 117–29.

Mapping Israel in grandiose terms can also be seen as a strategy of resistance, a refusal to admit to the diminishment of the nation amidst the loss of territory or to be defined solely by the maps of emperors and generals. Rather than accept a peripheral placement or no notice on someone else's map, Israel appropriated the maps of empire and put itself at the center. This kind of big thinking impacts the representational power of the Jordan. In the biblical maps, the Jordan stands on a par with the Nile and the Euphrates. Because the Nile and the Euphrates respectively signify Egypt and Babylonia, the correlation enables the Jordan and the People of Israel to assume symbolic import incommensurate with their size.

In addition to emulating empire, the maps speak to a complicated sense of origin. Alongside the tenet of a homeland west of the Jordan are concessions to ancestral beginnings east of the Euphrates, national burgeoning in Egypt, and the inevitable diasporic revisitation of both places. A late chronology could offer an easy detour by proposing that the maps, along with most of the Hebrew Bible, were written in exile in Babylon or during the homecoming sanctioned by Persia. According to this line of reasoning, the maps reflect the places to which Israelite/Judean communities and their scribes fled or were exiled rather than the empires with which Israel contended in an earlier stage.[54]

One can say with certainty that Egypt remains on the minds of biblical writers from different periods as an influence and a threat. Doubts have been raised about the degree of influence and even the existence of the Egyptian map so central to Mazar's and Weinfeld's argument.[55] Even if such a source document does not serve as the paradigm, the idea of an enslaving Egypt plays a role in the Priestly map as it appears in the book of Numbers. Just as the Reed Sea enacts a separation between Israel and Egypt, so the Jordan River distances Israel from the potential contamination of mixing with the strange peoples east of the Jordan (Num 25).[56] Nili Wazana advises understanding both the Priestly and the Deuteronomistic maps as literary creations, not accounts of where the People of Israel actually lived. I would add that because the borders shifted according to changing political and historical conditions, the Priestly and the Deuteronomistic maps are best seen in terms of distinct theories of the nation. By listing coordinates and emphasizing clear bound-

---

54. Along with the exile to Babylon, Judeans seem to have fled to Egypt in the wake of the Babylonian attacks of 597 and 586. Jeremiah refers to Judean communities in Egypt (Jer 40–43, 44:1).

55. For additional influences on the Priestly map, see Wazana, *All the Boundaries*, 127.

56. Wazana emphasizes the degree to which the Priestly map articulated in Ezekiel's final vision "changes the historical location of the two and a half (Transjordanian) tribes, moving them west from the east side of the Jordan. It thereby refuses to recognize the legitimacy of any settlement east of the Jordan." Wazana, *All the Boundaries*, 173.

aries, the Priestly map imagines Israel as upholding ritual categories in its pattern of settlement. Even when the strangers in Israel's midst command status and rights, Priestly texts emphasize Israel's distinction in their representation of space.

Israel in Terms of Babylonia

The Euphrates maps seem to reflect the political climate during the Neo-Babylonian period, when its threat to the kingdom of Judah peaked and alliance with the Transjordanian states was among the self-protecting tactics (Jer 27:2–3). The geographic schema within the Babylonian *Mappa Mundi* and Babylon Topography likely emerged from the Neo-Babylonian expansionist heyday under Nebuchadnezzar, who vaunted himself as "the protector of all humanity" and his capital as "the economic and administrative center of the world."[57] The Euphrates maps mimic such schemas and acknowledge Babylonian hegemony east of the river while situating Israel as a kind of mirror image just to the west. With Babylon on the rise, the idea of Israel reaching to the Babylonian shore as a mighty counterpart would both lessen the fear of the growing empire and strengthen, depending on the moment, either allegiance or oppositional resolve.

The Euphrates map can be explained as such a technique introduced by Deuteronomic (Dtr_1) scribes and reproduced in later versions by their successors, the Deuteronomists.[58] In terms of national sentiment, the Deuteronomistic maps present an Israel with military and territorial ambitions. Wazana suggests that while the Deuteronomistic writers recognize the Jordan as the border between Israel and the Transjordanian peoples of Edom, Moab, and Ammon (Deut 2), they also advance a utopian idea of God's eternal rule over a vast tract of land. On this count, she sees the adaptation of imperial language as a means of expressing divine sovereignty. A chapter like Joshua 1, in which the Jordan clearly marks a border at the same time that God promises Israel land extending to the Euphrates, according to Wazana, sanctions the dispossession of a limited number of people at the same time that it asserts a borderless future of divine rule. I would characterize it somewhat differently as the use of imperial geographic rhetoric in order to define the nation along with a concession to the nation's current restrictions.

---

57. Bill T. Arnold, *Who Were the Babylonians?* (Atlanta: Society of Biblical Literature, 2004), 96.

58. Wazana specifically relates the language of the Deuteronomistic maps to Neo-Assyrian imperial propaganda. Her account of a Neo-Assyrian origin is quite plausible, but she also speaks of a more general Mesopotamian orientation evident in the maps. Because Babylonia ultimately conveys more symbolic import in the edited Hebrew Bible, I would argue that the geographic rhetoric—which may indeed derive from Neo-Assyrian sources—adapts to the ascendancy of Babylonia. See Wazana, *All the Boundaries of the Land*, 119–22.

Where the Jordan maps appropriate the imperial terrain of Egypt, the Euphrates maps assimilate Israel into the Babylonian conception of the world. Neo-Babylonian imperial geography referred to the sweep of land west of the Euphrates as *Eber Nari* or Transeuphrates.[59] This geography did not account for all of the differing peoples situated between the Euphrates and the Mediterranean coast, but rather marked them as conquered people distinct from the "true" Babylonians east of the Euphrates. The biblical writers seem to pick up on this map and formulate it from their own perspective. The amorphous land west of Euphrates becomes the site of the great Hebrew nation and along the way its subjugation to Babylonia is effaced. In another twist, the right and the wrong side of the river are reconfigured. In the biblical maps, a position west of the Euphrates marks a position of blessing rather than shame. Not only is the Babylonian Empire scaled down in the biblical writings, but since "the vanquished wrote the history" they also "produced perhaps the most influential portrait of Babylon to survive antiquity."[60] Babylon is most remembered as Judah's archetypal foe. For the empires that inherited a biblical legacy, the Judean hierarchy of west over east certainly trumped the Babylonian hierarchy of east above west.

Israel between Egypt and Babylonia
The Euphrates maps of Deuteronomy 1:7 and 11:24, which fail to provide a southern border, perhaps leave the question of Egyptian influence open in the wake of Babylonia's rise. In other Euphrates maps as well as in Jordan maps, Egypt remains a point of reference abutting Israel at the southern border. Some maps reach to the Nile (Gen 15:18); some more specifically recall the Exodus by using the Reed Sea as a marker (Ex 23:31); and others halt more generally at the desert (Josh 1:4) or "the border of Egypt" (1 Kings 5:1), or more specifically at the wadi of Egypt (Josh 15:4, 47). It is possible that the omission of an Egyptian boundary in Deuteronomy speaks to perceived Egyptian quiescence.

With the rise of the Neo-Babylonian Empire and the Egyptian challengers of the Twenty-Sixth Dynasty, "the small state of Judah, located at the particularly sensitive crossroads linking Asia and Africa, was influenced more than ever before by the international power system, now that the kingdom's actual

---

59. This is the case, for example, in the Etemenanki cylinder that "delineates the cities and regions that contributed corvée laborers or raw materials for work on Marduk's ziggurat in Babylon." Vanderhooft, *The Neo-Babylonian Empire and Babylon in the Later Prophets*, 36, based on the edition of Weissbach, *Die Inschriften Nebuchadnezars II im Wâdî Brîsâ und am Nahr el-Kelb* (Leipzig: J. C. Hinrichs, 1906), 44–48.

60. Vanderhooft, *The Neo-Babylonian Empire and Babylon in the Later Prophets* (Atlanta: Scholars Press, 1999), 5.

existence was at stake."⁶¹ The enmity between Egypt and Babylonia arose because of Egyptian participation in the Assyrian challenge to Babylonian predominance.⁶² With the decline of Assyria, the antagonistic relationship played out on various fronts including the kingdom of Judah, which became a site of contestation as well as a bellwether for the two. Josiah, the Deuteronomic hero, lost his life in the battle to impede the Pharaoh Necho II from reaching the Euphrates (2 Kings 23:29–30; 2 Chr 35:20–27). Donald Redford interprets Josiah's actions as proof that he correctly understood the defeat of Assyria and Babylonian advance as "the wave of the future."⁶³ Necho's setback at Megiddo led to Egyptian hostility and an unchecked Babylonia. The pharaoh exacted his vengeance by imprisoning Jehoahaz and instating Eliakim, reinvented as vassal to the pharaoh rather than son to Josiah, with the name Jehoiakim (2 Chr 36:1–4).⁶⁴ Against prophetic admonition (Ezek 29:6–7, 17:17; Jer 37:5–11), the Judean leadership remained invested in the Egyptian alliance. Not even Nebuchadnezzar's successful march on Jerusalem convinced Jehioakim to resign himself to subordinate status. Following an unsuccessful Babylonian bid for Egypt, he led a successful rebellion that further imperiled Judah (2 Kings 24:1). Jehoiachin, the next king, surrendered to Babylon and allowed Nebuchadnezzar's troops to skim off the treasures and notables of Jerusalem and carry them into exile. At this time, Nebuchadnezzar had his opportunity to name and install a vassal, yet Zedekiah, nee Mattianiah, still looked to Egypt for signs of Babylonian weakness.⁶⁵ The Babylonian eclipse of the Egyptian Empire combined with the false sense of Judean security as an Egyptian protectorate led to the destruction of Jerusalem.

The struggle for hegemony between Egypt and Babylonia took form in the

---

61. Malamat, "Kingdom of Judah," 119. "The latter years of the Judean monarchy were dominated internationally by the collapse of the Assyrian empire and the emergence of a bipolar system of confrontation between the rising Neo-Babylonian empire and the Saite or 26th dynasty of Egypt." Greifenhagen, *Egypt on the Pentateuch's Ideological Map*, 249.

62. "In the late summer of 616 B.C. as Nabopolasser and his troops ravaged the land of the middle Euphrates, an Egyptian expeditionary force appeared and in concert with Assyrian forces pursued the retiring Babylonians partway down the Euphrates." Donald Redford, "Egypt and the Fall of Judah," *Egypt, Canaan, and Israel in Ancient Times* (Princeton: Princeton University Press, 1992), 430–69.

63. Redford, "Egypt and the Fall of Judah," 448.

64. The Judean population was itself divided along pro-Egyptian versus pro-Babylonian parties. "Egypt and the Fall of Judah," Redford, 449.

65. "Nebuchadnezzar's failure to invade Egypt in 601 only underscored the feeling that the supremacy of Babylon under the Chaldeans was a passing phenomenon . . . Consequently, the 'triumphal progress' of Psammetichus II to Palestine in 592, though basically a peaceful journey was intended . . . 'to galvanize his allies and subjects in hither Asia by his presence against the Babylonian menace.'" Mordechai Cogan and Hayim Tadmor, *II Kings: A New Translation with Introduction and Commentary* (Garden City, NY: Doubleday, 1988), 323.

battle for the land from the Egyptian Brook[66] to the River Euphrates (2 Kings 24:7), the very territory at stake in the Euphrates maps.[67] In the midst of imperial land grabs, the maps define Judah/Israel both in terms and in place of the dueling powers. These portraits do not express imperial aspiration so much as function as an ideological safeguard in the face of attenuating territory and autonomy. If the armies of Babylonia and Egypt could not be ousted from the land, then at least they could be confined behind their own waterways in symbolic renderings. The simultaneous dynamic of using its hegemonic tropes in order to negate empire has additional manifestations. Weinfeld recognizes the attributes of Assyrian emperors and characteristics of Mesopotamian royalty in the Isaianic depiction of the ideal king who will reign in the redemptive era to come.[68] Even visions of a divine kingdom arising from the humbled ground of annexed territory are spun of imperial language. Without images of an oppressor's glory, utopia has no legs on which to stand. The Euphrates maps offer such a utopia on a spatial plain. Beyond historical specifics, the two sets of maps situate Israel in the midst of a pull between Mesopotamia to the east and Egypt to the south and employ the standards of imperial cartography in order to map a nation.

The insistence on borders marking the land from neighboring peoples and engulfing empires brings into being a contiguous national space. Once the space becomes emblazoned in collective memory and enlists adherents, it need no longer correspond to actual dimensions of sovereignty. As I have discussed, symbolic potency and mythic allusion trump topographic accuracy from the beginning. In the case of Israel, geographic borders signified practices and rituals that maintained a manner of separation within an environment of interaction. The two maps communicate the complicated message that we are part of Egypt, Babylonia, and the empires yet to rise; however, we belong to a group whose uniqueness is indelible. One might think that the double message combined with the alternate identities outlined by two maps would be too complicated to remember and transmit or that the more definite the sense of home, the easier its preservation. Israel, however, attests to the opposite. Jan Assmann remarks that during the Persian period Israel alone

---

66. Its location, according to Paul Hooker, cannot be determined solely from biblical texts. Hooker, "The Location of the Brook of Egypt," in *History and Interpretation: Essays in Honour of John H. Hayes*, ed. M. P. Graham, W. P. Brown, and J. K. Kuan (Sheffield: JSOT Press, 2009), 203.

67. There is some evidence external to the Bible that suggests that Nebuchadnezzar may indeed have conquered this territory. Vanderhooft, *The Neo-Babylonian Empire and Babylon in the Latter Prophets*, 98; D. J. Wiseman, *Chronicles of the Chaldaean Kings* (London: British Museum, 1956), 69.

68. Moshe Weinfeld, "The Protest against Imperialism in Ancient Israelite Prophecy," in *The Origins and Diversity of Axial Age Civilizations*, ed. S. N. Eisenstadt (Albany: State University of New York Press, 1986), 181–82.

emerged as a "nation" "able to separate itself from the outside world and create an internal community entirely independently of political and territorial ties."[69] I suggest that this was the case in imperial epochs prior to the Persian. It seems then that the more fluid the sense of home, the easier it is to establish discrete community structures both at home and elsewhere.

---

69. Jan Assmann, "Five Stages on the Road to Canon: Tradition and Written Culture in Ancient Israel and Early Judaism," in *Religion and Cultural Memory*, trans. Rodney Livingstone (Stanford: Stanford University Press, 2006), 72.

Chapter Two

ISRAEL AND MOAB AS NATION AND ANTI-NATION

Moab shall become like Sodom and the People of Ammon like Gomorrah
—Zephaniah 2:9

It is not through maps that the banks of the Jordan become vivid. The opposing shores assume qualities first and foremost through the characters ascribed to them. Since it is a revolving cast, there is a sense in which east bank and west bank teams develop. The teams are premised on a spatial code—every one on each bank displays certain characteristics irrespective of time—in which location confers certain enduring "ethnic" qualities. The spatial code seems to trump genealogy because blood brothers find themselves on opposite sides of the Jordan. Every battle fought across the Jordan then figures as a civil war. I propose that by looking at biblical texts about Israel's neighbor nation Moab, we can see how location rather than filiation determines identity. I argue not only that one's spatial location is a stronger determinant than kinship or ethnic affiliation, but also that when location is represented in story, symbol, or map it bespeaks ideological position more than latitude and longitude. Said differently, people are associated with a place not necessarily because they are actually from that place but because they exhibit characteristics that support or underwrite what that place is supposed to be. In this chapter we will see how gender deviance in the Bible is categorized as Moabite behavior. This means that Moabites are depicted as deviants, and also that political passivity by men and political assertion by women lands them in that space the Bible calls Moab.

Abraham and Lot: The Founding Fathers

The stories of Abraham and Lot, Isaac and Ishmael, Jacob and Esau establish the mythic bond between rivals. Abraham is the archetypal crosser of rivers who walks across the Euphrates, the Orontes, the Jordan, and the River of Egypt in the name of his journey closer to God (Joshua 24:2–4). The name for the people he establishes, *Ivrim*/Hebrews, may itself be based on the verb עבר, meaning "to cross (a river)." Amidst his wandering, Abraham also founds a homeland in the mountains and deserts of Canaan, between the River of Egypt and the Jordan. He sets out on his travels from Haran (his father Terah having migrated there from Ur of the Chaldees with the family) with Sarah his wife and his nephew Lot. A border with Egypt comes into relief after Abraham seeks refuge there during one of the frequent Canaanite famines.

Abraham first signals the difference between Canaan and Egypt when he asks his wife to play the role of his sister. Since Sarah is so beautiful and Egypt a locale of fierce appetites, she must play a sororal role in order to spare his life. This is a setup in which Abraham prostitutes Sarah to the Egyptians: word of Sarah's beauty builds to the point that she is brought to the palace of Pharaoh with Abraham profiting to the tune of "sheep, cattle, donkeys, manservants and maidservants, mules and camels" (Gen 12:16). God plays a role in the ring by plaguing Pharaoh and his household for taking Abraham's wife, meaning that Pharaoh speaks correctly when he asks why Abraham deceived him and then let God exact vengeance. It is clear when Abraham goes back to the Negev, the desert north of Egypt, with his wife and his stockpile that he won't be going back to Egypt. The fact that he wronged the Pharaoh is what sets the line.[1]

The family trouble reverberates when Abraham returns home, giving rise to two schisms with geographic implications.[2] The first fraction transpires

---

1. F. Volker Greifenhagen suggests that the episode is part of a strategy of covering up Israel's origins in Egypt. See "The Pentateuch and the Origins of Israel: Ideological Linkage Around the Master Narrative," *Voyages in Uncharted Waters: Essays on the Theory and Practice of Biblical Interpretation in Honor of David Jobling*, ed. Wesley J. Bergen and Armin Siedlecki (Sheffield: Sheffield Phoenix Press, 2006), 110–22.

2. The fraught relationship between Sarah and Hagar is transmitted to their sons, Isaac and Ishmael. Since Isaac and Ishmael are separated by a more amorphous, frontier-like desert, their relationship will not be considered in the body of my text. The fact that there is no Jordan between them communicates something specific in literary and, likely, historical terms. Ishmael is not emplaced east of the Jordan because he is not resolutely outside of the covenant. As Abraham's son Ishmael is circumcised (Gen 17:25–26) and promised paternity of of twelve tribal chiefs and one large nation (17:20) at the same time that he is rejected as Abraham's heir, maternity is introduced as a key aspect

between Abraham and his nephew Lot, who has been with him since Abraham's father migrated from Chaldea to Haran. After the Egyptian adventure, an Abraham rich in cattle and coin returns to Beth-El, where he first built an altar. On the basis of his flocks and tents, Lot is also a chieftain of sorts. The text seems to speak euphemistically when it reports, "The land could not support them staying together" (Gen 13:6) and more bluntly when it confesses to a dispute between the shepherds working for Lot and Abraham (Gen 13:7). Rather than running away, Abraham makes a treaty with his kinsman. He offers Lot the land in a direction of his choosing, vowing to make do with the remainder. Lot follows his eyes to the lush Jordan Valley, well watered "like the Garden of God and the land of Egypt" (13:10). Through the analogy, the banks of the valley and the east bank in particular are associated with appetite. As the Garden of God was the place of ease and fertility and as Egypt offers a bounty due to the powers of the swelling Nile, so the Jordan Valley supplies bounty without demanding arduous labors. According to the typology, such ripeness has corrupted the inhabitants of the valley, particularly the citizens of Sodom and Gomorrah. Lot settles among them, so when Abraham turns and walks the other way, the land of God's promise is defined in contrast: "All the land that you see, I give to you and your descendants for eternity" (13:15). The eternal nature of this promise means that the split between Abraham and Lot will be reenacted in every act by which the descendants of Abraham define themselves.[3] The Jordan (or at least the Jordan Valley) comes to distinguish the founder of Israel from his closest relation.[4]

In spite of the geographic divide, Abraham and Lot remain connected. Abraham releases Lot from captivity following a war in which Sodom, Gomorrah, and the other Cities of the Plain are looted (Gen 14) and advocates on his behalf after God announces to Abraham His intention to annihilate the

---

of election. Neither quite in nor out of covenant, Ishmael's descendants wander in and out of the narrative (see, for example, Gen 37:25) without the symbolic drama of crossing the Jordan. Ishmael, by definition, is a man of the desert and indeed the eighth through sixth centuries B.C.E. saw the rise of Ishmaelite tribes in the Negev desert. Ronald Hendel, *Remembering Abraham: Culture, Memory and History in the Hebrew Bible* (New York: Oxford University Press, 2005), 47.

3. For the ways in which remembering Abraham involves acts that mark geography, the body, and genealogy, see Hendel, *Remembering Abraham*, 31–36,

4. David Jobling sees the partition of the east and west as part of a foundational Israelite self-definition. "According to the foundation myth of Genesis 13, Abram and Lot divided the Promised Land into west and east. The details of this division are not quite clear but in mythic logic it would tend to coincide with the best-known east-west division, between Cis and Transjordan. Lot's inheritance would then simply be (some part of) Transjordan, and would have passed to Moab and Ammon in the continuation of the myth in Gen. 19:30–38." Jobling, "'The Jordan a Boundary': Transjordan in Israel's Ideological Geography," *The Sense of Biblical Narrative II: Structural Analysis in the Hebrew Bible* (Sheffield: JSOT Press, 1986), 113.

cities of the Jordan Valley due to their wickedness (18:22–33). By this point, Abraham has been through his own share of troubles: Sarah's barrenness has been prolonged and a surrogacy engineered through Hagar the Egyptian servant has backfired. At the same time, Abraham has been brought closer to God by being party to a covenant. Marked through circumcision, this covenant designates Abraham as the progenitor of nations and kingdoms (Gen 17:6), establishes an eternal bond between Abraham's descendants and God (17:7), and transfers the site of Abraham's sojourns into a perpetual inheritance (17:8). The triangulation of God-person-land means that a descendant of Abraham such as Ishmael, son of Hagar, or a relative like Lot can be related by blood but not by territory. In order to be a party to the covenant, one must stem from the correct hereditary line, undergo circumcision, and live within a particular set of borders. Lot is already disqualified by not being in the right place when Abraham performs circumcision on the members of his household.

The ideal recipient of covenant comes on late in the scene after three angels appear at Abraham's tent to promise that Sarah will give birth to an heir. Following some negotiation between Sarah and God concerning the probability of a miraculous birth by a postmenopausal woman, two angels go on to bring destruction to Sodom while God tarries in order to inform Abraham of the plan. Abraham enters into a more elaborate negotiation with God in which he bargains in decrements of fives and tens to spare Sodom and Gomorrah in the name of ten righteous people. It would seem that a universal compassion for the people of the Jordan River Valley accompanies particular concern for his nephew.

Divine-human dialogue regarding the decremental minimum of justice in the world does not set the course of the plot. This is done rather by the visit of the two angels to Sodom in the evening.[5] As they approach, Lot, a figure of the margins, rushes to them from his position at Sodom's gate.[6] Like Abraham when the angels appeared at his tent, Lot extends hospitality to the strangers. Refusing to let them sleep in the street, he brings them home and invites them to a meal of drink and bread. Compared with Abraham's repast of cakes,

---

5. For the sense of danger inherent in this timing, see Weston W. Fields, "The Motif 'Night as Danger' Associated with Three Biblical Destruction Narratives," *Sha'arei Talmon: Studies in the Bible, Qumran, and the Ancient Near East Presented to Shemaryahu Talmon*, ed. E. Tov and M. Fishbane (Winona Lake, IN: Eisenbrauns, 1992), 17–32.

6. Lot appears as a marginal figure because he chooses to live in the borderlands of the Jordan River Valley. When he first arrives in Sodom, he dwells at the city's edge (Gen 13:12). In the realm of genealogy, Lot is marginal in terms of being outside of the covenant. On the marginality of Lot, see Lyn M. Bechtel, "A Feminist Reading of Genesis 19:1–11," *Genesis: A Feminist Companion to the Bible*. Second Series, ed. Athalya Brenner (Sheffield: Sheffield Academic Press, 1998), 108–29.

curds, and meat, Lot's offerings are sparse with the emphasis on strong drink communicating something about how they dine in Sodom. No sooner do the angels lay down their heads than the people of Sodom surround Lot's house, demanding that he turn over his guests to abuse.

One can't pass through Sodom without addressing the crime severe enough to warrant annihilation. The history of interpretation bears out the variant ways of understanding the transgression and since its nature lends characteristics to the east bank, it is worth exploring what goes wrong. The men of Sodom surround Lot's home, emphasizing how fragile a boundary it is between us and the sudden, perhaps latent, hostility of our neighbors.[7] "Bring the men out to us," they demand, "so that we may *know* them" (Gen 19:5). The knowledge that the men have in mind is really about power: they want to show their power over these strangers by humiliating them through penetration. The ancient equation of humiliation with penetration has the most to do with subjugating women and defining them as always passive even when in action, but it is here proposed as a mode of revenge for strangers who have penetrated the city.[8] As Lot goes out to face them, the door of his home slams behind him in a futile attempt to reinforce the boundary between his private house and the public arena. He proposes that the crowd leave his guests unmolested and make do with his virgin daughters. The suggested substitution only enflames the crowd—Lot's "judgment" of them emphasizes the degree to which he himself is a stranger—so they move to abuse him more violently than his guests (19:9). The logic by which Lot dispenses with his daughters in the name of protecting the guests who "have come under the shelter of my roof" (19:8, JPS) coheres with Mieke Bal's observation that men forge alliances by exchanging and conquering the bodies of women.

Next to Abraham's successful bartering with God in the name of justice, Lot's failed bargain reads as pathetic parody. It doesn't work—Lot's daughters' bodies are not the foreign ones over which the Sodomites would like to have power and, as is made clear when Lot later speaks to his sons-in-law, the virginity claim is a lie. The angels pull Lot back into the house and strike the men outside with blindness, leaving them groping for the door at this last moment in their lives. Current debates about the status of homosexuality in the Bible

---

7. Speiser translates נסבו as "surround" with the preposition על indicating "hostile intent" (*ABD Genesis*, 301). According to G. R. Driver, על may qualify נסבו (Jer 4:17; cp. Gen 19:4; 2 Chr 18:31; Isa 7:8; Job 16:13). Weston Fields observes that "there are only three places in the Hebrew Bible in which the niphal of סבב is used in the sense of 'surround, close round upon': Gen. 19:4; Judg. 19:22 and Joshua 7:9." Fields, *Sodom and Gomorrah*, 75.

8. Susan Niditch understands the proposed penetration as "an active, aggressive form of inhospitality." Niditch, "The 'Sodomite' Theme in Judges 19–20: Family, Community, and Social Disintegration," *Catholic Biblical Quarterly* 44 (1982): 369.

make it necessary to point out that the angels speak of an "outcry gone up to God" from Sodom that sent them to destroy the city in the first place (Gen 19:13). While the collective desire to subject foreigners to humiliation evinces the lack of ten righteous people, it is not the direct cause of Sodom's destruction. The cause instead lies in the outcry.

Although several Hebrew Bible texts reference the corruption and destruction of Sodom (Deut 29:23, 32:32; Isa 1:10, 3:9, 13:19; Jer 23:14, 49:18, 50:40; Amos 4:11; Zeph 2:9; Lam 4:6), only Ezekiel 16 speaks to what, specifically, brought about its downfall. "Only this was the sin of your sister Sodom: arrogance! She and her daughters had plenty of bread and untroubled tranquility; yet she did not support the poor and the needy" (Ezek 16:49). Sodom's abundance is understood as perverting, not in sexual terms (this, according to Ezekiel 16, is Israel's fatal flaw), but insofar as it leads to excess rather than to generosity. One can imagine how ascribing excess and selfishness to the peoples lucky enough to dwell in a river valley could provide the comfort of righteousness to the mountain-dwelling Israelites. Addressing the issue of interpretation, if we want to stay true to a biblical contextual meaning, then we would call the wealthy who do not help the poor, not practitioners of anal sex, Sodomites.

The history of early biblical interpretation supports such a reading. "In the Apocrypha, the sin of Sodom is understood to be the lack of hospitality."[9] The Jewish-Roman authors Josephus and Philo took the contrast between Abraham and the people of Sodom to indicate a position of Jewish nonparticipation in the practice of pederasty. Going a bit further in his imaginings, Philo writes that even pederastic norms in which an older, "active" man can only penetrate a younger, "passive" boy were violated in Sodom: "Not only in their mad lust for women did they violate the marriages of their neighbors, but men mounted males without respect for the sex nature which the active partner shares with the passive" (*On Abraham* 135–36). Midrash Bereshith Rabbah combines the Sodomites' sexual aggression with other sins of greed and violence. Perhaps the most vivid narrative about Sodomite selfishness tells of two girls who go to draw water from a well. When one girl explains that her family is out of food, the other exchanges her friend's water jug for a jug of flour. When the city hears of this, the girl who extended the gift is taken and burned to death (BR 49.3). Another account introduces the sexual element: "This is what the Sodomites had stipulated among themselves: 'As to any wayfarer who comes here, we shall have sexual relations with him and take away his money'" (BR 50.3, Neusner translation). This interpretation

---

9. Martti Nissinen, *Homoeroticism in the Biblical World: A Historical Perspective*, trans. Kirsi Stjerna (Minneapolis: Fortress Press, 1998), 47.

contends with the multiple registers of the Sodomites' proposed crime: it manifests hostility toward strangers, involves sexual subjugation, and springs from a rapacious appetite.

As traced by Mark Jordan, a similar interpretive trajectory is evident in early Christianity. "There is no text of the Christian Bible that determines the reading of Sodom as a story about same-sex copulation."[10] With provisions for categorizing correct and incorrect desires, the Church Fathers speak to other excesses in Sodom. Jerome lists the sins of Sodom as pride, bloatedness, and brazenness; Ambrose decries its luxury and lust, and Augustine accounts for the "debaucheries in men" as "a symptom of the madness of their fleshly appetites."[11] In Jordan's genealogy, the term "sodomy" and this species of sin is largely invented in Peter Damian's "Book of Gomorrah" addressed to Pope Leo IX and relates more to the atmosphere of monasteries than to the Jordan River Valley. The Sodom of Genesis 19 then is hardly a city of homosexuals, but rather a kind of shadowlands where all codes are inverted. This aspect is most important in regards to the Jordan River because it is from this context that Israel's rival neighbors, Moab and Ammon, emerge.

Lot's Daughters: Founding Mothers
With the press of imminent destruction, the escape of Lot and his family from Sodom is narrated at a hurried, accelerated pace. Lot's claim to his daughters' virginity is exposed when he tries to persuade his Sodomite "sons-in-law, who had taken his daughters" to flee with them (Gen 19:14). The sons-in-law laugh off the matter and even Lot moves reluctantly despite repeated admonition from the angels. Even in its final hour, there is something addictive about Sodom. Lot, his wife, and his daughters need to be dragged out by the angels; Lot begs them to let him find refuge in a nearby city rather than in the mountains and, most memorably, Lot's wife casts a backward glance only to be forever frozen as a pillar of salt. In the end, Lot is saved as a result of God's covenant with Abraham. Adhering to the reciprocal nature of covenant, God "remembers" Abraham amidst His fiery assault on the Jordan River Valley and sends Lot to a dreaded mountain destination (Gen 19:29–30).

Having witnessed chaos unleashed from above and finding themselves in a mountain cave, Lot's daughters logically reason that they are the remnant of humanity. After God wipes out the social order, why should taboos remain in place? The older daughter confers with her sister. "Our father is old and there is no man left in the world who can have sex with us in the natural way. Let us

---

10. Mark Jordan, *The Invention of Sodomy in Christian Theology* (Chicago: University of Chicago Press, 1997), 32

11. Jordan, *The Invention of Sodomy*, 35.

then get our father drunk on wine and sleep with him so that we can stay alive through our father's seed" (Gen 19:31–32). Abraham's covenant enables the legitimate concern of Lot's eldest daughter with the perpetuation of life, yet, because she stands outside of the covenant, her concern is cast as perverse.

Lot's passivity and love of drink, suggested in other episodes, is here a central theme. In the three spaces through which Lot passes during the destruction—home, road, cave—he is stripped of some aspect of his identity. When the angels save him from the crowd, Lot ceases to be the protective host; when his wife turns to salt on the road, he is no longer husband; and when his daughters lie with him, he can hardly be considered a father. Lot is thus entirely negated according to Israelite norms. The measure-for-measure justice at work in the daughters' actions should not be overlooked. Exactly as Lot tried to broker his daughters' bodies as a means of ensuring the well-being of his guests and likely his own survival, so the daughters employ their father's member as a means of perpetuating their names. The horror of the incest, both heightened and lessened by the fact that Lot remains unconscious of both unions with his daughters, is then portrayed as something of an inevitable outcome of life in Sodom.

The story of Sodom and its aftermath describes several features of ancient Israel's existence. On the level of natural history, it accounts for how it is that the Jordan River terminates at the Salt (Dead) Sea and that the southern portion of the lush valley is a vast desert. On the level of national formation, it attributes incestuous origins to Moab and Ammon, its neighbors east of the Jordan. The story of Lot and his daughters in the cave, which displays biblical Hebrew punning at its best, is on the order of the nasty lore disseminated by one group about their closest neighbors (figure 2).[12] There is nothing surprising about alleging immoral beginnings and questionable ethics to those on the other side of a border, yet there is something unexpected about also imagining these people as relatives. As stated, the covenant excludes Moab and Ammon—a fact realized and reinforced by their separation from Israel by the river and the river valley—and yet they are remembered as part of the ancestral beginnings of Israel.[13] Geographic location gives difference its meaning and the difference becomes readable in terms of location. Where

---

12. "Moab, particularly in its gentilic form *Mo'abi* (Moabite) can be parsed in Hebrew as *Me-* [from]; *'ab*—[father]; *i*—[my]. In other words, 'From my father.' Similarly, the eponymous ancestor of the *Benay-'ammon* (sons of Ammon; Ammonites) is here rendered as *Ben-'ammi*—'son of my (close patrilineal) kinsman.'" Bruce Routledge, *Moab in the Iron Age: Hegemony, Polity, Archaeology* (Philadelphia: University of Pennsylvania Press, 2004), 41.

13. Robert Alter infers the larger conclusion of the story: "these peoples will be somehow trapped in their own inward circuit, a curse and not a blessing to the nations of the earth, in consonance with their first begetting." Alter, "Sodom as Nexus," 154.

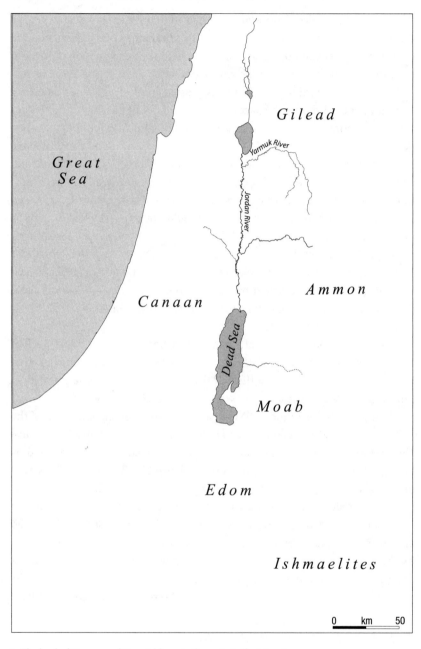

2. The land of Canaan and its neighbors to the east. Soffer Mapping.

a regional difference between a mountain range and a valley first separates Abraham and Lot, the writers impose the more definitive Jordan River between their descendants.

Depictions of Moab and Ammon confer the qualities of the foreign and the transgressive to the east bank largely through the figure of the Moabite woman. With the initiation of incest by Lot's older daughter in the background, Moabite women are portrayed as hypersexual, forward, and unmindful of necessary distinctions.[14] One might say that they are the focus of the danger and allure associated with bodies labeled Other. However Transjordanian, quasi-Moabite behavior can also be exhibited by Israelite women. Since the inclusion of Israelite women in the covenant forged with Abraham is ambiguous at best, they are something of free agents when it comes to ascribed location.[15] This means that they can keep their place within Israel by becoming wives and mothers who build families while remaining imperceptible on the national level. Those women who seek a place for themselves in national configurations tend to be transferred east—at least on the level of representation—and there discounted. Since wives and mothers are already domesticated in every sense of the word, the suspect category is "single" women, usually figured, in biblical terms, as daughters.

Moabite Daughters

The east-west split between Lot and Abraham eventually sets a definitional border at the Jordan. As I have intimated, the break is not quite clean since biblical narrative reports repeated Israelite-Moabite encounters. Thus the founding of the east and west banks by Abraham and Lot is but one level or stage of the myth. David Jobling discusses biblical narratives set in Transjordan in exactly this way, as structural transformations of the east-west binary. At the same time, historical stratification is important to Jobling insofar as "the diachrony of the text will be 'inscribed' in its final form not only through the clues of vocabulary, style, theological point of view, and so on, to which historical criticism attends, but also *structurally*, that is, at the level of deep assumptions."[16] Through an application of the principle, Jobling concludes not only that the ambivalent status of Transjordan jumps off the page in the received version of the Hebrew Bible, but that it is detectable in all of the iden-

---

14. Rashi says that a woman who does not exhibit modesty is called a "Moabitess."
15. On the question of whether or not Israelite women have a place in the covenant, see Shaye J. D. Cohen, *Why Aren't Jewish Women Circumcised? Gender and Covenant in Judaism* (Berkeley: University of California Press, 2005), and Havrelock, "The Myth of Birthing the Hero: Heroic Barrenness in the Hebrew Bible," *Biblical Interpretation* 16 (2008): 154–78.
16. Jobling, "The Jordan a Boundary," 91.

tified sources spliced into that final form.[17] In other words, there is a structure according to which the east-west binary is translated into other dichotomies, but such translations can also be read productively in diachronic terms.

This seems a solid method if two pitfalls can be avoided.[18] The first potential pitfall is to take the stories of Lot departing from Abraham and escaping the destruction of Sodom as "merely" legendary and claiming that it speaks to social relations in a fixed time period recoverable through other biblical texts like those describing the relationship between King David and his Moabite neighbors. The flaws with such an approach are many. To begin, a false hierarchy is established between one set of texts labeled "legendary" and another labeled "historical." Even if indeed the Lot texts are earlier than the David texts, determining that therefore the Lot texts are legend and the David texts history falls in line with a constructed theory of linear progress. The Lot texts and the David texts communicate different aspects of a symbolic Moab; these aspects may, in complicated ways, reflect the perceptions and mores of the times in which they were transmitted and eventually written down. Such differences can be acknowledged and then, as Jobling would have it, shown to interact on the level of deep structure. I would nuance this somewhat in order to see the ways in which various traditions accrue on the spatial plane and result in stable, although fluctuating, myths of place.

The second pitfall, related to the first, involves locking in on one moment that serves as *the* explanation for a range of ideological gestures. The moment in vogue has changed throughout the alternating eras of biblical scholarship and so we have seen monotheism, monarchy, and now, it seems, the conditions of Judean restoration under the Persian Empire as all-purpose explanatory agents. In such an approach, scholarly fictions about priesthood, kingship, and scribes provide the necessary backdrop for all explanations. In contrast, recognition of Moab as an ideological topos shows how narrated events are framed around this topos rather than reflecting things that historically transpired there or the political climate of a given moment. Still, I am persuaded enough by Jobling's conclusions to try to integrate his proposed division of materials according to source in order to see how Moab functions as a gender-deviant, anti-national site in the various strata of the Hebrew Bible.

Jobling proposes that the west-east split into Cisjordan and Transjordan is also coded in terms of female and male. After considering a string of women

---

17. Bruce Routledge works through the Moab motif while noting, "Moab is portrayed relatively consistently across the chronologically disparate units of the canonical text of the Bible." *Moab in the Iron Age*, 42.

18. It is not that Jobling falls into either of these pits, but that an application of his method easily could.

playing on the east bank "team," Jobling concludes that Transjordanian women "represent a logical problem in relation to Israelite legitimacy, and the best thing that can happen is for them to move to the west as virgins."[19] That this is not the usual outcome for Transjordanian women points to the degree to which Israel cannot assimilate the east bank. This resistance to assimilation further reveals how absorbing the foreign always poses a threat to the ideological systems embedded in the Hebrew Bible. Because the foreign is understood as a threat, Transjordan is a problematic place and its women are always cloaked in suspicion. The fact that according to archaeological finds,[20] the evidence of Israelite, Moabite, Ammonite, and Edomite epigraphy,[21] and the Bible's own admission (to be heard shortly) there are not really perceptible differences between the east and west banks in antiquity leads to the assessment that the natural dimension of the Jordan River as a border does not render it any less constructed.

The coding of Transjordan as female begins with the incestuous etiology regarding Moab and Ammon in which the daughters, rather than Lot, are the true founders of the nations. Lot, already rendered passive by the time he reaches the mountain cave, does not "know that" his daughters sleep with him, suggesting that he does not in fact "know" them in the sense of active male sexuality. Because they initiate sex and express abiding concern with the fate of humanity as well as their individual memories, the daughters can be labeled founding mothers. The two daughters may even be reflexes of Sarah and Hagar. However, by inaugurating a homeland—transforming the space of Transjordan into the places of Moab and Ammon—the daughters assume a male role. The perversion of incestuous origins is compounded by the reversal of gender roles.

Daughters of Moab in Priestly Texts

In the book of Numbers, Israel settles into the plains of Moab in preparation for the national homecoming across the Jordan River. The report of battles fought and territories conquered there will be of concern in a subsequent chapter. The people of Israel are accidentally blessed when Balak king of Moab hires a local prophet, Balaam, to place a paralyzing curse on their heads but

---

19. Jobling, "The Jordan a Boundary," 131.

20. There is "abundant evidence of close continuity between the material culture of the Moabites and that of their immediate neighbors, including Israel." "Moab," *Anchor Bible Dictionary* 4, 885.

21. "The language of the (Mesha) stela is a West Semitic variety so close to Hebrew that it is not entirely certain whether two of the three most marked differences (the masculine plural and feminine singular nominal forms) are not merely matters of spelling. The script is even closer, so much so that the Moabite inscription has been termed the first actual example of the Hebrew monumental style." Seth Sanders, *The Invention of Hebrew* (Urbana and Chicago: University of Illinois Press, 2009), 116.

God allows only praise to flow from his lips.[22] No sooner have Balaam's blessings concluded than the men of Israel engage in some of their worst behavior. Finding themselves in the vicinity of Moabite women, the men cross into the realm of taboo and engage in sex deviant enough to draw one of biblical Hebrew's more disdainful verbs (לזנות Num 25:1).[23] The Moabite women, it is implied, seduce the Israelites and actively lead them (to the greater chagrin of the Priestly writer) to a sacrificial feast dedicated to their gods.[24] In the metaphorical economy in which adultery and apostasy are one and the same crime, Israel amorously attaches itself to the god Baal-Peor.

Worthy of note is that the encounter with Moab is an encounter only with Moabite women. By beckoning the Israelite men to the altars of their gods, the women engage in subversion on a national level. So long as they are encamped on the east bank, Israel can fall prey to the irresistible Moabite women and thus be corrupted. This then necessitates Israel's crossing of the Jordan River as a technique of achieving distance from Moab. Because Moabite women will never but act in their own national interest, their bodies must be kept outside the community of Israel. The problem with Moabite women highlights the potential problem with all women: they can use their sexuality and reproductive capacity to further not an official male agenda, but their own ends.

This anxiety, so blatantly expressed through the Moabite orgy, is evident in more subtle terms in the association of Zelophehad's daughters, the only sanctioned land-owning women in the Hebrew Bible, with the east bank. A national census follows the raucous orgy and subsequent bloodbath in Numbers 25. It is as if the nation needs to be reconstituted through enumeration following apostasy. The census counts the able-bodied men, "everyone who will go out to war in Israel"; these warriors are said to make up "the whole Israelite community" (Num 26:2). The nation as defined here is male, and the gender performance that ensures their place in its roster is that of a soldier.

The census also introduces the category of daughters. Such daughters, should they mind their place and marry within their tribe, can legitimately

---

22. In this regard, the Jordan River Valley is a borderlands where outsiders like Balaam and Rahab are able to discern and communicate the will of the God of Israel. In the scheme of the wilderness trek, the two are reflexes of one another who respectively propel Israel forward and draw them across the Jordan.

23. Disdain is also expressed in Hosea 9:10b: "When they came to Baal-Peor, they turned aside to shamefulness, then they became as detested as they had been loved" (JPS translation).

24. Nathaniel Deutsch observes such charges leveled against religious groups in considerably later times. "Accusations of antinomian sexual behavior have long been leveled against religious sects in which women have played prominent roles, typically in order to attack their morality." Deutsch, *The Maiden of Ludmir: A Jewish Holy Woman and Her World* (Berkeley: University of California Press, 2003), 154.

own land in the absence of brothers (Num 27:8). The daughters of Zelophehad secure this right when they present their claim "before Moses, Eleazar the Priest, the chieftains, and all of the community at the entrance of the Tent of Meeting" (Num 27:2). Indeed, the initiative of the five daughters secures the right of inheritance for them and for any other daughter who has no brothers, and thus the category of a particular type of Israelite woman figures in the Priestly portrait of Promised Land holdings. Daughters then have the potential under certain circumstances to be numbered among the nation.

At the same time, however, there seems to be a strategy at work that distances the inheriting daughters of Zelophehad from the center or at least renders their patrimony ambivalent. The book of Numbers links Mahlah, Noah, Hoglah, Milcah, and Tirzah through their ancestors Gilead and Machir to the tribal territory of Manasseh east of the Jordan River (27:1, 36:1).[25] The territory in question falls outside of the Priestly parameters of the land (Num 34:12).[26] Although they receive the land in the right way according to the Priestly source, it is the wrong land. Seen in this context the five women gain a place for themselves in the census (Num 26.33) and in the land only to then be pushed beyond its boundary. Even a text like Numbers 27:1–11, in which five daughters secure inheritance rights, contains a strategy for marginalizing women insofar as the lands that they inherit are situated beyond the Jordan River, the legitimate boundary of the land in the Priestly definition. These legitimate daughters are thus rendered strange. Their strangeness in turn generates the association with the east bank, an association that makes them simultaneously distant and threatening.

The daughters can gain legitimacy by becoming the wives of their tribesman (Num 36). Jobling observes that although the tribe of Manasseh is split between the two riverbanks, the hope is sustained that the daughters will cross westward and thereby shed the taint of the foreign.[27] The fact that the marriage of Zelophehad's daughters to their cousins is the last act of the book

---

25. Jobling diagnoses the text's vacillation about whether the daughters are really Transjordanian or Cisjordanian as reflecting a deep ambivalence concerning Transjordanian women in general. Jobling, "The Jordan a Boundary," 119.

26. Josh 17 contains an alternate genealogy that positions the patrimony of the daughters of Zelophehad west of the Jordan. Called "daugthers of Manasseh," they claim their portion "in the midst of his sons" (Josh 17:6). This tradition does not marginalize them in the manner of the Priestly texts, and is substantiated by both extra- and intrabiblical evidence. Hoglah and Noah appear as place names in the Samaria Ostraca, and Tirzah, mentioned in other biblical passages as a place ostensibly in the Northern Kingdom (1 Kings 14:17; Song of Songs 6:4), is identified by scholars as Tell 'el-Far'ah near Nablus. Little can be said concerning the location of Milcah and Mahlah, although Tamara Cohn Eskenazi sees the site of Abel-meholah on the western edge of the Jordan as related to the name Mahlah. Eskenazi, "*Pinchas*: Legacy of Law, Leadership, and Land," *The Torah: A Women's Commentary*, 972.

27. This move could well be what is recorded in Joshua 17.

of Numbers (Num 36.10–12) shows how female ownership of land can be counteracted in order to not destabilize the male nation. By marrying correctly, the daughters of Zelophehad lose their land and with it their ambivalent status. As wives, the daughters become imperceptible in the national collective.

In Numbers 25, sex with Moabite women is an assault on national unity. The punitive killings of Israelite men that follow serve as an example of how grave an error it is to commingle with Moabites. In Numbers 27, five daughters of Israel secure the right to inherit land and inaugurate territory. They are then, in subtle terms, assigned to the east bank team. Numbers 36 implies that when these daughters marry their cousins, relinquish their land claims, and move westward, sex redeems them. "Moabites" no longer, they are neither particularly threatening nor particularly attractive and therefore warrant no further report. These examples show how any woman who challenges the Priestly notion of the nation can be described in Moabite terms and consigned to the east bank. To abandon the challenge is to be repatriated and consequently to become a woman of Israel. "Moabite," speaks not only to an ethno-national designation, but to gender performance other than that prescribed in the census in which men are soldiers and landowners and women are fleeting actors on the national stage. When the term "Moabite" is applied to a man, it suggests that the man is not a soldier and is more likely to be penetrated than to penetrate. When applied to a woman, "Moabite" implies one who refuses to be invisible, who attempts to claim and transmit territory, who tries to found a tribe or a cultic site, or who initiates action in order to be recognized as a member of the nation. The term does not necessarily express geographic or ethnic information, but instead operates as a code for whether one's behavior as a man or woman conforms to or deviates from national standards.

Daughters of Moab in the Deuteronomistic History
In legal language Deuteronomy makes its stand on Moabites clear:

> An Ammonite or Moabite shall not gain admittance into God's community; Up until the tenth generation they shall not enter into God's community, they shall not enter for eternity. This is because they did not greet you with bread and water on the road when you left Egypt and because they hired Balaam son of Beor from Pethor of Aram-naharaim to curse you. However Yahweh your God refused to listen to Balaam. He turned the curse into blessing for you, because Yahweh your God loves you. Do not seek their peace or their plenty all of the days of your life. (Deut 23:4–7)

Moabites and Ammonites are resolutely barred from Israel. What is primarily a religious prohibition—they cannot approach Israel's sanctuary or ritual space—also has social ramifications. Since participating in ritual is what defines Israel, Moabites and Ammonites are to have no contact with Israelites. No contact certainly includes no marriage. The mythic bad blood is further caught up in national memory through the command that Moabite misbehavior not be forgotten for eternity. How did the Moabites go wrong? It seems that Ammonites and Moabites offered no support to the Israelites after they escaped their Egyptian oppressors. Encoded in this is perhaps historical Israelite anger that their neighbor nations did not join them in an insurrection or go out with them to a particular battle. With the two-directional pressure of Egyptian and Assyrian Empires, alliances among the smaller nations of the Levant could be advantageous. Yet, ironically, Deuteronomy's prohibition itself would stand in the way of such alliances.

Separation is also maintained geographically. During its wandering in Transjordan, Israel is forbidden from attacking Moab and Ammon because their land is a God-given "inheritance for the descendants of Lot" (Deut 2:9, 19). Because Israel cannot claim Moabite and Ammonite land and Moabites and Ammonites cannot enter the community of Israel, the border is quite definite.[28] Although the prohibitions involve both Ammonites and Moabites, the relationship between Israel and Ammon lack the charge of that between Israel and Moab.[29] Despite the dividing line, Moabites and Israelites meet in war and in marriage.

Judges 3 narrates the political assassination of King Eglon of Moab by Ehud son of Gera of the tribe of Benjamin. According to the text, Eglon deserves his fate since he has crossed the Jordan and occupied "the City of Date Palms" (Jericho). Ehud goes to bring tribute to the corpulent king with a two-edged dagger stashed on his thigh beneath his clothes. Gaining private counsel with the king, Ehud draws close and stabs him in the stomach. Eglon's failure to act as a soldier means that he employs no strategy, expending all of his energy to simply "rise from his seat" (Judg 3:20). As Eglon's stomach swallows the

---

28. At the same time, the Moabites are not quite enemies. While Israel wanders through the wilderness and staves off a Canaanite-initiated battle (Num 21:1), an Amorite attack (Num 21:23), and the siege of Og, King of Bashan (21:33), the Moabites raise no arms against Israel.

29. Most mentions of Ammon enforce its position as military opponent. The Prophets rail against Moab and Ammon in the genre of "oracles against the nations" (Isa 15, 16, 25:10–12; Jer 48:1–47; Ezek 25:8–11; Amos 2:1–3; Zeph 2:8–11; Jer 49:1–6; Ezek 25:1–7, 10; Amos 1:13–15; Zeph 2:8–11.) Only Jeremiah concedes that a remnant of Moab and Ammon will be saved (Jer 48:47, 49:1–6). The apocryphal book of Judith recalls Achior, the Ammonite commander, who narrates Israelite history and attests to their divine protection to Holophernes. Achior defects to the Israelite camp, feasts with the leaders of Bethulia and marks his conversion through circumcision.

dagger and his innards flow out, an east bank male body is again penetrated.[30] Like Lot, Eglon is deceived and then feminized. This penetrated man who fails to behave as a warrior, like forward women, belongs to Moab.

In the days of the monarchy, Saul wages war against Moab (1 Sam 14:47) and David humiliates Moabite captives (2 Sam 8:2). The perverse nature of the Moabites manifests itself in a battle between King Jehoram of Israel, King Jehoshaphat of Judah, and the King of Edom against King Mesha of Moab, where the Moabite king sacrifices his oldest son in the eyes of the conquering armies (1 Kings 3:27). The Israelite soldiers are repelled in the sense of being driven back as well as in the sense of being disgusted, and so the geographic as well as the social border thickens.

Child sacrifice, as it turns out, is not quite so alien to Israel. Jephthah judge of Israel sacrifices his daughter in order to fulfill the oath that he made to God. Since Jephthah is from Gilead, the plain east of the Jordan River Valley and south of the Yarmouk River, the events in question transpire on the east bank. That Jephthah's father is Gilead's eponymous ancestor grants him status enough while only trouble is conferred by his mother, alternately called "an Other woman" and "a prostitute" (Judg 11:2, 1). Jephthah is thus doubly marked as a Transjordanian. Because his maternity poses a threat, Jephthah's brothers dispossess him. All he inherits is his mother's foreignness and so he flees to the Transjordanian land of Tob, where he gathers a gang of bandits around him. An Ammonite attack causes the people of Gilead, his brothers included, to reconsider the banishment and to beg Jephthah to come back as their leader. After leveraging a position of permanent leadership, Jephthah engages the King of Ammon in dialogue and then in war.

Along with the foreign woman motif, the motif of excess associated with the east bank is apparent when Jephthah makes a vow to God after he has already been animated by divine spirit. In the martial economy of the book of Judges, Jephthah need only be possessed by God's spirit in order win the war, yet he makes a pledge in order to receive further assurance. "If You deliver the Ammonites into my hands, then whatever comes out of the doors of my house to greet me when I return in peace from war with the Ammonites will be offered up to God as a sacrifice" (Judg 11:31). Oaths always present a problem in biblical literature—the future is too uncertain to lock in a necessary action and God alone can be aware of all contingencies—and further, as Mieke Bal has asked, what else could emerge from his doorway but his only

---

30. Susan Niditch observes that "the short sword worn on the thigh, a male erogenous zone" that is "also the seat of male fertility." "The short blade is a phallic image . . . the term used for Ehud's ample belly is the same as the term for womb, while the image of the fat closing around the blade is strongly vaginal." Niditch, *Judges: A Commentary* (Louisville: Westminster John Knox Press, 2008), 58.

child. As well as likely foreign, the daughter's mother is absent, so the girl seems to wait alone in the house.

Oath aside, Bal has argued that the daughter (whom she names Bath) is sacrificed because Jephthah cannot stand the thought of his own memory being erased when the daughter leaves his home to bear descendants for another man. Maybe she, as do Zelophehad's daughters in Joshua 17, would even cross westward in order to marry and thereby efface his Transjordanian patrimony along with his memory. The transgression for which she is punished, in Bal's estimation, is the inevitable act of leaving Jephthah's house, an act made certain by her exit to greet her father. Bal is interested in the space of the home, one to which women are confined at the same time that it serves as the site of their undoing. As I am interested in the space of the east bank, I see something different at work here. The mere existence of a daughter as an only child means that she could press a claim for territory as did the daughters of Zelophehad. As Jephthah has experienced, one's mother determines one's inheritance in the first place. In addition, the daughter is an east bank woman and therefore expected to seek territory and even perhaps the elevation of her own status at the expense of her father. Jephthah's sacrifice shows that when one cannot mark distance from a Transjordanian woman, that woman can be dispensed with.

Under a death sentence, Jephthah's daughter still initiates subversion. She asks her father to release her for two months in order to mourn her virginity in a manner of female pilgrimage to the mountains. After she is slain true to her father's words, the very hills where she bewailed her fate become a ritual site for the daughters of Israel: "for four days each year they lament the daughter of Jephthah the Gileadite" (Judg 11:40). Jobling casts the subversion in structural terms: east bank women can redeem themselves by crossing westward as virgins, yet the daughter dies an unredeemed virgin in the wrong place and further incites women on the west bank to cross eastward by establishing a festival. What the biblical text seems to decry in a clear case of blaming the victim, Bal celebrates. By being memorialized Jephthah's daughter receives "a form of afterlife, [that] replaces the life that she has been denied . . . the only alternative form of survival within the limits of what is left over of the father's power."[31] Like Bal, I appreciate mention of the festival and strain to imagine the versions of the daughter's story told there as the female oral history left out of the Bible. At the same time, I recognize the role of the festival in establishing the east bank as a site of incorrect, even rebellious memory. Things that should be suppressed, like incest and child sacrifice, are relegated there and cannot be expunged from memory because their residue

---

31. Bal, *Death and Dissymmetry*, 67.

lingers on the territory. By establishing a female festival and a pilgrimage in the wrong direction, Jephthah's daughter, according to the logic of the text, proves that her murder expunges corrosive elements from Israel.

In the Deuteronomistic history the degenerate nature of the east bank brings the good land of Israel into relief. When performed east of the Jordan, nation building acts like founding, inheriting and memorializing appear as perverse; when women perform such nation-building acts, they tend to occur in Transjordan. According to this symbolic channel, female attempts to participate in the activities of the nation cannot be other than perversions. Women who would be members of Israel are instead granted perpetual citizenship in the zone of the anti-nation. Moab is the place other that enables the definition of Israel.

Moabites: The First "Shiksas"

Later biblical texts, written in the Persian period, are more permissive toward female heroes. Women like Esther and Ruth have their own books in which they perform the roles of savior and founder. In these books, the inherent foreignness of women is shown to be an asset for the Jews. Esther's indeterminate nationality lands her in the Persian palace, a queen amongst a harem of women, from whence she foils an extermination plot launched by an enemy of the Jews. Her "uncle" Mordecai prostitutes her just like Abraham pimped Sarah, but this time sleeping with a foreign ruler does not provoke God's wrath. The key is that Mordecai is hungry for power rather than for wealth and is able to use this power to the benefit of his people, now scattered to the far corners of the Persian Empire. Esther is an instrument to these ends and also a heroic precursor of Jewish political power in the diaspora. The diaspora appears more conducive to female participation than the land of Israel because, in the absence of land, ethnic borders have relaxed.

The second heroine comes straight from the other side, as emphasized by the oft-repeated gentilic, *ha-Moaviah*, the Moabite.[32] Ruth the Moabite, as a widow escorting her bereft mother-in-law, actually crosses over from Moab to Judah, seduces a prominent man of Bethlehem, and secures territory to be transmitted with her name. According to the book of Ruth, the land, not the women, of Moab is the problem. It is the destructive bounty of Transjordan that kills off Ruth's husband, his brother, and his father and not the fact that they mingled with Moabites. Since the men die without sons, the legal precedent of Zelophehad's daughters authorizes any daughters to inherit the property of the deceased. The plot thickens when it is revealed that the only

---

32. There is no need to strain to translate this as "the Moabitess," since, as I have shown, "Moabite" functions as a feminizing term.

survivors are Naomi of Bethlehem and two daughters of Moab. It would seem that the Moabites are once again successful in bringing down Israel and that the grieving Jewish mother will have to return home, shamed and alone. On Naomi's advice, Orpah, the other daughter of Moab, returns to her "mother's house" (Ruth 1:8), but Ruth insists on sticking with Naomi and crossing the border at her side. The Jordan crossing, according to Ruth's witness, transforms her entirely:

> Wherever you go, I will go; wherever you stay, I will stay. Your people are my people and your God my God. Where you die, I will die and there I will be buried. May God do thus to me, and even more, if anything but death separate me from you. (Ruth 1:16–17)

By pledging to be Naomi's other half, Ruth breaks down the divide between Moab and Israel. However, as I have already suggested, the gender code implicit in the appellation "Moabite" trumps the ethnic one. On the level of gender, the affinity between Ruth and Naomi is natural and thus we see how the boundary between Israel and Moab in fact works to separate men and women or, said differently, men who represent the nation and women who must be kept out.

Scholars have correctly characterized the book of Ruth as a narrative polemic against the measures described in the books of Ezra and Nehemiah.[33] In the name of the restoration of the Temple in Jerusalem, Ezra and Nehemiah order the expulsion of all foreign wives and half-breed children in the name of purifying "the holy seed" (Ezra 9:2).[34] Although the returnees from exile have long mixed in with the diaspora, they must refine their bloodlines in the name of reestablishing a homeland.[35] The border of blood is enforced through the ritual proscription of male "peoples of the land" from building the Temple (Ezra 4:4) and physical separation from women (Ezra 10: 10–11). The contaminating bodies in question include "the peoples of the land whose abhorrent

---

33. Still, Ilana Pardes' literary reading that "the Book of Ruth manages to challenge the biblical tendency to exclude the other" is more supple. Pardes, *Countertraditions in the Bible: A Feminist Approach* (Cambridge: Harvard University Press, 1992), 99.

34. For an account of how biblical narratives contest "the ideology of the 'holy seed,'" see Mark Brett, *Genesis: Procreation and the Politics of Identity* (London: Routledge, 2000).

35. Daniel Smith-Christopher seeks to temper the outrage of Bible scholars regarding the treatment of foreign wives by pointing out that oppressed groups may well have self-preservation in mind when they look askance at mixed marriages. Smith-Christopher reaches this place of understanding by considering concerns expressed by Native Americans about non-Natives adopting Native children. See Smith-Christopher, "The Mixed Marriage Crisis of Ezra 9–10 and Nehemiah 12: A study of the Sociology of the Post-Exilic Judean Community," *Temple and Community in the Persian Periods*, eds. Tamara Eskenazi and K. Richards (Sheffield: JSOT Press, 1994), 243–65.

practices resemble the Canaanites, Hittites, Perizzites, Jebusites, Ammonites, Moabites, Egyptians, and Amorites" (Ezra 9:1). It is worth noting that the corrupting peoples are not actually Canaanites, Ammonites, Moabites, etc., but rather resemble them through the practice of abominable acts. How one behaves, particularly in regard to sex, exerts influence on one's ethnic label.[36]

Nehemiah focuses in more closely on the offending Ammonites and Moabites. In the midst of Temple rebuilding, a version of Deuteronomy ("the book of Moses") is read aloud and the eternal prohibition of Ammonites and Moabites from "God's community" is discovered to have contemporary relevance (Neh 13:1). The reasons are familiar—the two nations did not bring bread and water to Israel in the desert and instead hired Balaam to put a curse on them—but the ramifications are novel. As a result of hearing the law anew, it is reported that Israel disengages from all of the mixed-bloods in its midst (Neh 13:2).[37] How then is it possible for a Moabite to be singled out as the heroine of a biblical book? Certainly, as many have noticed, the book of Ruth contests the ruling and concedes to the possible advantages of marrying a Moabite. Yet the book proposes something more radical than that forbidden sexuality can be enjoyed under the radar. The heroine as Moabite is an extreme example used to make the point that women can and should be included in the activities of the national collective. If the most subversive of women can cross the boundary and enter Israel, then perhaps a place can be opened for all the women of Israel.

Ruth is indeed an exceptional character—the women of Bethlehem praise her as "better than seven sons" (4:15)—but, all the same, she displays the very qualities that readers expect from a Moabite. Recently arrived in Bethlehem, she goes out to work in the fields in order to sustain Naomi and herself. This shows that she gladly accepts the responsibility as the head of a family and feeds Naomi the food grown on her own land, something that Naomi's husband and sons failed to do. There are plenty of clues that Naomi and Ruth are in the throes of a conspiracy to reclaim the land rightly due to them. Ruth just "happens" to end up gathering barley in the field of Boaz, the overseer of Naomi's land (Ruth 2:3) and, with perfectly orchestrated timing, Ruth goes

---

36. Smith-Christopher, suggesting that Ezra may really object to the status of those Jews who did not undergo exile, questions, "whether the Ezra documents are really talking about 'foreigners' at all." I don't think that the designations pivot on exile, but rather on relationship to the power structure put into place by Ezra and Nehemiah. See Smith-Christopher, "Between Ezra and Isaiah: Exclusion, Transformation, and Inclusion of the 'Foreigner' in Post-Exilic Biblical Theology," *Ethnicity and the Bible*, ed. Mark G. Brett (Boston: Brill Academic Publishers, 2002), 126.

37. I would even suggest that these "mixed-bloods" as well as those so designated at other points in the Hebrew Bible are not necessarily ethnic hybrids, but rather those who assume an oppositional stance toward the powers that be.

down to the threshing floor where she seduces Boaz. Subsequently, when Boaz negotiates his marriage to Ruth with the elders of Bethlehem, the land in question becomes the heart of the matter. Boaz employs the ruse that Naomi wishes to sell her late husband's fields as a way to secure his ownership of the land and marriage to Ruth. Although Boaz would imagine himself the master, he is more the instrument with which Ruth and Naomi secure land and an heir. Fulfilling the duty of levirate marriage, Boaz speaks to the need "to establish the name of the dead on his patrimony" (Ruth 4:10), but after the child is born, the women of Bethlehem name him Obed, "a son born to Naomi" (4:17). Ruth successfully gives birth to Naomi's child.

Contemporary exegetes have perceived lesbian undertones in the relationship between Ruth and Naomi.[38] Without aiming to contradict this interpretation, I suggest that female partnership, whatever its nature, is part and parcel of Moabite behavior. Like Lot's two daughters, the women of Moab at their sacrificial feast, the daughters of Zelophehad, and the women who lament Jephthah's daughter, Ruth and Naomi work together. Ruth's rhetoric always functions to protect Naomi. In her report to Naomi, Boaz's initial invitation to "stick close to my girls" (Ruth 2:8) becomes "stick close to the boys who work for me" (2:21). Perhaps the substitution expresses Ruth's preference for male labor rather than female alms seeking; perhaps she wants to play down the suggestiveness of Boaz's every proposal. When Ruth comes back to Naomi loaded with grain after her roll in the wheat with Boaz, she tells her that Boaz said, "don't go back empty-handed to your mother-in-law," a concern far from Boaz's mind in the midst of Ruth's seduction (Ruth 3:17).

Resembling Abraham and Mordecai, Naomi prostitutes Ruth to Boaz. Her pimping is perhaps the most laudable since it is done not for money or power, but for an heir to keep her name alive. Naomi instructs Ruth explicitly:

> Wash, anoint yourself, put on a dress and go down to the threshing floor. Do not let the man know that you are there until after he has eaten and drunk. Pay attention to where he lies down and then go, uncover his legs and lie down. He will tell you what to do. (Ruth 3:3–4)

Ruth again makes it clear whom she is serving when she responds to Naomi, "All that you tell me, I will do." Despite the fact that Naomi orchestrates every

---

38. See Rebecca Alpert, "Finding Our Past: A Lesbian Interpretation of the Book of Ruth," *Reading Ruth: Contemporary Women Reclaim a Sacred Story*, ed. Judith A. Kates and Gail Twersky Reimer (New York: Ballatine Books, 1994). Lesleigh Cushing Stahlberg calls the book of Ruth "the prooftext the religious left needs for sanctioning forbidden marriages." Stahlberg, "Modern Day Moabites: The Bible and the Debate About Same-Sex Marriage," *Biblical Interpretation* 16 (2008): 474.

aspect of the seduction, Ruth is shown to be characteristically Moabite when she slides up to a sleeping Boaz and provides him with pleasure "until the morning" (4:14). Although rendered in the laconic prose of biblical narrative, this represents the Moabite fantasy at its most full-fledged. That it does not end with a massacre, like the orgy with Moabites east of the Jordan, means that the attraction is explored without the overhanging sense of dread. By the time Ruth lands one of the richest men of Bethlehem, the readers do not begrudge her the comfort. After all, the book indulges the fantasies of mothers-in-law more than those of Israelite men. It seems that Ruth could take or leave Boaz with his pomp and puff and, in case one is still concerned about how a Moabite infiltrated Israel, the text ends with a genealogy that traces King David to Ruth.

Commentary has long latched onto Ruth as the self-effacing, humble woman whose actions are so genuinely good that they warrant her inclusion in the community of Israel. This pious reading along with the by-now traditional notion that the book of Ruth is a story about conversion has, in fact, obscured the power of what it communicates and how it operates in the Bible as a whole. Ruth is a woman who presses a land claim and, to the degree that a woman can own land, is awarded it. Where Boaz would have her be satisfied with the occasional repast, gleanings, and the wages dispensed by God (Ruth 2:14, 12), Ruth expands the parameters of a Moabite's—read woman's—place in the nation. The initial disparity between how Boaz would treat her and how Ruth intends to be treated comes out beautifully in the always mistranslated verse when Ruth thanks Boaz for his kindness and then lets him know, "I will not be like one of your female servants" (2:13).[39] Ruth's seduction then should be understood as sexual initiative rather than submission. The son "born to Naomi" and begotten by Ruth inherits the land reclaimed by Naomi and Ruth. A Moabite is the progenitor of the messianic line; the Bethlehem where David is first discovered (1 Sam 16) becomes inflected, through the book of Ruth, with Moabite memory. Ruth's name is embedded in the heartland of Judah.

The book has also been read as a propaganda piece disseminated by monarchs in the Davidic line in order to shore up pastoral, populist credibility. One could even imagine it as an overture made in times of reconciliation or occupation of Moab. The Moabite motif may also work as a kind of editorializing on David's acute sexual appetite, which ultimately proves his downfall.[40]

---

39. Translated by the JPS as "though I am not so much as one of your maidservants" and the NRSV as "even though I am not one of your servants."

40. David professes that his love for Jonathan was "stronger than the love of women," seduces and impregnates Bathsheva while she is married to one of his soldiers, and picks up wives and concubines at every turn of his journey.

The link to Ruth as an ancestress may speak to something foreign and destabilizing about the golden king's sexuality. However, the chronicles of Davidic kingship do not shy away from reporting the marriages between Israelites and Moabites. Solomon brings Moabites along with Egyptians, Ammonites, Edomites, Sidonians, and Hittites into his harem (1 Kings 11:1); a genealogy of Judah's lineage records marriage with Moabites (1 Chr 4:22). In this regard, the book of Ruth only exposes the persistent case of Moabite-Israelite intermixing.

In the Hebrew Bible, "Moabite" functions more as a charge leveled against gender deviants than an ethnic designation; uses of the term reveal the gendered dimension of national boundaries. This is particularly true when it comes to women who seek political power or representation. These women are either called Moabites or linked to the east bank, the terrain of Moab. When Ruth—with praise and in public—joins the community of Bethlehem, engineers a female territorial claim and founds the Davidic line, a Moabite makes it into the center of the Israelite power constellation of God, land, and heredity. The book of Ruth advocates not only for the stranger at the gates to be ushered in, but also for women's access to the centers of power. As a variant myth of the promised land, Ruth revises prior myths and contends with the conventional wisdom in its own day. When the relationship in question transpires between a Moabite woman and a woman, rather than a man, of Israel, the opposition between them morphs into alliance. By simply changing the gender of one of the players, the stark difference begins to fade. In this way, the book of Ruth can be seen as a mythic variant that refigures the border between Moab and Israel and thereby imagines the Jordan River as a place of crossing rather than of division. For the writers to allow a Moabite across the Jordan border is a radical act with revolutionary consequence. It suggests that incorporation of the foreign does not necessarily compromise the national body and that there is a place for women in the politics and economy of land ownership.

Chapter Three

TWO CAMPS

Ancient Israel Between Homeland and Diaspora

A wandering Aramean was my ancestor.
—Deuteronomy 26:5

If place in its represented form is ideology rendered as location, then what can one say of the place that Israel calls home? The question drives this chapter about Jacob, a wanderer who becomes the eponymous Israel just before a westward crossing of the Jordan. Where Abraham's split from Lot determines the Jordan as a border with Ammon and Moab, north and along the coast of the Salt (Dead) Sea, Jacob's dealings with his relatives yield a frontier with Edom south of the Salt Sea and a borderland with Aram (Syria) along the Jordan north of the Jabbok River. In terms of the map, Jacob's borders frame Abraham's. Yet the places where Jacob parts from his relatives do not form borders as deep as those with Moab and Ammon. The cycle of Jacob stories in Genesis involves several dualities including the fact that the stories about Jacob and Esau likely reflect traditions connected to the Southern Kingdom of Judah and those about Jacob and Laban are related to the Northern Kingdom of Israel's relations with Aram.

On the level of composition, the characteristics of a Judean hero are merged with those of an Israelite ancestor. In the Northern stories about Jacob and his uncle Laban, the mountain range east of the Jordan functions as a frontier and in the Southern stories, geographic distance protects Jacob from dangerous intimacy with his brother Esau. These more muted, regional distinctions bespeak the fluid processes of sedentarization through which nomadic Aramean and Edomite tribes became states abutting Israel.[1] The amorphous borders point to prolonged processes of state formation and differentiation. Ironically, the greater distance marked between Jacob and Esau seems a technique of distinction amidst a situation of significant overlap between Judah and Edom, while the story of Jacob's negotiations with Laban admits to interaction and occasional dependence upon Aram. This more open depiction of borders, I argue, is part of a distinct Northern national myth that acknowledges and even at times advocates alliance and interpenetration with neighbors.

Where tales of Moabites prescribe the correct gender behavior of Israelites, the stories of Jacob's contact and flight from Esau and Laban admit that the homeland depends upon the dreams of exiles. With admitted anachronism, I am prone to call the space beyond the Jordan a diaspora because biblical narratives already convey the tension of reference/invention and resistance/assimilation that characterizes communities consciously living outside of where they "belong." The Jacob stories not only concede that Israel's place is to be found somewhere on the road between home and diaspora, but also suggest that Israel's identity is produced in this space between. If place bespeaks ideology, then the ideology of ancient Israel is a composite that encompasses opposition.

While many biblical traditions convey a sense of Israel's outsider status, the Northern traditions seem most comfortable with marginality. The tales of Northern forerunners speak to the inevitability of blending with other peoples and to a fluctuating affiliation with Israel. As Jacob travels in and out of the land, so the Northern tribes seem to splinter off and rejoin the national collective. In this light, I read the recurring accusation that the Northern tribes that reside east of the Jordan are half-breeds as pointing toward a dynamic in which these tribes at times ally with Israel and at others with Aram, Sidon, or different neighboring groups. The Northern traditions embedded in the biblical canon present a different version of national identity that involves a particular orientation toward the Jordan River.

Jacob: Without a Place to Call His Own

The recurrent struggle between Jacob and Esau reflects a sense of competition between Israel and Edom; that the two figure as twins heightens the stakes of

---

1. On Edom, see I. Beit-Arieh, "The Edomites in Cisjordan," *You Shall Not Abhor an Edomite for He Is Your Brother: Edom and Seir in History and Tradition* ed. Piotr Bienkowski (Atlanta: Scholars Press, 1995), 33–40. On Aram, see William Schniedewind, "The Rise of the Aramean States," *Mesopotamia and the Bible: Comparative Explorations,* ed. Mark W. Chavalas and K. Lawon Younger, Jr. (Grand Rapids, MI: Baker Academic, 2002), 276–87.

the contest. Esau's status as the elder signals that he will be humbled in the reversal of primogeniture—that favorite of biblical plotlines—and suggests an operative timeline in which Transjordan boasts older, more established nations. Bible scholars have long bandied the idea that proto-Israelite tribes may have migrated from Transjordan, meaning that the ambivalence about the place speaks to the difficult conditions that precipitated the migration or to a kind of embarrassment about such local origins.

When he is born, Esau is already a place. His physical description reads as an etiology: "the first one emerged red, covered in hair like a cloak. They named him Esau" (Gen 25:25). The red baby is really the red rock of southern Jordan (Seir) and his covering of hair, *Sear* שער, is both a synecdoche and synonym of Edom.[2] Jacob is not yet a place. The best he can do is to grasp Esau's ankle עקב עשו and be named Jacob, "grasper" יעקב, as a result.[3] The grasping it seems is an attempt to be a territorial body like his brother. The identification of Esau with land continues in the description of the boys' youth when Esau is a hunting "man of the fields" and Jacob "a dweller in tents" (Gen 25:27). Not only is Esau associated with the out of doors while Jacob remains inside, but Jacob's tent also suggests a kind of uncertainty about the place where he belongs. Jacob's next attempt at territorialization is to buy Esau's birthright for the price of bread and a pot of lentils; although Esau swears off his birthright in order to eat Jacob's food, the deal doesn't get Jacob anywhere.

Rebecca, Jacob's mother, puts Jacob in Esau's place when her husband indicates his readiness to bless Esau with the transmission of his legacy. Jacob and Rebecca cook and scheme while Esau hunts in the fields and so, after eavesdropping on Isaac's request that Esau bring him game as prelude to the blessing, Rebecca prepares a favorite meal for Isaac, dresses Jacob in Esau's clothing, and makes the smooth Jacob hairy by covering his arms and neck in the fur of goats. Playing the role of Esau, Jacob is described for the first time in topographical terms. Picking up the scent of Esau's clothing, Isaac says unwittingly of Jacob, "the smell of my son is like the scent of a field blessed by God" (Gen 27:27). The ensuing blessing bestows agricultural bounty and

---

2. Gen 25:30 has Esau's mouth watering over Jacob's red lentils, which he calls "this red, red stuff." This is a more deprecatory etymology that faults Esau for his appetites. Puns advance the idea that as the men, so their territories: "Esau's being 'hairy' (שער; 27.11, 23) is a pun on Seir (שעיר). Jacob's being 'smooth' (חלק) is a pun on Mount Halak (חלק), 'that rises toward Seir' (Josh. 11.17, 12.7)." Bert Dicou, *Edom, Israel's Brother and Antagonist: The Role of Edom in Biblical Prophecy and Story* (Sheffield: Sheffield Academic Press, 1994), 141.

3. For the probable historical form and origin of Jacob's name, see P. Kyle McCarter, Jr., "The Patriarchal Age: Abraham, Isaac, and Jacob," *Ancient Israel: A Short History from Abraham to the Roman Destruction of the Temple*, ed. Hershel Shanks (Washington, D.C.: Biblical Archaeology Society, 1988), 24.

political supremacy on Jacob. Although he is not quite blessed to be a nation, Jacob is told that "peoples will serve you and nations will bow to you" (27:29). Yet in opposition to what Jacob actually seems to want, the blessing leads him out on the road.

Esau returns to receive trauma where he had expected blessing. The scene in which Isaac trembles and stumbles over his words while Esau is seized with bitter grief holds the mirror of consequence up to Rebecca and Jacob's deception. Esau's repeated cry to "bless me too, father" and the paltry promise that one day Edom will be redeemed through insurrection (27:40) confer sympathy on the wronged brother. In between his supplications for paternal blessing, Esau reiterates Jacob's name: "was he named Jacob so that he might supplant me these two times? He took my birthright and now he has taken my blessing" (Gen 27:36). In Esau's estimation, Jacob remains the man without a place who must use cunning to make his way in the world. Esau is the man of Edom, where Jacob is the trickster forced to shape-shift in order to gain benefit.[4] This reiteration of Jacob's name and its implicit contrast with Esau's show that neither birthright nor blessing confers land.

Only when Isaac dispatches Jacob to Paddan-Aram in order to find a wife does he speak to him of patrimony.

> Do not take a wife from among the daughters of Canaan. Get up and go to Paddan-Aram, to the house of Bethuel, your mother's father, and take a wife there from among the daughters of Laban, your mother's brother. May El Shaddai bless you, make you fertile and numerous, so that you become an assembly of peoples. May He grant the blessing of Abraham to you and your offspring, that you may possess the land where you are sojourning, which God assigned to Abraham. (28:1–4, JPS)

In forswearing Canaanite women, Isaac finds a justification for blessing Jacob in place of Esau. Esau is married to two Hittite women, while Jacob remains unmarked by association with women. In another tragic scene in which Esau misses the point only to find himself further discounted, he finally understands that "the daughters of Canaan" are unacceptable to his father so marries Mahalat, daughter of Ishmael (28:8–9). Esau tries to behave correctly, yet by marrying into the wrong line of Abraham finds himself further removed from the covenant. Jacob, in contrast, is sent back to the family's place of origin in order to marry properly; only in this dispatch does Isaac speak to

---

4. "Tales of the trickster Jacob are, indeed, central to Israelite identity and self-image" Susan Niditch, *Underdogs and Tricksters: A Prelude to Biblical Folklore* (Urbana: University of Illinois Press, 1987), 117.

Jacob of fertility, of the founding of nations and of the land. A further irony is that the land of Jacob's sojourning is granted to him only when he departs from it. "The blessing of Abraham" then means that one acquires a place of one's own only through the act of leaving it. Here we begin to see the homeland-diaspora environment from which a place called Israel emerges. Jacob's territory becomes meaningful in its absence.

When Jacob sets out from his hometown of Beersheva, he has no map and no sense of what awaits him on the other side of the Jordan. The way in which Jacob walks down the road forms the lived dimension of space. Michel de Certeau calls the ways in which people negotiate the boundaries that define them "everyday life" and sees living in the everyday as a mode of redefining while adapting to one's world. Stories work in much the same way, according to his estimation, by traversing space in ways often impossible in the here and now. Legends, like those of Jacob's wanderings, "permit exits, ways of going out and coming back in, and thus habitable spaces."[5] The legend is a practice that opens up a metaphoric level of space within ideological geographies that is unconstrained by ideology. The characters of legend navigate their path according to adventure's caprice; heroes transgress hierarchies and wander beyond boundaries into uncharted terrain. These wanderings represent a reordering of space that sets new limits authorized by the footsteps of an ancestor. De Certeau equates the legend with walking, an individual improvisation that makes space one's own; both establish new orientations and redefine relationships of difference. The story of Jacob, a legend about walking, would doubly please the French theorist. Like that of de Certeau's urban pedestrians, Jacob's story—which is Israel's story—"begins on ground level, with footsteps."[6]

• • •

Jacob's place materializes through his movements; these movements in turn open up Israelite practices. The practices are both spatial (a variant map that bridges the east and west banks comes into relief) and religious (Jacob is "the only patriarch to erect massebas [memorial stones] to establish the legitimacy of various cult centers").[7] Jacob's wanderings, like Abraham's, set the contours

---

5. Michel de Certeau, *The Practice of Everyday Life*, trans. Steven Rendall (Berkeley: University of California Press, 1984), 106.

6. De Certeau, *The Practice*, 97. Bob Marley correlates Jacob's walking with the story he sings in "Talkin' Blues."

7. Magnus Ottosson, *Gilead: Tradition and History*, trans. Jean Gray (Lund, Sweden: CWK Gleerup, 1969), 43.

of the land, yet Jacob's epiphanies more precisely mark sacred places. Jacob's first tactile experience of place occurs on his first night away from home.[8] Carrying only the intangible baggage of blessing, Jacob demonstrates youthful adaptability when he uses "one of the stones of the place" for a pillow (Gen 28:11). His first significant contact with land also provides the occasion for his first dream, in which he sees a ladder that reaches to heaven and on which angels ascend and descend. The dream points to the dual nature of Jacob's journey: a dream journey toward increasing proximity to God parallels his physical wanderings across the horizontal plane of space. The stations between home and exile have a corollary in the rungs between heaven and earth.

Although the ladder appears in a dream, Jacob's vision of God is physical in nature. God "stands over" Jacob and speaks to him:

> I am the God of your father Abraham and God of Isaac: the land on which you are lying I will give to you and to your descendants. And your descendants shall be like the dust of the earth, and you shall spread out to the west and the east and the north and the south. Through you and your descendants all the families of the earth shall find blessing. And here I am, with you: I will watch over you wherever you go, and I will bring you back to this soil. I will not abandon you as long as I have yet to do what I have promised you. (Gen 28:13–15) (David Stein trans. with one emendation)

The central theme of God's speech is land: God bequeaths the site of revelation to Jacob and his descendants and establishes it as a sacred center. Unlike the maps from the Deuteronomistic and Priestly sources discussed in chapter one, the cardinal directions are mentioned without specific coordinates.[9] God further vows to accompany Jacob on his way and to restore him "to this soil," an element not before associated with him. As in the case of Isaac's blessing, God speaks to Jacob of promised territory only as Jacob prepares to depart from it.

In parallel to God's standing over Jacob while he slept, Jacob takes his stone pillow and erects it as "a standing stone" memorial to his vision. Jacob thus marks a connection to his site of epiphany that he renames *Beth El*, House of

---

8. His arrival is described by a verb פגע that stresses the physical nature of Jacob's encounter with the place. The valence of this verb, in my reading, emphasizes that this is Jacob's first real encounter with land as well as one that begins his transformation into an eponymous founder of a nation.

9. Scholars believe Genesis 28 to derive largely from the Elohistic (E) source associated with the Northern Kingdom of Israel.

God, but he is not yet the man to found a nation. That he remains Jacob, the grasper, becomes manifest when he answers God's blessing with a bargain.

> If God will be with me, watch over me on the road that I am walking, give me bread to eat and clothes to wear and I return in peace to my father's house; then God will be my God. And this stone that I have set up as a memorial will be a house of God and all that you give me, I will present to you a tenth. (Gen 28:20–22)

Jacob wants real-life indicators that God will watch over him wherever he goes, and will return him to his native soil. He pledges that if he meets with success on his journey, then the God of his father's household will be his God, his pillow will be the cornerstone of a temple, and he will sacrifice a tenth of his bounty. Jacob views his epiphany as yet another opportunity to derive benefit. Because the land cannot yet be his, he tries to secure an assurance that this will one day be the case. The very self-assurance that helps Jacob achieve blessing also necessitates his emigration.

The distance between Beth El and Haran exists in a vacuum of unpracticed space. This lacuna means that Jacob's initial crossing of the Jordan goes unrecorded as if no boundary is at this point acknowledged. Still, the two Jordan crossings serve as the framework for this leg of Jacob's travels. The vision of angels moving up and down on a ladder will be balanced and extended by Jacob's nocturnal wrestling with a divine being prior to his return home. His transformation becomes evident in the change of tone in which he addresses God from either side of the Jordan. The hero pattern provides a relevant lens because the phases of Jacob's development are expressed in geographical terms. Jacob's movements are also phases of self.[10]

> Jacob's two direct encounters with Yahweh/Elohim are arranged symmetrically, the first on the way from Beersheba to Haran, the second on the homeward journey. Both encounters can be viewed as rites of passage in a geographical and spiritual sense: Jacob is passing through

---

10. For the hero pattern as an interpretive tool, see Alan Dundes, "The Hero Pattern and the Life of Jesus," *In Quest of the Hero* (Princeton: Princeton University Press, 1990), 179–223; Lord Raglan, *The Hero* (New York: Vintage, 1956); Otto Rank, "The Myth of the Birth of the Hero," in *Myth of the Birth of the Hero and Other Writings by Otto Rank*, ed. Phillip Freund (New York: Vintage, 1964). For the application of the hero pattern to the Jacob story, see Ronald S. Hendel, "The Jacob Cycle and Israelite Epic," *The Epic of the Patriarch: The Jacob Cycle and the Narrative Traditions of Canaan and Israel* (Atlanta: Scholars Press, 1987), 99–165, and Niditch, *Underdogs and Tricksters*, 93–125. For the hero pattern as it relates to Rachel and Leah, see Havrelock, "The Myth of Birthing the Hero," *Biblical Interpretation* 16 (2008): 154–78.

the threshold of his native land, while at the same time entering into a new phase of his heroic identity.[11]

The nature of Jacob's identity depends upon his orientation toward the land. Jacob's departure indicates that he is not yet fused with place, while his return shows his readiness to assume the role of national founder.[12] The disparity between these two orientations shows the territory east of the Jordan (and perhaps the diaspora in general) to be a politically generative space. In Aram, Jacob teams with four women to establish a clan and realizes the value of having a place during the long sojourn in a place not his own.

The problem with Aram is Laban, a character whose deception and eye for profit show that he is kin to Rebecca and Jacob. Rebecca and Jacob's actions, however morally ambiguous, ultimately serve a covenantal ideal while Laban's actions find no such justification. As Jacob labors for Rachel only to be tricked into marrying Leah, works an additional seven years to rightly call Rachel his wife, and then builds up his flocks while allegedly making Laban rich, it becomes clear that Jacob will never claim land belonging to Laban. When Rachel, after a prolonged period of barrenness, finally gives birth to Joseph, Jacob asks Laban to let him go back to his land. After implicit and explicit contests over what Jacob can actually claim as his own, Jacob flees Laban's household. The mention of Laban as "the Aramean" at this juncture points to the national implications of the conflict between kinsmen (Gen 31:20).[13] Just as the primacy of Esau's birth signifies that Israel claims something rightly belonging to Edom, so the statement that both Rachel and Jacob "steal" from Laban suggests the stealth acquisition of important assets from Aram (Gen 31:19–20).[14]

When Laban catches up to Jacob in the mountains of Gilead, their property dispute involves whether Laban's flocks can count as Jacob's wages, if Jacob's wives and children fall under Laban's jurisdiction, and the whereabouts of Laban's household gods. The treaty between Jacob and Laban addresses

---

11. Ronald Hendel, *The Epic of the Patriarch*, 63. "According to the directions of Jacob's route, Bethel and Penuel would roughly correspond to the high points from which Jacob would descend into the Jordan rift valley on his exit from and return to the promised land." Hendel, *The Epic of the Patriarch*, 63.

12. "Land functions in this Cycle as subject of the binary pair *exile/homeland*. The actions of Genesis 25:19–35:22 can thus be viewed along a spatial axis. Jacob flees from Canaan and has an encounter with the divine at the border shrine of Beth-el (28:10ff.); he stays in Aram until Rachel gives birth, whereupon he returns and encounters the divine at the border shrines of Mahanayim and Penuel (chapter 32). The shrines mark the transition of action from sacred to profane space, and back." Michael Fishbane, *Biblical Text and Texture: A Literary Reading of Selected Texts* (Oxford: Oneworld Publications, 1998), 6.

13. God similarly speaks to "Laban the Aramean" when He instructs Laban in a dream to watch what he says to Jacob.

14. Rachel "steals" Laban's household gods where Jacob, more metaphorically, "steals Laban's heart."

none of these issues, but instead divides territory. Jacob again raises a *matseva*, a memorial stone, whose force as a barrier thickens when the men accompanying him pile additional stones into a mound. This mound indicates a border reinforced by the language of a treaty. The tension that suffuses the dialogue between Jacob and Laban will erupt into future hostilities between Israel and Aram, but the treaty confers equality upon them.[15] The two narratives that come into contact only to be divided by the stones are signaled by the series of doublets that comprises the treaty scene. There are two stones, two meals, two maps, two gods, and two levels to the treaty—one familial and one political.

Gilead: No Direction Home

The hauntingly prescient scene in Genesis 31 shows treaties as deepening rather than reconciling conflicting accounts. Gilead, a site of continual contest, gains definition through the treaty between Laban and Jacob (figure 3). This may reflect material conditions when the texts were written, but on the literary level its conflicting meanings further multiply.[16] The Gilead disputed by Israel and Aram also has a disputed status within Israel. The Hebrew Bible reflects the dispute internal to Israel insofar as some accounts present Gilead as an inseparable region of Israel and others place it resolutely outside of Israel. The question of whether Gilead is a part of or apart from Israel renders it ambivalent ground even in those texts that unequivocally include it. Thus Gilead is sometimes an auxiliary site of home, a proximate diaspora, or the place in between homeland and exile where the concepts lose their integrity exactly as they come into being.[17] The third possibility most pertains to the

---

15. "The story of Jacob and Laban who settle a dispute and mark a border between them reflects a time in which the Aramean siege of Samaria (2 Kings 6:25–7:20) and occupation of Gilead (Amos 1:3) still rankled as recent memories." Karel Van Der Toorn, *Family Religion in Babylonia, Syria, and Israel: Continuity and Change in the Forms of Religious Life* (Leiden: E. J. Brill, 1996), 262.

16. "During Iron I and II Gilead, or at least its northern region toward Bashan, was a point of controversy between the Israelites and the Aramaeans because of its importance as a passage for trade via the Kings' Highway. The Old Testament sources have countless references to this fact, while no Aramaic material mentioning this has so far been discovered . . . The first reference to 'Gilead' is in connection with Tiglath Pileser III (745–727), although not in the sense of an area but as a frontier town against Aram in a fragmentary royal inscription on a stone from Nimrud . . . Nonetheless it is clear in this context that the 'town of Gilead' in Tiglath Pileser's inscriptions must be identical with Ramoth Gilead, the main center in northern Gilead, which during the period of the Divided Kingdom was a disputed frontier town between Aram and Northern Israel." Magnus Ottosson, *Gilead: Tradition and History*, trans. Jean Gray Lund (Sweden: CWK Gleerup, 1969), 19–21.

17. Syrian-Jewish communities of the second and third centuries are portrayed with similar ambivalence. Syria is "a middle case between that in the Land of Israel and outside the Land." Richard Sarason, "The Significance of the Land of Israel in the Mishnah," *The Land of Israel: Jewish Perspectives* (Notre Dame: University of Notre Dame Press, 1986), 122.

3. The region of Gilead and the kingdoms of Israel, Aram, Ammon, Moab, Edom, and Judah. Soffer Mapping.

stories of Jacob in Gilead. If Canaan is home and Aram diaspora, then Gilead is a borderlands where Jacob severs his Aramean ties and reencounters his Edomite brother. If all he did in Gilead were negotiate with relatives not embraced by covenant, then it would simply be a foreign place. But since Jacob finally becomes the eponymous Israel in Gilead, it figures as a generative, inextricable part of Israel. Gilead, the place under perpetual negotiation, gains its name through the terms of a treaty.

The Treaty

As anyone who has tried to reconcile individuals or nations can attest, little progress can be made as long as the two parties insist on airing their grievances. On this count, the ancestors of Syria and Israel are portrayed as particularly histrionic and intractable. Laban catches up with Jacob in the mountains of Gilead, enumerates the wrongs done to him, and searches in vain for the household gods concealed by the guilty Rachel.[18] When the search turns up no gods, Jacob demands to know why he has been made to suffer such indignity at Laban's hand. No doubt relishing the last word that "the daughters are my daughters, the sons are my sons, the flocks are my flocks, and all that you see is mine," Laban then proposes a treaty (Gen 31:43–44). Laban's insistence that all that Jacob beholds belongs to him raises the question of whether the treaty will confer any territory on Jacob and by extension whether Gilead rightly belongs to Aram. In any event, Jacob acts upon Laban's suggestion by raising his memorial stone, instructing his men to build a rock mound, and inaugurating the site through a feast (31:45–46). The splitting of symbols ensues.

The differentiation that often accompanies translation becomes apparent as Laban names the cairn *Yegar-sahadutha*, "the mound of witness" in Aramaic, and Jacob calls it in Hebrew *Gal-ed*, "mound of witness," the given etymological origin of the name "Gilead." Laban's echo of the appellation *Gal-ed* produces an additional name: "And [it was called] Mizpah, because he said, 'May God watch between you and me, when we are out of sight of each other'" (Gen 31:49, JPS). Laban stipulates the terms of the treaty, which Jacob ratifies through ceremonial acts. Laban seeks assurance that Jacob will not replace his daughters with newer wives (Gen 31:50) and that Jacob will not transgress the border now defined by the stones in the name of raids or vengeance (31:52). The mound of witness then functions as a border that keeps Laban on one side, vowing to never "cross over to you past this mound" as long as Jacob similarly abides by the limits of the stones (31:52). The names of two gods or "two epithets for God," the God of Abraham and the God of Nahor, underwrite the treaty (31:53).[19] Jacob accepts the terms by evoking "the Fear of

---

18. Genesis 31:25 has Jacob encamped "on the mountain" and Laban "on the mountain of Gilead." Biblical scholars have proposed a range of solutions to the crux. Ottosson qualifies Noth's idea that we have a J and an E tradition that both refer to Mount Gilead through de Vaux's observation that the story bears out the compound place name Mizpeh Gilead. "The complete form is Mizpeh Gilead (Jg. 11:29) and the shortened form Ha-mizpeh (Jg. 11:11, 34)." See Roland de Vaux, "Notes d'histoire et de topographie Transjordaniennes," *Vivre et Penser: recherches d'exégès et d'histoire* 1 (1941): 25–29. Ottosson goes on to read Gen 31 as indicating, "that the cult site Gal'ed-Mizpah was situated on a mountain in the hill country of Gilead and that this is identical with Mizpeh-Gilead in Jg. 11." Ottosson, *Gilead*, 23.

19. Magnus Ottosson reads necessity in Laban's verbosity. "It is a peace treaty in which Laban lays down the conditions, but he is forced to offer the treaty. In fact if possession of the household gods

Isaac his father," an additional name for the God of Abraham (31:53), and setting out the bread and meat of a sacrificial meal. As night falls, the distinction between Jacob and Laban reverberates in the two stone configurations, two place names, two Gods, and two meals. The duality of the armistice anticipates a war prevented only for the time being. The terms of Jacob and Laban's agreement alone make the rocks a border among the mountains of Gilead, a frontier between Israel and Aram and also between Israel and Ammon (Judg 11:29). Although Jacob separates from Laban here, it is also their most significant point of contact and translation. What the Jordan should distinguish becomes blended in Gilead.

Two Camps

With a contained father-in-law behind him and an unpredictable brother ahead, Jacob considers his return home. The encounters of Gilead multiply as "angels of God" meet him on the road. The anticipated encounter with Esau balances that with Laban and the angels met on the way recall those he saw moving up and down the ladder in his first dream.[20] By approaching the Jordan, Jacob enters a zone of conflict as well as vision. In this respect, Gilead functions as a literary device, "as the link joining the two legends. It is on the 'border' of Gilead that Jacob consents to a treaty with Laban and just before he enters Canaan that he settles matters with Esau."[21] But at the moment in question, nothing is settled with Esau. While the borders of Edom may have already taken shape, where is Jacob's place? What land can he claim as his? Other than a pile of stones that keep Laban at bay, what borders protect him? None of this is certain. The vulnerability of being without a place becomes stark when he receives word that Esau is on his way with four hundred men. It seems the correct moment to remind God of Jacob's request to "return in peace to my father's house" (Gen 28:21).

The humble tone of Jacob's supplication shows a change in the man who once drove a hard bargain with God. He is a second Jacob, a different person from the one who used a stone for a pillow and crossed the Jordan without event. He looks back on his past from the vantage point of the other side.

---

brings with it the legal right to Laban's property and if Laban is aware that in spite of this intensive search Jacob has the gods with him then Laban can use the treaty to nullify further property claims by Jacob." Ottoson, *Gilead*, 41. For the two epithets for God, see ibid., 42.

20. Angels function here, as they do in considerably later Jewish and Christian texts, as "frontiersmen ... [who] live on both sides of a border that they cross at all times." Beatrice Caseau, "Crossing the Impenetrable Frontier between Earth and Heaven," *Shifting Frontiers in Late Antiquity*, ed. Ralph W. Mathisen and Hagith S. Sivan (Aldershot, UK: Variorum, 1996), 335.

21. Ottosson, *Gilead*, 37. When Jacob dispatches tribute to "the land of Seir, the fields of Edom, " Esau's rootedness is emphasized (Gen 32:4).

> God of my father Abraham and God of my father Isaac, Yahweh who said to me, "return to your land and to your birthplace and I will take care of you," I am unworthy of all your kindness and of the truth that you have performed for your servant. I crossed this Jordan with only my staff and now I have become two camps. Save me please from the hand of my brother, from the hand of Esau, because I am afraid that he will come and strike me, the mother together with her sons. You are the one who said to me, "I will take care of you and make your descendants like the sand on the shore that cannot be counted because there is so much of it." (Gen 32:10–13)

Jacob couches his story between citations, remembered fragments of divine promises. The form of the address speaks to the tenuous nature of a life lived between blessings and how Jacob's homelessness means that God alone protects him. Of himself Jacob says little. Where once he crossed the Jordan unencumbered, "with only my staff," now he has become "two camps." However sparse, the evocative self-description demands interpretation.

Quite literally, Jacob has divided his family into two camps. In the name of protecting them from the onslaught of Esau, a panicked Jacob "divided the people with him, the sheep, cattle, and camels into two camps" (Gen 32:8). With the division Jacob hopes to protect a remnant of his clan should Esau overtake the rest. The sense of his family as divided into two camps also results from the fact that his two wives are sisters who have struggled with one another for the attention and affection of both Jacob and God. They have given birth to Jacob's children while also claiming these children as well as those of their servants as members of either a Leah or a Rachel "camp."[22]

Amidst the anxiety about his family's survival, Jacob also takes stock of himself. With staff alone, he has grown his blessing into a fortune. The two camps of wives and children, sheep and cattle are the result of his own labor conducted in the atmosphere of blessing. Jacob is also of two camps in that he has a kind of dual citizenship in the camp of people and the camp of angels. From the moment he dreamed the ladder, a dream journey toward heaven has paralleled his physical wanderings across the plane of space. The encounter

---

22. The basic division of Israel into Leah tribes and Rachel tribes becomes clear in Numbers 1–4, the initial census of the tribes taken during the wilderness wandering. See Havrelock, "B'midbar." Israel's configuration as a kingdom divided into north and south can also be accounted for matriarchally. All Israel may descend from Jacob, but Rachel is the mother of the Northern Kingdom (Jer 31:15) and Leah of the Southern. If, as Martin Noth argues, the ultimate written form of the Jacob legends is intended to promote reconciliation between the once and soon to be again distinct Northern Kingdom of Israel and Southern Kingdom of Judah, then the division into two camps on the basis of the matriarchs might be a strategy of admitting and protecting difference.

with angels always precedes the encounter with opponents. Between facing Laban and Esau, Jacob runs into angels on the road. Upon seeing them, he says, "this is the camp of God" (Gen 32:3).[23] Naming the place *Mahanaim*, Two Camps, Jacob defies biblical grammar by turning מחנה אלהים, "the camp of God," a construct in the singular, into מחנים, Two Camps, a noun in the dual form. To Geoffrey Hartman, this form points to the "betwixt-and-between status of Jacob: he is a wanderer; he dwells in the space of Mahanaim, the double camp on this and the other side of the river."[24] Jacob's life has been divided between the land and the other side of the Jordan.[25] The experience of being inside and out, in the land and much in the diaspora has formed Jacob and made God's blessing both potent and necessary. "O God," Jacob seems to say, "I am not of this land you speak of. I am a man of split identity." If God wants the man of two camps to become landed, then He must protect and deliver Jacob. The creation of homeland in this case requires miracle.

Jacob's admission of his split self operates on a national as well as an individual level. The individual prayer doubles as Israel's confession that division is stitched into its identity; like its ancestor, the People Israel will continually traverse the space between home and diaspora. It is this in-between space that generates the identity of ancient Israel. Homeland is a motivating factor, yet only half of the picture. In order to discern the historical implications, I avail myself of biblical source criticism and try to plumb the mind of the Northern writer.[26] This Northern writer or school of scribes assumes the perspective of the Kingdom of Israel, ten Northern tribes said to split from Judah following the reign of Solomon. The Northern Kingdom follows its own, sometimes intersecting, political trajectory of succession, insurrection, and alliance until 722 B.C.E., when the Assyrian army defeats the Kingdom of Israel and renders its ten tribes "lost" in the diaspora. What source critics have termed the

---

23. Yitzhak Peleg observes that the phrase "Angels of God" appears in the Hebrew Bible only when Jacob dreams the ladder (Gen 28:12) as he prepares to leave the land and when he encounters them in Mahanaim prior to his return (Gen 32:2). Peleg, "Going Up and Going Down: A Key to Interpreting Jacob's Dream (Gen 28, 10–22)," *Zeitschrift für die Alttestamentliche Wissenschaft* 116 (2004): 7.

24. Geoffrey H. Hartman, "The Struggle for the Text," in *Midrash and Literature*, ed. Geoffrey Hartman and Sanford Budick (New Haven: Yale University Press, 1986), 10.

25. "Jacob's/Israel's having to leave the country that YHWH has destined for him, to stay for many years in Mesopotamia, and his subsequent return under God's promise, are elements that occur not only in Genesis, but also in Isaiah, Jeremiah and Ezekiel." Dicou, *Edom, Israel's Brother and Antagonist*, 162.

26. These Northern traditions are associated with the Elohist writer (E). Although E has fallen out of fashion of late, I uphold it because the political sensibilities expressed in Northern traditions are so different from Southern and Jerusalem-oriented sensibilities. For other defenses of E, see Joel S. Baden, *J, E, and the Redaction of the Pentateuch* (Tübingen: Mohr Siebeck, 2009).

E source preserves distinctly Northern traditions that include a positive sense of Gilead and Transjordan in general as a vital, inseparable part of Israel.

According to biblical texts, considered subsequently, two and a half tribes claim their patrimony east of the Jordan River. In the logic of the texts, these tribes constitute an essential part of the Northern Kingdom but also form part of the United Kingdom prior to the split. Israel then, like Jacob, is divided into the "two camps" of the east and west bank tribes. As Moshe Weinfeld writes, "It is also possible that the story of Mahanaim in the Valley of Sukkoth, in which Jacob divides his people into two camps (Gen 32:8), reflects the ancient reality of the existence of Israelite camps east of the Jordan analogous to those on the west bank."[27]

According to Weinfeld, the story of Jacob's two camps sanctions the early territorial configuration in which the tribes of Israel were divided into east and west bank factions. Genesis 32, a Northern text, implies that the east bank constitutes an integral site of Israelite identity. The legend of ancestral passage opens up the space of the east bank as a legitimate component of Israel's identity. So Gilead, from the point of view of a Northern scribal school, is a site sanctified by theophany as well as by Jacob's footsteps. According to this line of thinking, the tension surrounding Gilead is a product of Northern traditions coming into contact with Southern traditions that vilify Moab and Ammon as well as Priestly traditions that draw a self-defining line at the Jordan.[28] Ambivalence in this case results from the compilation of variant traditions and their lingering incongruity.

However, source criticism cannot explain away the ideological valence of Gilead. In the Jacob legends themselves, Gilead is a place of uncertainty, anxiety, and encounter. Even if Jacob's actions inaugurate its cultic sites, Gilead is still where Jacob finds himself exposed and without solid ground to stand on. I propose that even the Northern writer who defends the primacy and sanctity of Gilead admits that its nebulous borders allow other peoples to seep in and disrupt what Israel is trying to become.[29] There is yet a third way to think about how the texts may relate to political processes. Let us imagine, for example, that the two and a half Transjordanian tribes—Reuben, Gad,

---

27. Weinfeld, *The Promise of the Land*, 41.

28. Weinfeld accounts for the tension as the result of different sources collated in the Hebrew Bible. Thus the Shilonite priests, interested in protecting "the exclusive sanctity of the camp of Shiloh, declared the area east of the river to be unclean (Josh. 22:19)." Weinfeld, *Promise*, 46. The E writer, also referred to by Weinfeld as "the folk tradition," displays a more reconciled position about the two camps of Israel through the story of Jacob.

29. Gilead was territory disputed by the Israelites and the Arameans (1 Kings 22:3; 2 Kings 9:14), and a later tradition suggests that the founding father of Gilead, Machir, was an Israelite-Aramean half-breed (1 Chr 7:14).

and half Manasseh—were initially discrete tribes or peoples brought into an Israelite confederation through marriage, treaty, or defeat.[30] Truly absorbing these groups would require a manner of folkloristic syncretism whereby their local traditions would be incorporated into a national narrative. This means that their founding father Jacob would need to be given a status similar to that of Abraham, the founder of the Southern Kingdom. This would explain why Jacob doesn't seem at all at home in Isaac's house—he has simply been spliced into Abraham's family to represent the union of initially disparate traditions.

Although anxious while in Gilead, Jacob is there closest to God, which suggests that his holy land lies east of the Jordan. Building on these points, let me try to add a new twist on this analysis. The component parts of nationalist narrative when held up to scrutiny ultimately erode its coercive unity. For example, looking at the variant traditions that accrue at border sites calls the distinction of those things divided by the border into question. More pointedly, the incorporation of traditions in which Gilead is both beyond the border and still connected shows how strange places can easily become home and how home includes foreign elements.

Encounter

I have left Jacob east of the Jordan, which is not where Genesis leaves him. After all, the point of Jacob's appeal is for God to save one, perhaps both of his camps from Esau and then to usher them into the land. Not quite confident that God will protect him from Esau's revenge, Jacob sends waves of gifts in the hopes that the livestock can compensate Esau for his stolen blessing as the lentils paid him off for his birthright. Jacob then escorts his family across the Jabbok and crosses back in order to be alone on the other side.

On this riverbank threshold, Jacob is resolutely in the camp of angels. Full body contact is the closest that Jacob will ever get to God and, according to his own testimony, as close as anyone can get without dying. Jacob's attacker is a shifter of names and perhaps of shapes who plays dirty when down by wrenching Jacob's hip from its socket and then demanding release as dawn breaks. Jacob, who knows how to wrest a blessing from every struggle, demands one this time as well. The blessing takes the form of the name "Israel," "because you have struggled with God and with people and prevailed" (Gen 32:29). The injured Jacob will not again wander so vastly on his own two legs, but he is finally made into a man that is also a place. His opponent will not di-

---

30. Such imagining on my part is influenced by Noth's ideas about canon formation as well as the mention of "the men of Gad" as a discrete group in the Mesha stele. This notion of the incorporation rather than primal presence of the east bank tribes in Israel helps me to read the ambivalent status of Transjordan throughout the biblical canon.

vulge a name, almost taunting Jacob, "Why would you ask my name" (32:30). Unable to call his adversary anything, Jacob names the place "Peniel, The Face of God, because I have seen God face to face and I'm still alive" (Gen 32:31). Jacob, who has been dropping names all along—Beth El (The House of God) for the place where he dreamed the ladder, Mahanaim (Two Camps) for the road back home, and Peniel (Face of God) for the place of contest—now too has a place name.

Wounded by the encounter, Jacob limps across the river.[31] His dual residence in heaven and earth, at home and in diaspora is encoded in his new name and evident in the ambulatory injury that results from shuttling back and forth between the two.[32] The two crossings of the Jabbok and the two names of Israel reinforce the dual nature of events in Gilead, where two rivers seem to be at issue. When he speaks to God, Jacob remembers crossing the Jordan with his staff. Is this not the river he should cross and the one where he should contend with an opponent in order to gain passage? This is among the "abrasive frictions" that lead Roland Barthes to try to read the text. "Jacob's purpose is to return home, to enter the land of Canaan: given this, the crossing of the River Jordan would be easier to understand than that of the Jabbok."[33] Furthermore if Jacob can freely cross the River Jabbok, then why does he face an opponent there?

Possibly, the episode at the Jabbok contests the Jordan as the definitive national border endowed by God. Having just established a boundary with

---

31. Richard Friedman observes that the word for inquiry (ותקע) used in Genesis 32:26 is used only here and in Numbers 25:4 (in the Hiphil—והוקע), where the men of Israel are punished for their participation in the Moabite orgy. Richard Elliott Friedman, *Commentary on the Torah* (San Francisco: Harper San Francisco, 2001), 112. In both cases, "Israel" is defined through a dangerous encounter that scars the "body" of Israel before it crosses the Jordan. According to de Certeau, border struggles are the processes through which bodies gain definition. De Certeau, *The Practice*, 126. The notion of contact between Jacob and representatives of other nations as well as a divine representative ties in to his stature as a "limping hero." In a comprehensive study of the figure, Peter Hays understands the limp as a negotiation of virility and sterility. Without the leisure to take on all of his examples, I would add that the "limping hero" seems always to be caught between realms and it may be the negotiation of categories in general that leads to the scar. Peter L. Hays, *The Limping Hero: Grotesques in Literature* (New York: New York University Press, 1971). Claude Lévi-Strauss sees limping or otherwise impaired heroes as performing a mediating function between the categories of life and death. This characterization would certainly apply to Jacob on his way to meet the man who once vowed fratricide, although I see the heaven-earth mediation at the Jordan as equally forceful.

32. In some sense, Jacob is destined for a leg injury. His name, Jacob, means "heels" or "one who grabs at the heels" and reflects that he emerged from the womb grabbing at his brother's heels (Gen 25:26). After Esau is duped by Jacob a second time, he affirms that Jacob lives up to his name (Gen 27:36). As Jacob grabs at his brother's feet, the angel wounds his thigh.

33. Roland Barthes, *Image—Music—Text*, trans. Stephen Heath (New York: Hill and Wang, 1977), 130.

Aram in the mountains, Jacob's struggle with an angel could then be the legend that draws an additional border of Gilead at the Jabbok. In Judges 11, the Jabbok figures as such a border between Gilead and Ammon. Yet Ammonites play no role in Jacob's story. The subsequent encounter with Esau would suggest that the Jabbok serves as the dividing line between Gilead/Israel and Edom. Other texts, however, place Edom far south of the Jabbok. The problem is compounded when Jacob arrives in the West Bank city of "Shechem, which is in the land of Canaan, having just come from Paddam Aram" (Gen 33:18). This version, it seems, ushers Jacob into Canaan directly from Aram. Gilead, the ambivalent place of encounter, cannot quite be situated.

In order to avoid the charge that I too have gotten stuck in Gilead without orientation or sense of where the borders lie, let me try to resolve some of this while leaving, as Barthes suggests, the text's "significance fully open." Barthes correctly calls the naming scene a *"metonymic montage"* in which the themes (Crossing, Struggle, Naming, Alimentary Rite) are *combined*, not "developed." Let us tease out the individual elements and then reckon with their juxtaposition in the montage. The ease with which Jacob crosses the Jabbok either shows it to be a border of no significance or rehearses the crossing of the Jordan in a manner similar to how the encounter with the angel serves as prelude to that with Esau.[34] On the spatial level, Jacob's successful match with a divine being sanctifies Penuel, Mahanaim, and perhaps the east bank in general as legitimate cultic sites. The polemical force of this claim will soon become apparent. Jacob's acquisition of the new name "Israel" plays an essential function in his heroic biography—at long last Jacob is one with a place—but the naming does not, as Barthes would have it, create something new. In Barthes' reading, "By marking Jacob (Israel), God (or the Narrative) permits an anagogical development of meaning, creates the formal operational conditions of a new 'language', the election of Israel being its 'message'. God is a logothete, a founder of a language, and Jacob is here a 'morpheme' of the new language."[35] This reading fails not because of its structuralist orientation, but because a Christian structural trope—typology—leads to a misreading in which continuity suddenly figures as a radical break from the past. Other than the fact that Jacob's opponent turns out to be a divine manifestation, there is no uniquely "anagogical" or spiritualized meaning at work. "God" has been speaking the very language since creation, and the covenant (Barthes' "election") is a message familiar to readers of Genesis. The additional name conferred on Jacob is no founding moment; hereafter, the names "Jacob" and

---

34. Barthes, *Image—Music—Text*, 131. "In the 'night encounter' Jacob wrestles with the 'Esau' he carried within him." Michael Fishbane, *Biblical Text and Texture*, 53.

35. Barthes, *Image—Music—Text*, 135.

"Israel" are interchangeable and serve as synonyms in the parallelism of biblical poetry. Jacob's victory that compromises his body simply makes him like Esau/Edom, a body with a place.

Still, despite Jacob's strivings, he remains not quite like Edom. First of all, Jacob's acquisition of the name is not something natural with which he is born, but rather something he must struggle to achieve. Second, the acquisition results from having been outside of the land for some twenty years. Jacob's status as national progenitor is continuous with, rather than a radical break from, the experience of diaspora. Israel is unlike Edom because he gains a place through contention outside of the land and then crosses into the land to face yet more trouble (Gen 34). The experience of being outside of the land creates an interior and the experience of being inside necessitates boundaries in order to keep things out. This point wrests the wrestling scene from a reading that would see the extra name as having "the value of a spiritual rebirth (of 'baptism')."[36] Rather than a rebirth of any sort (when Jacob meets Esau he is the same old trickster), the events around the Jordan borderlands encode something important about Israel—its development depends upon the dialectic of homeland and diaspora; they encode something about nations as well—they are always amalgamations of prior elements. Barthes knows this to be true of language, yet in his reading of the Bible holds to a kind of Christian futurity in which a divine manifestation brings something new into the world.

Many lines of continuity intersect in the borderlands east of the Jordan. Angels that Jacob once witnessed in transit touch and mark him; Leah and Rachel, like their aunt Rebecca, migrate to Canaan; and diaspora laps up on homeland while some symbolic stones, for the time being, keep Aram at bay. As with Abraham and Lot, Jacob divides his land from Esau's in the Jordan River Valley. Where Abraham's actions were performed with a certain clarity, Jacob's are more murky. Reunion with Esau marks the third and final encounter in Transjordan. Although Esau comes backed up by four hundred men, he seems to bring the more sincere intentions. Jacob engages in no small amount of bowing, flattering, and bribing, yet Esau disarmingly runs to his brother with hugs and kisses. Esau assumes that they will travel together and that Jacob will remain at least for awhile in Edom.

Unaware of Jacob's second name, Esau assumes that he still lacks a place and will need to make himself at home in Edom. Since Jacob spent their childhood grasping for what was Esau's, the assumption has validity. Jacob alone knows at this point that he is not only the grasper at his brother's heels, but also a man that is a nation like Esau/Edom. Unlike Abraham when he sepa-

---

36. Barthes, *Image—Music—Text*, 131.

rated from Lot, Jacob does not stipulate that since Esau is going southeast that he intends to go northwest.[37] Instead he dissembles, faulting young children and animals for the delayed pace well behind his brother. Further deflecting Esau's help and the watchful eyes of his men, Jacob turns in the other direction and sets up a wanderer's household in Sukkot.

The encounter with Esau would seem another scene of border drawing. Yet it is only a drawing if what transpires is a border at the Jabbok River between Gilead, the land of Jacob/Israel, and Edom, the land of Esau/Seir.[38] This is geographically strange, since Edom is located south of the Jabbok and even south of the Salt Sea, and thematically strange, because such a move would efface Ammon and Moab.[39] It certainly makes sense as a Northern tradition that situates Gilead north of the Jabbok, but opens up a whole question of where Edom belongs. The scene has Esau traveling to and from the Jabbok, so perhaps it is important only that Jacob settle accounts with his Transjordanian brother here.[40] On this count, the scene does double work. With its Northern provenance, it shows Gilead to be the place where Jacob garnered divine strength and beat his brother through cunning once and for all. In the contexts of Genesis and perhaps of regional politics, it shows how Jacob left his brother behind in the east before crossing the Jordan and settling in the land. Seen in this light, Edom then resolutely belongs, like Moab and Ammon, to the other side. When read in conjunction with the stories of their youth,

---

37. Or simply north if we accept the Northern view in which Jacob's "home" is at Sukkot in Gilead (Gen 33:16). I would then distinguish Gen 33:18–20, in which Jacob travels from Paddan-aram to Shechem, as belonging to a different source interested in having Jacob settle west of the Jordan.

38. Ottosson sees this as the dynamic: "Israel takes Gilead while Esau inherits Edom. There is as yet no question of a 'distribution theology' but the justification of the heritage of Gilead depends on the blessing which Israel wins by force." Ottoson, *Gilead*, 53.

39. This murkiness perhaps bespeaks the fluidity of division between Israel and Edom (later Judea and Idumea) and the ongoing Edomite migration westward. See Bartlett, *Edom and Edomites*, 140–43, and A. Kasher, *Jews, Idumaeans, and Ancient Arabs: Relations of the Jews in Eretz-Israel with the Nations of the Frontier and the Desert during the Hellenistic and Roman Era (332 BCE–70 CE)* (Tübingen: Mohr, 1988), 1–6.

40. Noth understood the geographic problem in terms of the development of traditions. The first stage involved the tension between pastoralists on the west side of the Jordan and hunters to the east. Later when a Southern, Judean group joined with the Northern tribes, the story expressed "nationalistic antipathy between Judah and Edom." P. Kyle McCarter, Jr., "The Patriarchal Age: Abraham, Isaac, and Jacob," *Ancient Israel: A Short History from Abraham to the Roman Destruction of the Temple*, ed. Hershel Shanks (Washington, D.C.: Biblical Archaeology Society, 1988), 16. Noth's identification of the East Jordan Jacob tradition helps to solve the crux of Jacob's double burial. The East Jordan tradition records a ceremony of lament for Jacob by his sons at Goren ha-Atad east of the Jordan (Gen 50:10–12). Noth reasons that the ceremony was once part of a burial, but that Priestly writers had the sons carry Jacob's body to Hebron and bury it at the cave of Machpelah (Gen 50:13–14). Martin Noth, *Überlieferungsgeschichte des Pentateuch* (Stuttgart: Kohlhammer, 1948), 95.

this scene makes the Jordan a dividing line between Esau's wild "nature" and Jacob's "culture."

Frank Crüsemann attributes more peace to the region—seeing the brothers thereafter living "as equals and in freedom next to each other"—than I think the Bible allows.[41] After all, Jacob sets the paradigm of how to avoid Edom and not venture toward alliance (Num 20:18–21; Deut 2:5). Edom, contra Crüsemann, is an ideological topos much like Moab: a mirror kingdom to Israel and Judah that it is best not to get too close to and that it is permissible to oppress. Gilead is more of a practice than a topos and Jacob something of an ideal Transjordanian subject who shuttles back and forth across the river while negotiating Aram and Edom in the form of Laban and Esau. Gilead is a site of reconciliation as well as one of anxiety and injury. Unlike Moab and Ammon, Gilead is not blatantly Other; it is the place where Israel takes form through contact and negotiation.

---

41. Frank Crüsemann, "Dominion, Guilt, and Reconciliation: The Contribution of the Jacob Narrative in Genesis to Political Ethics," *Semeia* 66 (1995): 69. Noted, however, is the tension between "the generous conciliatory presents" and the prenatal prophecy determining Jacob's dominance over Esau. "A reconciliation with the fraternal nation does not merely take place against the background of a history saturated with guilt. On the contrary, it happens in light of an existing divine promise, which has assured Jacob/Israel permanent dominion over Esau/Edom. Precisely what the narrative does *not* show is such a divine promise being carried out irrespective of human behavior. On the contrary, the divine pledge is *not* realized, and the reason why events happened differently is explicated." Ibid., 74.

Chapter Four

THE BOOK OF JOSHUA AND THE IDEOLOGY OF HOMELAND

Rabbi Judah bar Simon in the name of Rabbi Yohanan: In the Torah, in the Prophets, and in the Writings we find proof that the Israelites were able to cross the Jordan only on account of the merit achieved by Jacob.

In the Torah: With my staff I crossed this Jordan and now I have become two camps (Gen 32:11).

In the Prophets: You will tell your children, "Israel crossed this Jordan on dry land" (Josh 4:22).

In the Writings: What alarmed you, O Sea, that you fled? O Jordan, that you reversed course... from the presence of the God of Jacob (Ps 114:5, 7).

—Genesis Rabbah 76:5

One cannot approach the biblical book of Joshua without contending with the ideology of homeland. Discerning the ideology of homeland, in fact, may be the very reason to read Joshua, as the book dramatizes immigration, conquest, and the establishment of territorial boundaries. Joshua is a disciplined general, an exemplar of Deuteronomistic virtue who sways "neither to the right nor to the left" (Josh 1:7). As a straightforward national hero, Joshua lacks the complexity of Moses, the anger of Samuel, the paradox of David. The book of Joshua narrates the fulfillment of the Deuteronomistic national myth through the phases of destruction, extermination, and the partitioning of land. Only the collective character of Israel rescues the book from the realm of complete propaganda. Even amid wars to conquer the homeland, the collective Israel refuses totalizing unity and engages instead in rebellion, protest, and subversion. Where biblical scholars have tended to smooth over such conflict by appeal to different scribal schools whose positions clash because they reflect different historical periods juxtaposed in canon, a folkloristic approach best enables readings of Joshua's homeland ideology.

Biblical scholars have long puzzled over the text of Israel's Jordan crossing in Joshua 3–5. The chapters present overlapping, contradictory versions of the same events. This cacophony of story has tortured scholars who have labored with elaborate codes and arcane paradigms in order to account for each utterance in an identifiable historical source.[1] This method of reading misses in certain important ways what the text is trying to do. It is difficult, and ultimately may not make sense, to fully separate the different sources spliced together in Joshua 3–5 because they constitute a genre and are collectively preserved in the same textual environment. Setting definitive boundaries between such stories is near impossible despite clear evidence of repetition and distinct vocabulary. While alerting us to the patchwork character of the text, the tools of redaction and source criticism may never really succeed in explaining the overall picture of the tradition because they do not consider what the form of the text has to do with its theme and its place in the larger structure of the Tanakh.

A folkloristic reading can explain the function of repetition in Joshua 3–5 and show how the composition of the text is integrally related to its theme through the trope of storytelling. The central role that narrative plays in Israel's ability to cross the Jordan becomes strikingly apparent in the story of its crossing, a story that itself thematizes the telling of story. Crossing the Jordan is a new beginning for Israel at the same time that it closes the national journey begun with the crossing of the Red (Reed) Sea.[2] In this way, it reverses the stories of the wilderness while fulfilling the story of the Exodus. By telling a story while they cross the Jordan, Israel memorializes its actions as they occur. The act of commemoration involves forward and backward glances that create a unified group by imagining a collective future and reconstitute the past in light of the present. In narrative time, the ancestors and descendants of Israel cross the Jordan together.

The Jordan crossing is a multiphonic text composed of coexistent, competing narratives. There is a discernible plot sequence,[3] yet the text is structured like a storytelling event in which multiple speakers are roused to recount their

---

1. For a summary of divergent source-critical opinions, see Paul P. Saydon, "The Crossing of the Jordan: Josue 3;4," *CBQ* 12 (1950): 194; for an example, see Jan A. Wagenaar, "Crossing the Sea of Reeds (Exod 13–14) and the Jordan (Josh 3–4): A Priestly Framework for the Wilderness Wandering," *Studies in the Book of Exodus*, ed. Marc Vervenne (Leuven: University Press, 1996), 466. Frustrated attempts at source-critical reconciliation include G. A. Cooke, *The Book of Joshua* (Cambridge: Cambridge Bible for Schools and Colleges, 1918), 17; S. R. Driver, *Introduction to the Literature of the Old Testament* (New York: C. Scribner's Sons, 1900); Martin Noth, *Das Buch Josua* (Tübingen: J. C. B. Mohr, 1953), 9.

2. George W. Coats acknowledges that "the general tendency of the redactional framework is to show the end of one age and the beginning of another." See Coats, "An Exposition for the Conquest Theme," *CBQ* 47 (1985): 49.

3. Robert Polzin lists eleven events that structure the plot. Polzin, *Moses and the Deuteronomist: A Literary Study of the Deuteronomic History* (New York: Seabury Press, 1980), 95.

versions. The variants are structured so that the content of one legend recalls another, which in turn recalls another. In this way, the text is ordered like a performance in which various narrators follow one another's stories by filling in details, revising elements, and engaging with lacunae. Perhaps the Deuteronomistic editor of Joshua looked upon the available narratives as vital homeland traditions, all of which deserved preservation and so maintained the unique features of individual stories while splicing them together. The impulse toward preservation would help to explain why the editor allows contradiction to stand and includes transgressive traditions along with normative ones. By preserving multiple tales, the editor shows Israel's continued adherence to the instruction to "tell the story" to future generations (Josh 4:7, 22).[4]

Crossing the Jordan

At the beginning of the book of Joshua, Canaan, the land promised to Israel, does not seem so much a place as a stream of heterogeneous narratives. The Promised Land is an overdetermined site in which conflict is made inevitable by the contact and clash of stories. A comparison with the books on either "side" of Joshua, Deuteronomy and Judges, already shows how the land and the collective are multiply defined. Where Deuteronomy forges a silent collective as the addressee of law and Judges depicts the utter dissolution of civic bonds, Joshua airs the array of local and tribal traditions with the assurance that each one has its place.[5] This collation of disparate traditions, I argue, constitutes the national territory represented in the book of Joshua. However, this very territory already bears traces of tribe, clan, and family as well as of those displaced. In fact, the amorphous and leaky quality of these traditions makes the clear demarcation of a border necessary. There simply has to be some clear line that facilitates the claim and denial of traditions, for without it the edifice of nation would simply collapse like levees in a flood. With some ironic consistency, the definitional border is a flowing river that can be made to halt when God wants to confer distinction.

The book of Joshua is the primary, although hardly singular, site where the Jordan River functions as a border. Before enumerating all of the oppositions that it clarifies, it is worth asking what the Jordan *doesn't* separate. The dichotomies line up (the old era and the new, the foreign and the familiar, other and

---

4. Thompson reads Josh 24:11, Ps 114, and Micah 6:5 as variants of the Jordan crossing story. "The story was recited in the Jerusalem temple, in an oracle of Micah, and in an exilic setting." Leonard Thompson, "The Jordan Crossing: *Sidqot* Yahweh and World Building," *Journal of Biblical Literature* 100, no. 3 (1981): 358.

5. See Martin Noth, *Überlieferungsgeschichte des Pentateuch*, translated as *A History of Pentateuchal Traditions*, trans. Bernhard W. Anderson (Chico, CA: Scholars Press, 1981).

self, chaos and law, death and life, exile and homeland) so that even Moses and Joshua, the protagonists to either side, cannot really be brought into comparison. On the representational level, territorial bifurcation enables the social imagining and political assertion of the nation. Without knowing what (and where) the nation is not, one cannot grasp what the nation is. At the same time, the nation is unstable—each border crossing changes it in some way. When Israel crosses the Jordan into its land, estrangement and exile also come home.

When Joshua leads the twelve tribes of Israel across the Jordan, the men of Israel's bodies are recast from the flesh of slavery into brigades of soldiers marching in formation. By activating Israel as an army, the Jordan crossing is meant to cleanse the People of its leanings toward rebellion and foreign adventures. Internal rebellion is to be sequestered through the order of the military and the foreign is to be persistently removed from Israel's path. The central goals of the conquest—the marking of territory, the synthesizing of memory, the forging of rituals—are all part of a national-colonial rhetoric that devalues the lives of the current inhabitants as mere obstacles to the realization of a grand dream.[6] However this discourse is but the official register of Joshua, the grand dream is already fractured when Israel crosses the Jordan.

The fracture is most evident on the formal level or, better said, by the nature of the scene's composition. Rather than a definitive plot of crossing, demarcating, memorializing, the scene at the Jordan unfolds in a kind of temporal vacuum characterized by repetitions and gaps. The import of the crossing as national inauguration is betrayed by the fact that the event in question does not involve action, but storytelling. In other words, what we have at the beginning of the book is not the record of Israel's crossing of the Jordan, but a compendium of stories about Joshua and Israel at the Jordan. This collection of material regarding the Jordan may, as the venerable Bible scholars Martin Noth and Frank Moore Cross posited, reflect the existence of a festival at the Jordan staged in ancient Israel.[7] Such a festival, it seems, would have been

---

6. It may still be necessary to note that the conquest of Canaan as reported in Joshua most likely did not transpire in history. No other record corroborates its claims. We should be most disturbed, I think, by the application of Joshua to real national-colonial projects where the murders in question seek sanction from the Bible.

7. Noth believes that the Jordan stories are regional etiologies gathered together by scribes at the Benjaminite sanctuary at Gilgal. Martin Noth, *Das Buch Joshua* (Tübingen: Mohr, 1953), 11, 21. J. Alberto Soggin supports this idea: Soggin, "Gilgal, Passah und Landnahme: Eine neue Untersuchung des kultischen Zusammenhangs der Kap. III–VI des Josuabuches," *VT Sup* 15 (1966): 263–77, and Soggin, *Joshua* (Philadelphia: Westminster, 1972), 9. On the cultic dimension of the crossing, see Jay Wilcoxen, "Narrative Structure and Cult Legend: A Study of Joshua 1–6," *Transitions in Biblical Scholarship* (Chicago: University of Chicago Press, 1968), 56–59. Cross reconstructs a spring New Year's festival at Gilgal based on Josh 3–5. "The ark was born in solemn procession from the battle-camp across the Jordan at Abel-shittim to the river and from thence to the shrine at Gilgal where a covenant-renewal ceremony was consummated. The crossing of the Jordan which was 'divided,' that

of a Northern Israelite provenance and likely involved a ritualized forging of the river. Although the evidence is negligible, such a festival or its ritual relic would explain what gave charismatics like Theudas and John the Baptist their start. Whether or not the Jordan River crossing myth had an accompanying ritual, its variations exist in compendium form in Joshua chapters 1–5. The variations may reflect the various scripts used at different convenings of the festival at Gilgal or derive from one occasion much like the extant records of pre-Islamic poetry festivals at which different tribes gathered to sing their stories.[8] The effect of the compendium in its canonized form is that the border as well as the territory it delineates depends upon who is telling the story.

Leaving aside the frame stories in Joshua 1–2 and 5 for the moment, let us focus on tales of the crossing itself in chapters 3 and 4. I divide the text of 3:1–4:24 into five stories: 3:1–17, 4:1–8, 4:9, 4:10–14, and 4:15–24.[9] My divisions are based on proximity of verses, logically progressive actions, and repetition (for example, repetition of the same action is a cause for separation). Myopic focus has led Bible scholars to consistently misread the repetition that characterizes the opening of Joshua. Repetition does not distract the reader from some true event that transpired in history, but rather makes the point that Israel's political reality hangs in the balance of how well it executes the command to "tell the story" to future generations (Josh 4:7, 22).

• • •

The first story (3:1–17) emphasizes the miraculous dimension of the crossing with anticipatory directives that are then fulfilled as actions. Within the predictions of what will occur are instructions about how the events should be interpreted. Israel departs from the east bank site of Shittim with the morn-

---

is, dammed, so that Israel in battle array could pass over on dry ground, was understood as dramatic reenactment of the crossing of the sea, and as well the 'crossing over' to the new land in the conquest. Exodus and entrance, the sea-crossing from Egypt and the river-crossing of the conquest were ritually fused in theses cultic acts." Frank Moore Cross, *Canaanite Myth and Hebrew Epic: Essays in the History of the Religion of Israel* (Cambridge: Harvard University Press, 1973), 138.

8. It is notable that although a festival at Gilgal is suggested by the Jordan crossing story, the suggestion is quickly curtailed by the claim that what transpires at Gilgal is a second Passover. A hypothetical explanation is that editors with Judean biases let the Joshua stories stand while effacing the Gilgal festival through its merger with Passover.

9. Boling and Wright divide the text as such: 3:1–16, 3:17–4:8, the parenthetical transition of 4:9, 4:10–14. Robert G. Boling and G. Ernest Wright, *Joshua: A New Translation with Notes and Commentary* (Garden City, N.Y.: Doubleday, 1982), 170–71. Saydon divides the text into four different parts: 3:1–13, 3:14–16, 3:17, and chapter 4. Saydon, "The Crossing of the Jordan," 197. Soggin favors this division: "The preparations for the crossing of the Jordan (3.1–13); The miracle of the waters (3.14–16a); The crossing (3.16b-17; 4.10–11, 14–18); The twelve stones (4.1–9); The crossing of groups from the east of the Jordan (4.12–13); The arrival at Gilgal (4.19–5.1)." Soggin, *Joshua*, 51.

ing sun as if making a definitive break from the recent Transjordanian past and their predecessors' sexual commingling with Moabite women. Once they are encamped on the Jordan shore, preparations are made for revelation. A period of three days is observed (3:2), a distinguishing distance is imposed between the Ark of the Covenant and the People (3:4), and a mass purification rite serves as the prelude to miracle (3:5).[10] All of these acts echo the ritual script of Sinai and suggest that the land is a second gift to Israel bound up with the first gift of Torah. Joshua identifies the crossing as the collective miracle in which this generation will participate while at the same time bearing witness. The distance to be observed from the Ark not only demarcates the holy, but also serves as guide and compass on a "road that you have never gone down before" (3:4). Similarly, the purification is both prelude and sign that "tomorrow Yahweh will perform wonders in your midst" (3:5).

The impending miracle is certain to transform Israel, making them victorious west of the Jordan, so the story hovers around the question of who will hold power after the Jordan crossing. Some of the confusion about the forms of authority in the story may also be the result of the merging of different traditions in the text or of different editors writing themselves into history. The Kohanim, first-order priests with ultimate authority over Israel's sancta, are the agents of miracle from whose feet the Jordan River retracts (3:15–16). Although secular officials transmit instructions to the People, the Kohanim are mentioned often enough to overcome the abiding sense that there are other agents involved (3:6, 8, 13, 14, 15, 17). All the same, the officials have their say: "When you see the Ark of the Covenant of Yahweh your God carried by levitical priests, rise up from your positions and follow it" (3:3). In this instance, contestation arises from mention of Levites ("levitical priests") bearing the Ark. The Levites most often figure as a second-order Priestly caste associated with regional sanctuaries rather than central cultic sites.[11] Priestly writings that mention Levites represent them as a subsidiary service class singled out for sacred duties from among Israel yet at a remove from the sanctified Kohanim. This hierarchy of Levites and Kohanim seems to be

---

10. I am simplifying a temporal jumble. Assis charts the confusing timeline: "Verse 1 deals with the day before the crossing ('they spent the night there before crossing'); verse 2 treats the day of the crossing ('three days later') and indicates the completion of the three days mentioned in Joshua 1:11 ('in another three days you are to cross this Jordan'). However verse 5 implies that it refers to the day before the crossing ('tomorrow Yahweh will perform wonders in your midst'). Verse 6 returns to the day of the crossing ('today I will increase your stature in the eyes of all Israel')." Assis, "From Moses to Joshua," 84. I maintain that the multiple temporal markers speak to the folkloristic nature of the text. Joshua 3:1–7 serves as a kind of compendium of possible beginnings for the Jordan crossing story. Another way to look at the overlaps and contradictions is as a preservation of recorded or recalled beginnings. A storyteller could choose one of the openings or reconcile them in a sequence as I have done above.

11. Karel Van Der Toorn suggests that the Levites began as administrators serving in the state sanctuaries of the Northern Kingdom. Van Der Toorn, *Family Religion in Babylonia, Syria, and Israel*, 305.

a retrospective product of Temple culture as well as Priestly literature. The Kohanim likely emerge as a class with the establishment of a central shrine; they coexist with the Levites, who administer other ritual sites then deemed marginal. The verse in which the Levites are said to bear the Ark indicates a contesting tradition because Priestly texts expressly forbid the Levites from touching or transporting the Ark.[12] Alternately, the verse places the Levites at a kind of center—carrying the Ark as the People cross the Jordan—or at least puts them on par with the Kohanim.

The Jordan crossing confers supreme authority on Joshua, who is shown commanding the Kohanim as well as the military (3:8). By helping God choreograph the miracle, Joshua assumes the position of Moses. "God said to Joshua, 'This day, for the first time, I will exalt you in the sight of all Israel, so that they shall know that I will be with you as I was with Moses'" (3:7). The crossing elevates Joshua as the central figure of authority while creating a space—the land—in which his authority can take shape. In order to perform his new role, Joshua gathers the People to "hear the words of Yahweh your God" as if he were, like Moses, a prophetic medium (3:9). However, the laconic nature of his message betrays that he is a general rather than a prophet: "By this you will know that a living God is in your midst who will dispossess and expel from before you the Canaanite, the Hittite, the Hivite, the Perizzite, the Girgashite, the Amorite, and the Jebusite" (3:10). The sign in question—the "this" (בזאת) in the verse—seems to be the very events that are about to transpire, meaning that Joshua understands the cutting off of the Jordan's flow for Israel's footpath as a precursor to God's clearing of nations from Israel's path of settlement.

Prediction gives way to action in verse fourteen, when the Kohanim begin the movement across the Jordan.

> When the bearers of the Ark reached the Jordan and the feet of the Kohanim, the bearers of the Ark, touched the edge of the water (the Jordan overflows its banks all the days of the harvest), the water flowing down stopped and piled up in a single heap at considerable distance at Adam, the town next to Zarethan. The downstream waters that flow into the Sea of Arabah/the Salt Sea were completely cut off and so the nation crossed near Jericho. The Kohanim, who bore the Ark of God's Covenant, stood on dry land at exactly the middle of the Jordan, while

---

12. This may be a product of divergent Priestly and Deuteronomistic traditions: "Deuteronomy emphasizes that the entire tribe of Levi is to serve God, in contrast to Numbers where they are to serve Aaron and his sons ... [The Levites] are in charge of the written Torah of Moses and shall teach its judgments (Deut. 33:10). Yet curiously, on first crossing over the Jordan, it is the *people*, not the Levites, who are commanded to offer sacrifices." Adriane Leveen, *Memory and Tradition in the Book of Numbers* (Cambridge: Cambridge University Press, 2007), 57.

all Israel crossed over on dry land, until the entire nation had finished crossing the Jordan. (Josh 3:15–17)

The Kohanim, as stated above, are singled out during the crossing of the Jordan, a miracle amplified by the river's overflow. When their feet dip in the river, the downstream waters freeze and halt at attention while the waters on route to the Salt Sea cease to flow (3:16). The "single heap" (נד אחד) into which the upstream waters pile recalls the "flowing heap" (נד נזלים) of waters at the parting of the Reed Sea (Exod 15:8), thereby closing the Exodus with a symmetrical coda.[13] The place names—Adam, Zaretan, and Jericho—which have generated midrash and vexed historical geographic scholars, are best thought of as indicators that the Joshua generation is marching out of the wilderness and into marked geography. At the same time, a mythic framework characterizes the description. The imagery of congealed waters works to "mythicize the conquest—insofar as the Canaanite enemy is now rendered helpless by God's mighty arm, like the waters years before, when the nation of Israel 'passes' through them into the promised land."[14] God's subduing of waters, a mythic theme, signifies the annihilation of Israel's (future) neighbors. The dry land on which the People walk across the Jordan then symbolizes the land that Israel will conquer and settle. The motif of God's dominance over nature provides the foundation of nationalist myth.[15]

• • •

The second story, Joshua 4:1–8, concerns the tribal structure from which Joshua derives his legitimacy as a leader.[16] An emphasis on a league of twelve

---

13. Also Psalms 78:13b.

14. Michael Fishbane, *Biblical Myth and Rabbinic Mythmaking* (New York: Oxford University Press, 2003), 51. This is an instance in which mythic parallelism is both maintained and transformed. In the Ba'al Epic, Ba'al (Ba'lu) crushes an opponent called both Prince Sea and Judge River. In the narrative transformation of this myth, Yahweh defeats the Sea and Egypt during the course of one human generation and the Jordan River and the Canaanites during another. See Propp, *Exodus 1–18*, 560.

15. Imagining possible historical contexts in which the story was told further highlights its mythic nature. If we imagine the recitation of the story before or between the exile of Israel by the Assyrians and the exile of Judah by the Babylonians, then it functions to provide an idealized past in which Israel, empowered by God, subdues its neighbors as a means of drawing distinguishing social boundaries between Israel and these neighbors (or others occupying their symbolic place) in the present. Conquest is then a narrative spun in the name of group cohesion. If we imagine the story told in an exilic context, then it functions as a charter for group unity amidst dispersion as well as a promise that God will again lead the People across the Jordan. The dry ground beneath the Jordan in every case expresses territorial aspirations.

16. The story is structured according to command (4:1b–7) and action (4:8) with the twist that God directs Joshua (4:1b–3) and then Joshua transmits the plan to the twelve chosen rep-

tribes whose territory lies west of the Jordan comes across through the theme of memorialization. Unlike the story in Joshua 3, this one does not stress the difference from other peoples but rather the unity of the twelve tribes. Each and every tribe has a single representative designated to lift a stone from the riverbank and place it in a memorial configuration west of the Jordan. The memorial indicates the equal inclusion of the twelve tribes while setting a limit on membership. The twelve stones lifted onto Israelite shoulders, shuttled across the Jordan, and planted in the land are analogous to the twelve tribes lifted out of bondage, brought across the Jordan, and planted in the land by God. By analogy the stones tell the story of Israel. They also recall the miraculous retreat of the Jordan's waters and Israel's procession across a dry riverbed. The stones have the external function of declaring intended permanence to their witnesses. Positioned at Israel's initial west bank campground, the memorial is its first territorial marker. As Joshua instructs the twelve men in the bearing and placement of the stones, he also schools them in the necessary symbolism:

> This will be a sign among you. In the future when your children ask, 'what do these stones mean to you?' You will tell them, 'the waters of the Jordan were cut off before the Ark of God's Covenant. When it passed through the Jordan, the waters of the Jordan were cut off.' These stones will be an eternal memorial for the People of Israel. (Josh 4:6–7)

The ritual script that Joshua provides does not concern the current miracle, but rather its future commemoration. Joshua's story dispenses with the ancestral past and replaces it with a present resonant with meaning for the future. The sign (אות) provokes the dialogue that enables the stones to be a legitimate and abiding memorial (זכרון).[17] In the story, deferential children curiously inquire about their parents' perspective—"what do these stones mean to you?"—and the parents respond with the chiastic remembrance of how the

---

resentatives while appending an accompanying narrative (4:4–8). In verse 3:12 Joshua instructs the People to choose twelve tribal representatives. In 4:2–3 God tells Joshua to perform a similar selection and for these men to lift stones from the Jordan. Joshua calls upon twelve men in 4:4, an action that Assis sees as a continuation of the previous verses as well as 3:12. Assis, "From Moses to Joshua," 84.

17. Babylonian kings and Assyrian imperialists were known to erect stones proclaiming conquered lands and establishing new borders. An important aspect of these stones or stelae, as Seth Sanders points out, is that they tended to bear writing that spoke of "the king's conquests in his voice." Sanders, *The Invention of Hebrew* (Urbana: University of Illinois Press, 2009), 120. It is then noteworthy that both sets of stones described in the crossing stories come with transmitted oral traditions rather than written ones.

Jordan waters were cut off (Josh 4:7).[18] Joshua's story in effect creates a past for Israel in the land.[19] This past, which begins at the Jordan, seems to allow for no reminiscence of previous, contending notions of the land. The memorial reinforces the border at the Jordan while creating a legitimate place and time for the nation Israel. The story concludes with the perfect fulfillment of Joshua's instruction, suggesting the tacit consent of all Israel.

After one set of stones is lifted from the Jordan and set down at the encampment (4:8), another brief story recounts how Joshua set up an additional twelve stones in the middle of the Jordan to mark where the Kohanim stood.[20] These stones "are there to this day" (4:9), a temporal deictic different from the "forever" of verse 4:7 (עד־עולם). The stones set upon the riverbed signal Joshua's authority rather than that of Israel's constitutive tribes. His mark is made on ground beneath the river, which may correlate with the first story to imply Joshua's central role in defeating the peoples of Canaan. Joshua's twelve stones still speak to the nature of the people he leads, but the fact that he is the only agent of memorialization emphasizes his singular hegemony.[21]

---

18. Note that the children here ask, "what do these stones mean to you?" (4:6), where the children in 4:21 ask, "what are these stones?" Jan Wagenaar observes that the "twofold reference to the questions the children put to their fathers" in Joshua mirrors the two sets of questions posed by imagined children concerning the Exodus (Ex 12:26–27 and Ex 13:14–16). The parallel is precise in that the children in the first case ask "what is this ritual to you?" (Ex 12:26) while the children in the second case pose the more laconic "what is this?" (Ex 13:14). This shows the conscious shaping of these traditions by an editorial hand. Jan A. Wagenaar, "Crossing the Sea of Reeds (Exod 13–14) and the Jordan (Josh 3–4): A Priestly Framework for the Wilderness Wandering," *Studies in the Book of Exodus*, ed. Marc Vervenne (Leuven: Leuven University Press, 1996), 461. Aaron Sherwood sees the parallelism between the two episodes as part of a larger mirroring pattern "wherein the elements of a broad-stroke view of Exodus play out in reverse to form the skeleton of Joshua 1–12." Sherwood, "A Leader's Misleading and a Prostitute's Profession: A Re-examination of Joshua 2," *JSOT* 31.1 (2006): 59.

19. Dan Ben-Amos refers to this episode as "monumental testimony (that) was used for historical commemoration (Josh 4:6–7)." Ben-Amos, "Folklore in the Ancient Near East," *Anchor Bible Dictionary*, 823.

20. Saydon presents the possibility that "the two groups of stones represent two traditions amalgamated together into one narrative," as a solution to the two sets of stones at the encampment and in the river. Saydon, "The Crossing of the Jordan," 201.

21. Boling and Wright divide Joshua 3:1–4:18 into "two blocks of material." The first block narrates the crossing and say nothing of the twelve-stone memorial (3:1–16 + 4:10–18). They attribute this block to Dtr 1 and describes the omission of the memorial as motivated by the desire to forget regional Israelite centers in the name of recognizing Jerusalem as Israel's sole center. The second block, 3:17–4:8, focuses on the construction of the memorial. The account of Joshua placing stones in the river (4:9), "which is supposed by many critics to be a rival etiological tradition," reads to Boling and Wright like a seam or "an explanatory parenthesis which makes sense because the redactor was working with largely preformed units. Robert G. Boling and G. Ernest Wright, *Joshua: A New Translation with Notes and Commentary* (Garden City, N.Y.: Doubleday, 1982), 170–71. I argue that the stones placed within the Jordan must be taken seriously as a distinct tradition that belongs with the thematic constellation relating to Joshua's succession of Moses at the Jordan.

The next story, in 4:10–18, echoes elements of 3:1–16 while giving them a militaristic spin. The Kohanim cross "in front of the People" (לפני העם) (4:11/ 3:6, 14); "God exalts Joshua in the eyes of all Israel" (4:14) as promised in 3:7; and the "feet of the priests" (4:18/3:13) stand on "dry land" (4:18/3:17). The story begins in medias res with the Kohanim standing in the middle of the Jordan while "everything that God commanded Joshua to say to the People was fulfilled" (4:10). The People "hurry" across the Jordan as if eager to take up the battles of conquest. The two and a half tribes of Reuben, Gad, and half Manasseh (the subject of subsequent analysis) march forward arrayed as an infantry 40,000 strong prepared for battle in Jericho (4:13). The fact that the two and a half tribes "cross" before God on their way to war represents another device of correlating the crossing of the river and the battles to come. At the same time, the confident image of a vanguard marching in formation works to cover up the challenge that the two and a half tribes pose to the very definition of homeland. The portrait of the two and a half tribes as infantry suppresses their problematic nature and instead writes them into the homeland mythology. Through an equation in which the army mirrors the infantry, the portrait of the vanguard further suggests the degree to which the Jordan crossing transforms the People into perfect soldiers commanded by Joshua.

Again the overlap of story generates questions: Who in fact marches in front of Israel? Is it God? The Kohanim carrying the Ark? Joshua? The infantry? At one juncture or another, each is placed at the forefront. This can be explained as the work of different narrators endowing the groups they represent with authority by inscribing them in the moment of national rebirth and thus may speak to the competition among factions. It could likewise indicate an anxiety underlying the conquest traditions in general: who is ultimately responsible for the migration and the wars and who for their incompleteness and failure? Ambivalence and anxiety emerge not from any single story, but from the context of multiple stories. The story in question moves with assurance from the infantry marching before God to the leadership of Joshua. "On that day God elevated Joshua in the eyes of all Israel so that they looked upon him as they had Moses all the days of his life" (4:14). The certain tone in which the miracle of retreating waters and the elevation of Joshua are narrated attempts to compensate for any sense of loss involved with crossing the Jordan. What is lost of course is Israel's actual past, now effaced and covered over by the creation of a new past in the land. At the same time, the actual past cannot be entirely disregarded since Israel must remember Moses and the Exodus. As a figure, Moses authorizes Joshua while also serving as a persistent reminder that Israel comes to its land from the outside. Therefore Moses's figure must be handled correctly at this important juncture of change.

The story conveys the incompatibility of longing for Moses with the new era of Joshua.

• • •

The synthesis of national territory, memory, and futurism becomes apparent in the last of the chain of river crossing stories (4:15–24). In this conclusive account, Israel's temporal and spatial coordinates become rooted in the historical. The first step towards this is that the Kohanim carry the Ark (ארון העדות) out of the river. Since they have been holding up the Ark in the middle of the Jordan in order to repel its waters and enable the People to cross over on dry land, when the Kohanim touch the western shore the miracle is suspended and the Jordan resumes its regular flow. When the People (who ostensibly are already across when the Kohanim arrive) complete the Jordan crossing, they enter chronological time, "the tenth day of the first month," and indexed space, "Gilgal at the eastern edge of Jericho" (Josh 4:19). National territory and history begin west of the Jordan, so Joshua attends to its mapping and chronicle. Joshua lifts twelve stones from the Jordan into a formation that signals the onset of a nation composed of twelve tribes and thereby domesticates the unknown land through a memorial. Stone mounds or other improvised structures tend to mark territory as claimed space.[22] The stone memorial allows Israel, heretofore forbidden from registering the places of its wandering and marking the graves of its dead, to move in a linear fashion across the grid of time and space.[23]

The stones generate additional meaning through an accompanying story told by Joshua. The device of a story within a story further thematizes the degree to which the crossing is a storytelling event.

> In the future when your children ask their parents, 'what are these stones?' You will tell your children, 'Israel crossed this Jordan on dry land.' For Yahweh your God dried up the waters of the Jordan from before you until you crossed as Yahweh your God did to the Red Sea when he dried it up from before us until we had crossed. Therefore all

---

22. "In the Greek tradition of colonization, we find that the first settlers used to erect pillars of stone and monuments at the conclusion of their journey (e.g., Strabo 3:171, and cf. Herodotus 2:102–3), a custom which is paralleled in the setting up of stones at Gilgal by the people of Israel at the end of their migration (Josh 3–4)." Weinfeld, *The Promise of the Land*, 36. Also see Greenblatt, *Marvelous Possessions*, 56.

23. For the lack of proper burial as a means of forgetting the exodus generation, see Adriane Leveen, "Falling in the Wilderness: Death Reports in the Book of Numbers," *Prooftexts* 22 (2002): 262.

of the nations of the land will know that the hand of Yahweh is mighty and you will fear Yahweh your God forever. (Josh 4:21–24)

The didactic, coercive dimension of the story constitutes a kind of official register in which a multivalent symbol is forced to point in one direction: the stones recall the crossing, which points to the miracle of the Exodus, which justifies the conquest. The novel memory intended to domesticate the land is shored up by an event in the more distant past, the Red (Reed) Sea crossing. At the same time, this story within a story is a rare moment in the Hebrew Bible in which the future is more instrumental than the past. All that has transpired and is to take place happens in the name of the "children"; they are the population for whom stones are raised and nations are extinguished. This then is not only a moment of memorial, as Yerushalmi notes, but also one of futurism.[24] Futurism, rather than memory, seems to be the foremost motivation for installing a narrative in the stone structure. The syntax of the final verse opens a question: will the nations know of Yahweh's might and Israel's election because of the story told to children or because of the event itself? It seems that the parents endowing the children with the story is the very cause of knowledge among "the nations of the land" and so the concern for the future establishes another sort of boundary around Israel.[25]

Along with the official register, there are both populist and revisionist registers to the story. The populist register becomes apparent through comparison with the story in Joshua 4:1–8, in which a ritual narrative is also appended to the stone structure. In Joshua 4:1–8, the story that accompanies the stone memorial is transmitted as specialist knowledge to the twelve tribal representatives. Because the twelve are the custodians of knowledge, the question "what are these stones *to you*?" has particular force. That the stones signify simply the fact that the waters of the Jordan stopped shows how symbolic value often depends upon its exclusivity rather than its complexity. The force of the stones as memorial makes the children of the wilderness into the par-

---

24. "Not the stone, but the memory transmitted by the fathers, is decisive if the memory embedded in the stone is to be conjured out of it to live again for subsequent generations." Yosef Hayim Yerushalmi, *Zakhor: Jewish History and Jewish Memory* (Seattle: University of Washington Press, 1982), 10.

25. Robert Polzin, who reads the repetitions in Joshua 3–5 as shifts in spatial, temporal, or psychological perspective, believes that "the crossing of the Jordan is narrated first from the spatial and psychological point of view of participants in the procession (3:1–4:14) and then from the spatial and psychological point of view of those who witnessed this miraculous event from afar: the non-Israelites whose land was being invaded (4:15–5:1)." Polzin, *Moses and the Deuteronomist*, 103. Thus the memorial function described in Josh 4:7 is internal to Israel and the symbolic function described in 4:24 is externally directed to non-Israelites.

ents of the land and simultaneously casts them as a pioneer generation whose acts of inscription cannot be separated from the land itself. The twelve are the particular, although anonymous, heroes that a pioneer generation requires for its values to be readable as biography. The external recognition of Israel and its God by the nations is not at issue in this story. Instead, its theme revolves around the production of an elite who are the custodians of national memory and thereby the defenders of homeland. The populist element in the story at hand (4:21–24) arises from the fact that it is addressed to all Israel rather than a select elite and thus imagines the People as a collective in the future as well as in the present.

The revisionist aspect of our story (4:21–24) also comes into relief through a comparison with Joshua 4:6–7. Remember that in 4:23 Joshua equates the Jordan with the Red (Reed) Sea crossing: "For Yahweh your God dried up the waters of the Jordan from before you until you crossed as Yahweh your God did to the Red Sea when he dried it up from before us until we had crossed." Joshua equates the two crossings, yet places himself in the wilderness generation for whom God dried up the Red (Reed) Sea rather than in the new generation for which the Jordan is retracted. At the moment of the Jordan crossing, Joshua appears as the exemplary representative of the Exodus generation. Now that Joshua represents the past, Moses is conveniently forgotten and Joshua can thus narrate the past in a manner that suits his program. Since only he and his fellow spy Caleb have survived the wilderness, Joshua can claim a virtue that exceeds that of Moses, who dies just east of the Jordan.

In the same way that Joshua's survival reframes the life of Moses, so Joshua's reading of the Jordan crossing in light of the parting of the Red (Reed) Sea erases the wandering and rebellions of the wilderness from memory. In other words, memory is reconstructed through omission of the potential crossing into the land that was averted by the resistance and counternarrative generated by ten of the twelve spies that Moses sent into Canaan (Num 13–14).[26] On this count, Joshua's memorialization is partisan. He and Caleb were the two dissenting spies who contradicted the majority report of "a land that devours its inhabitants" (Num 13:32). Instead, they insisted that the peoples of the land were "food" to be consumed by Israel if only they would trust God (Num 14:9). Since the spies' story was effectively an opposition to the plan of conquest, Joshua circulates an edited story with no mention of an averted crossing. The stones thus establish a limit on what can be remembered.

---

26. See Havrelock, "Sh'lach-L'cha (Numbers 13:1–15:41)," in *The Torah: A Women's Commentary*, ed. Tamara Cohn Eskenazi and Andrea Weiss (New York: Union of Reform Judaism Press, 2008), 886, and Ilana Pardes, *The Biography of Ancient Israel: National Narratives in the Bible* (Berkeley: University of California Press, 2000), 100–126.

Joshua's commemoration that imagines the future is a national charter that establishes boundaries of territory and ethnicity.[27] Although the territory in question awaits further demarcation and memorial, it is defined by the limit of the Jordan River. Any land traversed during the wandering is excluded from claim and likewise the events of the wilderness are best forgotten in the new era west of the Jordan. The very purpose of commemoration is collective unity in the present: the unified memory of the commemorated unifies the commemorators and, inevitably, the commemorated are remade in the commemorators' image. The first narration of the Exodus at the Jordan River changes the story as it launches a process of storytelling required by the festival of Passover. The injunction to "tell your children the story" generates an endless array of variations.

The first celebration of Passover is constructed as a kind of inverse or even "anti-Exodus" in which each ritual action alludes to a specific event of the Exodus and reverses its effects.[28] Stones play a further role in the establishment of Gilgal as a memorial site. Joshua forges stone knives in order to circumcise the men of Israel. With some apologetic rationale the narrator explains that the generation born in the wilderness went uncircumcised by their circumcised fathers (Josh 5:4, 5, 7). For this reason, Joshua must serve as the national father who brings all of his sons into the covenant through circumcision. This "second circumcision" of Israel marks a new beginning through ritual (5:2); since circumcision legally takes place on the eighth day of a boy's life, never before have all the living men of Israel undergone collective circumcision. The process leaves behind the somewhat grotesque place name, "the Hill of Foreskins" and, it seems, a more ephemeral mound of skin (5:3). The narrative accompanying the ritual lambastes the Exodus generation for their failure to engage in conquest, but also does its share of rewriting the past in light of the present. The members of the Exodus generation are twice referred to as "warriors" (5:4, 6). Immediately after, they are disparaged for the usual sins of "dying in the wilderness" (5:4) and "not listening to the voice of God . . . that promised their ancestors that He would give us a land flowing with milk and honey" (5:6), but all the same they are spoken of in military terms. The most plausible explanation is that because the term "warrior" is equated with male-

---

27. Israelite ethnicity, according to Van Der Toorn, is constructed through terms of valuation. "'Israelite' was merely an honorific self-designation, whereas 'Canaanite' was a derogatory adjective applied to others." Van Der Toorn, *Family Religion in Babylonia, Syria and Israel*, 328.

28. William H. C. Propp lists a series of chiastic parallels between the Exodus and the "anti-Exodus" in Joshua 3–5 including angelic visions (Exod 3–4/Josh 5:13–15), Paschal celebrations (Exod 12:1–42/Josh 5:10–11), requirement of circumcision (Exod 12:43–49/Josh 5:2–9), and crossings (Exod 13:17–15:21/Josh 3:1–5:1). Propp, *Exodus 19–40: A New Translation with Introduction and Commentary* (New York: Doubleday, 2006), 795–96.

ness in biblical Hebrew, the text stresses their identity as soldiers in order to emphasize the male membership of the nation. This plausibility is weakened somewhat by the fact that verse 5:4 goes out of its way to speak of "the nation that left Egypt, the males, all of the warriors." On the basis of the repetition of "males" and "warriors" here, it seems that the warrior designation is intended to give the Joshua generation a martial pedigree. Instead of descending from a generation that resisted the battles of the land, they are born of soldiers. It is true that the Exodus generation engaged in battles, but the only territory they conquered lies problematically east of the Jordan.

Following the ritual, the place name Gilgal is explained: "God said to Joshua, 'Today I have rolled the shame of Egypt from them' and so the name of that place was called Gilgal until now" (5:9). Gilgal, whose name is a reduplicative form of a noun used for things rolled up like a wave or a heap, is the place where Israel removed the taint of exile from its collective body. Interestingly, the generation that is circumcised has never known slavery in Egypt and so they stand in as bodies in transition from exile to homeland. The name "Gilgal" also suggests a circle of stones and is etymologically related to "Gilead," Jacob's stone mound of witness. Gilgal, like Gilead, is a point of contact and passage between margin and center. Once the men of Israel are healed from their circumcisions, the People conduct the Passover meal with produce of the Promised Land. The first commemoration of Passover reinterprets the events of the Exodus in light of its own experience and speaks to how the meaning of the Exodus rests in its use and application. The same can be said for the Promised Land; its location and character are determined by commemorative narratives.

The existence of many stories and several stones frustrates the attempt to establish a stable location for homeland or a coherent national memory. This means that where and what the land is depend upon narration. The variation of narrative at the beginning of Joshua should serve as an index to how to read the book as a whole. This is why the unceasing labors of Bible scholars to date the strata of Joshua 1–4 miss the point on a political as well as on a literary level. The point is that the land is not an object or even really a place that can be fixed in time, but rather an act of narration with the power to determine where Israel belongs. Admittedly, a most disturbing narration of conquest and extermination follows the Jordan crossing, but even this is shown to be naught but narration in the settlement stories that follow those of the conquest. The settlement stories (and border lists) in Joshua confess that, as it turns out, the peoples of the land are not exterminated or even cleared from Israel's path. Instead, a host of nations continue to populate the land and constitute its character (Josh 6:25, 8:33, 35, 9:21, 26–27, 11:22, 13:1–7, 13,

15:63, 16:10, 17:11–13). For example, Joshua 15:63 states, "As for the Jebusites, the inhabitants of Jerusalem, the tribe of Judah was unable to dispossess them and so the Jebusites and the people of Judah dwell together in Jerusalem until today."[29] Although it is not canonized, the narrative of Jebusite Jerusalem is recognized by mention of their long past and continued present in Jerusalem. The verse serves as a record of a parallel, variant, and perhaps competing narrative that shapes the character of the land.

Not Crossing the Jordan

In Robert Polzin's formulation, "The Book of Joshua's constant theme [is that] of the outsider in Israel." The first two sections of Joshua deal with "outsiders inside the land" and the third "insiders outside the land."[30] The insiders who stay out stage their full-blown resistance and then negotiate their relative autonomy in Joshua 22, where they contest the Jordan as a border and show how the ascription of memory can quickly domesticate any terrain as homeland. These inside-outsiders are the two and a half tribes seen in the Jordan crossing story marching as infantry. The spin in that story comes into relief through comparison with every other story about them, where their very function seems to be the troubling of national boundaries.

When Israel is encamped on the east bank, the tribes of Reuben and Gad voice their desire to call it home (Num 32:1–5).[31] The tribes bring their proposal before all the figures of authority, "Moses, Eleazar the Priest, and the leaders of the community," by requesting that "this land be given to your servants as a grant; don't make us cross the Jordan" (Num 32:5). The tribes tag a second, more subversive request onto the petition for east bank lands; by asking that they not be made to cross the Jordan, the tribes point toward the coercive dimension of the impending crossing. Not only do they not want to dwell within the sanctioned territory, but they also wish to be excused from the act of return. The entreaty shakes the pillars of the Israelite identity narrative by suggesting that the period of wandering may not have a definitive end, that Israel need not be identified with a fixed territorial entity or be distinct from other peoples, and that perhaps the People themselves don't want to cross the Jordan. In other words, the tribes propose that the Jordan may not

---

29. This verse, as well as Judges 1:21, records the presence of Jebusites in Jerusalem.
30. Polzin, *Moses and the Deuteronomist*, 134.
31. The tribes' "seeing" of the land (ויראו) alludes to Lot's "lifting of the eyes and seeing" (וירא) and suggests that the land claim east of the Jordan, like Lot's, is predicated upon a seductive visual appeal. The narrator in both cases tries to temper the innate appeal of the lush Jordan Valley by explaining that both Lot and the tribes are motivated initially by the potential grazing land for their numerous flocks (Gen 13:5; Num 32:1).

be essential as a divider or as a signifier. With their request, "the Transjordanians break the *integrity* of Israel. One people about to enter one land becomes two groups with different territorial intentions."³²

Moses interprets the request through another narrative, the story of the land told by the ten spies (Num 13:27–33), and therefore understands it as an attempt to thwart arrival at the intended destination.

> Shall your brothers go to war while you sit here? Why would you turn the People of Israel against crossing into the land that God gave them? This is what your fathers did when I sent them from Kadesh Barnea to investigate the land. They went up to the Wadi Eshkol, saw the land, and turned the People of Israel against entering the very land that God gave to them. On that day God was furious and swore that 'these people who came out of Egypt, from twenty years old and up, will not see the land that I promised to Abraham, Isaac and Jacob because they have not listened to me. Only Caleb the son of Yefunneh the Kenizzite and Joshua the son of Nun [will see it] because they have listened to God.' God was furious with Israel and made them wander in the wilderness for forty years until the entire generation that transgressed against God expired. Now you have established a culture of transgression in place of your parents' that only adds to the fury of God toward Israel. If you turn away from Him and He sets them out in the wilderness again, you will have destroyed this entire nation. (Num 32:6–15)

The prophet begins by unpacking the implications of not crossing the Jordan: the abstention of the two tribes from war will dissolve the tribal federation and influence the other tribes to abandon the land and its battles. The theme of influence connects the scenario with that of the spies (Num 13–14). After scouting the land, ten of the twelve spies (all except Joshua and Caleb) produced a story that overturned each trope of God's story of the land and transformed "the land flowing with milk and honey" into "a land that devours its inhabitants." The story told by the spies and embraced by the People caused a generation to lose the land and wander forty years in the wilderness. According to Moses, the tribes of Reuben and Gad practice the wrong kind of memory. Where Israel should remember its ancestors, but forget its parents, the two tribes reproduce the previous generation's "culture of transgression."

Jobling locates a flaw in Moses's "reading together" of the two incidents.

---

32. Jobling, "The Jordan a Boundary," 103.

"By far the most tendentious omission is of the People's *allowing themselves to be discouraged* by the discouraging report; this was an integral feature of the Num. 13–14 story, and its absence from Num. 32 makes Moses' argument founder."[33] Another important difference between the two scenarios lies in the territorial implications of the counternarratives. Influenced by the spies' horror story, the People ask "is it not better for us to return to Egypt?" (Num 14:3). Reuben and Gad, however, neither want to go back to Egypt nor remain in limbo; instead, they reason that any land could be designated as *the* land and thus the journey towards home could cease at any point. The two tribes do not want to go backwards, they just don't want to press forward. Desire, in their proposal, can turn the very terrain of wandering into home.

The tribes appease the prophet through a rhetorical strategy in which they maintain the request for east bank land while conceding to crossing the Jordan.

> They approached him and said: We'll build pens for our flocks here and cities for our dependents. We will march as infantry in front of Israel until we have brought them to their dwellings, while our dependents remain in cities fortified against the inhabitants of the land. We will not return to our homes until each man in Israel receives his allotted portion of land, but we won't receive our portion with them beyond the Jordan, on the other side, because we received our allotment on the eastern side of the Jordan. (Num 32:16–19)

The two tribes reassert their intention to settle Transjordan, while recognizing that it unsettles the master narrative in which the land is categorically opposed to wandering. They propose to compensate for the destabilization by placing themselves in the most vulnerable of military positions. Although the two tribes call other what the rest call home, they consent to uphold the binary system with the Jordan at the center. Thus, the river continues to separate home and exile, but their exile is located on the west bank and their home on the east. For Reuben and Gad, the national homecoming is an act of leaving home behind.

The negotiation through which the Jordan comes to separate an "us" and a "them" within Israel—two and a half tribes on the east bank and the other nine and a half on the west—is elaborate. The dialogue between Moses and the tribes is formalized and transmitted to the future leadership in an oath-taking ceremony. The volley of compromises structures the chapter as a

---

33. Jobling, "The Jordan a Boundary," 94.

chiastic "alternation of positive and negative statements."[34] Moses stipulates conditions for retaining the Transjordanian lands: "After you return, you must be clear before God and Israel in order for this land to be your divinely recognized allotment" (Num 32:22).[35] Deviation from the oath will constitute sin. Yet when Moses transmits the oath to Eleazar the Priest, Joshua son of Nun, and the tribal leaders, he imposes another condition. Should the tribes not cross the Jordan and fight, they will be forced to settle among the other tribes in Canaan proper (32:30).

As Joshua prepares to lead the People across the Jordan, he finds himself troubled by the inheritance of Moses's pledge to the two and a half tribes. The suspense of crossing the border halts when Joshua turns to confront the tribes of Reuben, Gad, and half Manasseh. That his first word to them is זכור "remember" suggests latent suspicion about the nature of their memory. The very national memory that Joshua is in the process of forging is threatened by the fact that the two and a half tribes remember the land beyond the Jordan through their narrative of claim.[36] By settling on the other side of the Jordan, they upset the dichotomy of exile and homeland and show instead an interpenetrating continuity. At the border, where exile and homeland should be powerfully differentiated, they are shown to be arbitrary.

Joshua attempts to reconcile the paradox by renegotiating the terms under which the tribes can lay claim to the other side of the Jordan. After enjoining them to "remember what Moses the servant of God commanded you" (Josh 1:13), Joshua then introduces a new dyad into the east-west dichotomy: the land granted by Moses versus the land given by God (Josh 1:15b). The national problem, one could say, is given a religious spin—the land beyond the border is unsanctified in the sense that it is promised by a person rather than by the Divine.[37] In order to compensate for the fact that the tribes plan to live on unsanctioned land, the men must serve as the infantry that "crosses into" the homeland before the other regiments. In other words, they must perform

---

34. Jacob Milgrom, *The New JPS Torah Commentary: Numbers* (Philadelphia: Jewish Publication Society, 1990), 493.

35. For the sacred status of this land, see Horst Seebass, "Erwägungen zu Numeri 32:1–38," *Journal of Biblical Literature* 90 (1999): 33–48. I disagree that the mention of Yahweh conquering Transjordan in Numbers 32:4 confers holiness upon it. Rather, it seems a note that Yahweh once again defeats Israel's enemies.

36. Since memory is a mode of authorization, alternate recollections disrupt extant authority by conferring it elsewhere. Maurice Halbwachs, *On Collective Memory*, trans. Lewis A. Coser (Chicago: University of Chicago Press, 1992), 182–83.

37. The sacred nature of the land west of the Jordan is made clear by Joshua's heavenly counterpart, "the captain of God's celestial army," who tells him to remove his shoes since "the place where you stand is holy" (Josh 5:15). In contrast, the east bank lands where the two and a half tribes dwell are called "unclean" (Josh 22:19).

as exemplary fighters for the homeland in order to prove an enduring allegiance. Attentive to the innovation at work in Joshua's memory, the two and a half tribes introduce their own condition into the agreement. They pledge obedience to Joshua as long as "Yahweh *your* God is with you as He was with Moses" (Josh 1:17). With the turn of phrase, the tribes distance themselves from the invoked deity who fails to recognize their lands.

The east bank tribes stand accused of a transgressive memory that runs counter to the normative national memory formulated during the Jordan River crossing. Their faulty memory is also charged with reminding Israel of the very things that it would like to forget. Moses accuses them of emulating the rebellious behavior of the spies (Num 34:14), and a suspicious tone underlies Joshua's exhortation to remember the agreement with Moses. The assertion of an east bank Israelite identity becomes a troublesome reminder that Israel is contiguous with rather than separate from the foreign. The two and a half tribes problematize the locative identity of Israel by claiming the other side of the river as home. Although they persistently attest that locations outside of the land constitute an inextricable component of Israel, the two and a half tribes continue to be recorded as a challenging variation on the subject of the land. The most homeland-oriented biblical book concedes to its proximate diaspora and to competing memories that suspend rather than fixing the border. Neither the Jordan nor the twelve stones quite draws a dividing line.

Chapter Five

THE OTHER SIDE

Sooner or later this will involve our telling the Zionists that the ancient territory of Reuben, Gilead and Manasseh is not to form part of the National Home.
—W. Ormsby-Gore minutes on General Clayton's telegram to the British Foreign Office, 2 November 1918[1]

Having considered the ramifications of crossing or not crossing the Jordan, we now turn for an immersive visit to Transjordan, the other side. Keeping in mind that location also carries the meaning of topos, we will notice how little the representation of place has to do with where people actually are and how much it has to do with the projection and organization of ideological positions. In order to have force, national myths must draw their opponents in bold colors and locate them in a problematic space. When the opponents are foreign enemies, their colors can be as vivid and their home as disordered as the myth needs them to be; when the opponents are co-nationals, the trick is to highlight the contrasts at certain moments and to mute them at others and to cast their space as sometimes marginal and sometimes iconic. As an illustration, one can think of the role played by New York City in pre- and post-9-11 American national mythology.

A spectrum of tension that spans from alliance to civil war characterizes the role of co-national rivals in national mythology. Because the case of the

---

1. BNA FO 371/3384/181689; see also Yitzhak Gil-Har, "Boundaries Delimitation: Palestine and Transjordan," *Middle Eastern Studies* 36, no. 1 (2000): 69.

two and a half east bank tribes exhibits this whole range, we will follow their depiction in a plot woven through the books of Joshua and Judges that moves from suspicion to treaty to a war fought over the pronunciation of a single consonant. The Transjordanian tribes trouble Israel—they perennially upset nationalist myth by blending the border, they live in a rich land that seems rightly to belong to someone else, and they find all the loopholes of ethnonationalism. Yet they never disappear from the national chronicle.[2]

There are several ways to interpret this. We can observe how national myths air variant and even contesting versions exactly as they assert themselves and work toward a theory of canonization in which national myth—codified in terms of historical beginnings—preserves the memory of disenfranchised people. Recent scholarship has decried such disenfranchisement and endeavored to tell history differently as a result, but if we think in mythic rather than historical terms we can find additional uses for unwitting mentions that seem to record not only the ways things once were but also how they could be different. What if, for example, we thought of the two and a half tribes as the founders of Jewish diaspora? This would not only make them heroes to some and cowards to others, but would also imply an ancient Jewish nationalism in which homeland and diaspora develop in tandem. If we think of the tribes Reuben, Gad, and half Manasseh as the founders of diaspora, then we see that a diasporic stance on homeland is an integral part of Jewish national mythology.

Calling Home the Diaspora

As the Jordan River crossing came to a halt when Joshua confronted the two and a half tribes, so the national myth spun in the book of Joshua confronts its own limitations when these tribes settle beyond the parameters of homeland. Every acknowledgment of Israelites east of the Jordan ends up unsettling the exile-homeland dichotomy upon which the Deuteronomistic national myth depends.

The dominant plot line involves twelve tribes that escape servitude in Egypt, wander in the desert, defeat Transjordanian kings, and then cross the Jordan in order to conquer their homeland. Since the tribe of Levi serves as a

---

2. Perhaps in a manifestation of zeitgeist, Jeremy Hutton and I wrote near-simultaneous dissertations concerning the Jordan River as a border and the status of Transjordan. We did not learn of one another's projects until we both had completed our dissertations. While our studies stand out as distinctly individual, we reach several shared conclusions. For example, Hutton speaks of the "two-pronged ideology" of "the final form" of Deuteronomistic texts: "(a) Transjordan is an Israelite territory, not only ethnically, but politically and economically; and (b) Transjordan is *not* the site of God's dwelling on earth." Jeremy Hutton, *The Transjordanian Palimpsest: The Overwritten Texts of Personal Exile and Transformation in the Deuteronomistic History* (Berlin: Walter de Gruyter, 2009), 43.

largely nonterritorial clerical group, the split of the Joseph tribe into Ephraim and Manasseh preserves the structure of twelve land-owning tribes. This myth of origins is deployed as the common memory of various tribes and clans in order to incorporate and enlist them in the national project of Israel. At the same time, enlistment means that the heroes and ancestors of the tribes and clans gain mention in the national history. The concept of alternating generations often accommodates such varied traditions in genealogical terms.[3] This means that different territorial conceptions or regional traditions often appear in the Bible as stages of settlement. As a result, texts of the Hebrew Bible involve a unifying narrative as well as significant overlap and contradiction. Because Israel's national myth comprises multiple traditions and the loss of any one of these traditions would indicate the defection of some part of Israel, biblical texts preserve them despite their contradictions. The Exodus story further generates competing narratives by portraying Israel as the nation that arrives at its land from outside. The position of being outside thus appears as the quintessential position of Israel, which means that those somehow outside of the primary national framework are outsiders among outsiders or, in other words, Israel par excellence.

With different groups always moving between margins and centers, however, no particular party remains Israel par excellence for very long. This constitutes part of the trouble and part of the fun of trying to define a group that, at various junctures, looks like a nation, an ethnicity, a philosophy, and a religion. Along with my argument that Israel maintains the Jordan River as a fundamental national border for a very, very long time, I propose that it sees the east bank as the zone of proximate exile, a place for half-breeds, renegades, and refugees.[4] Sometimes east bank tribes are tacitly included in Israel, but most of the time they appear as distressed hybrids out at a perimeter. Such portrayals reinforce the border by raising the stakes of living beyond it and show the corresponding spatial and social definitions of the marginal.

After Israel duly commemorates its Jordan crossing, an action-packed campaign to conquer the land ensues. The miraculous victories, open violations, and unwitting alliances halt in Joshua 11, as it is declared:

---

3. Sometimes difference is recorded in terms of the simultaneity of brothers rather than the inheritance of sons. In both cases, "the genealogical idiom" enables the incorporation of outsiders. See Saul Olyan, *Rites and Rank*, 73.

4. Hutton views the figuration in historical terms: "In all likelihood, therefore, the population of Transjordan by the time of the divided Israelite monarchies was one of mixed ethnicity, which recognized itself as only nominally Israelite at most, and which, throughout its long association with the Cisjordanian Kingdoms (ca. 1250–742 BCE), attempted to maintain some degree of autonomy." Hutton, *The Transjordanian Palimpsest*, 52.

> Joshua conquered all of the land exactly as God had promised Moses. Joshua distributed territory to Israel according to their tribal divisions and the land grew quiet from the din of war. (Josh 11:23)

After the battles come inventories of kings defeated and punctiliously detailed lists of regional boundaries. In retrospect, the campaign to colonize Canaan appears mostly to have been a question of borders. I would name the boundary lists in Joshua 14–21 regional maps that collectively and not unproblematically signal contiguous national territory and further argue that they represent local traditions canonized for nationalist purposes (see figure 4). Because Transjordan was the first territory conquered by Israel in the Exodus narrative, it is the first to be mapped. At the same time, one hears a compensatory nervousness in the repeated legitimization of territories east of the Jordan (Josh 12:1b–6, 13:8–32, 14:3a, 17:1, 18:7). This is compounded by the fact that the settlement of Transjordan almost proves an exception to the claim that after fighting the Canaanites, Israel heard no sounds of war.

As Joshua dispatches the tribes of Reuben, Gad, and half Manasseh to the east side of the Jordan, he upholds the tenets of previous negotiations as well as his distinction between the divine and the prophetic lease:

> Now that Yahweh your *God has given peace to your brothers* as he told them, you can turn and go to your tents in the land of your possession that *Moses the servant of God gave you* on the other side of the Jordan. (Josh 22:4)

Joshua then admonishes them with a sermon spared all the other tribes. "But be vigilant in performing the commandments and the teaching that Moses the servant of God transmitted to you: Love Yahweh your God, walk along His paths, abide by His commandments, cling to Him, and worship Him with all your heart and all your soul" (22:5).[5] In Joshua's view, not only does the Jordan define the territory of Israel, but it also sets a geographic limit on the feasibility of performing the law. The two and a half tribes become a test case of the law's spatial bounds.

---

5. Elie Assis misreads this verse by drawing a parallel between Moses's injunctions to the People prior to their entrance to the land and Joshua's exhortations to the Transjordanian tribes. Elie Assis, "'For It Shall Be a Witness Between Us': A Literary Reading of Joshua 22," *Scandinavian Journal of Old Testament* 18 (2004): 211. Assis misses the fact that all the other tribes are spared the repetition of national expectations as they go to settle their allotments. Since Joshua holds his next audience with the half-tribe of Manasseh alone, one can imagine the tribes of Reuben and Gad receding beyond the river as the Israelite leader rattles off the well-established tenets.

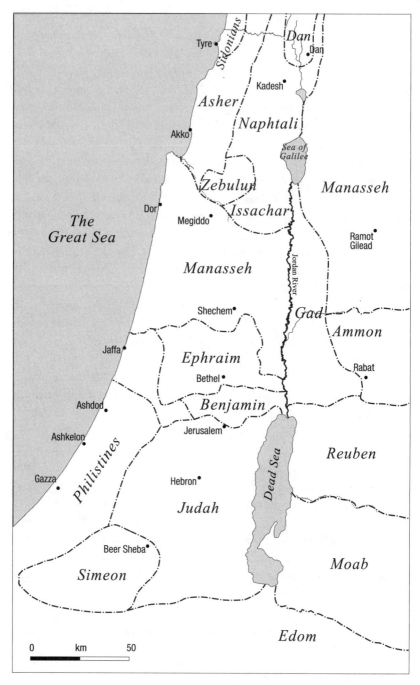

4. The territory of the twelve tribes according to the book of Joshua. Soffer Mapping.

The fears voiced by Joshua seem to belong to the anxieties of diaspora and have led many scholars to read this late episode in the book of Joshua as a postexilic parable of deterritorialization.[6] The question that permeates each twist in the plot is how group cohesion can survive dislocation. Although Transjordan is just barely exile, it involves the problems and the alternatives posed by any diaspora. Joshua also fears that the wealth of the diaspora may not benefit the homeland: "Return to your tents with great wealth, with abundant livestock, with silver, gold, copper and iron and a large wardrobe, but be sure to share the spoil of your enemies with your brothers" (Josh 22:8). As they embark on their own homecoming, the two and a half tribes are subject to suspicion and scrutiny. Yet in contrast to Joshua, the narrator confers divine approval by speaking of "the land of their possession granted *by God* acting through Moses" (Josh 22:9).

After the tribes of Reuben, Gad, and half Manasseh cross the Jordan and reach their land, they repeat the sequence of the other Jordan crossing by building a memorial by the river's edge.[7] Yet, their version of a memorial is an altar with cultic intent. This is no humble pile of rocks, but "a highly visible altar" (Josh 22:10). The altar is problematic on two counts: its magnitude implies the disproportion of an Israelite cultic place outside of the land, and its visibility demands notice.[8] The east bank tribes build a memorial so prominent that it doesn't allow the others to forget they exist. The elaborate dimensions also seem to play on Joshua's concerns about the tribes' great wealth.

---

6. Douglas Knight observes that "the Easternerers [what I call the east bank tribes] are not even called 'Israelites' in the vast majority of occurrences of the word in this chapter." Knight, "Joshua 22 and the Ideology of Space," *'Imagining' Biblical Worlds: Studies in Spatial, Social, and Historical Constructs in Honor of James W. Flanagan*, ed. David M. Gunn and Paula M. McNutt (Sheffield: Sheffield Academic Press, 2002), 61.

7. There is considerable ambiguity concerning the precise location of this altar. The two and a half tribes build the altar at *Gelilot* of the Jordan. The LXX and Syriac versions read *Gelilot* as *Gilgal*. Another problem arises from the discrepancy between "*Gelilot* of the Jordan that is in the land of Canaan" (Josh 22:10) and "the altar across from the land of Canaan in *Gelilot* of the Jordan, on the other side of the People of Israel (22:11). The Anchor Bible commentators note that "lack of clarity belongs to the structural integrity of this story." Robert G. Boling and G. Ernest Wright, *Joshua: A New Translation with Notes and Commentary* (Garden City, N.Y.: Doubleday, 1982), 505. Another possible reading is that *Gelilot* of the Jordan, on the east bank, belongs to the land of Canaan from the point of view of the two and a half tribes, the subject of v. 10, but opposes the land of Canaan from the point of view of the "People of Israel," that is, the nine and a half tribes, the subject of v. 11. In any case, I follow v. 11 in locating the altar on the east bank. This contradicts Noth's assessment that the altar was west of the Jordan and that verse 11 is an addition, Noth, *Das Buch Josua*, 134.

8. Frank Moore Cross picks up on the ambivalence surrounding the tribe of Reuben and the east bank in general. However, his suggestion of a functioning Israelite shrine at Nebo is a conclusion too literal to be drawn from the conflicted representations of Transjordan. F. M. Cross, "Reuben, First-Born of Jacob," *Zeitschrift für die Alttestamentliche Wissenschaft* 100 supp. (1988): 46–66.

Perhaps most unsettling of all, the altar is a simulation of the official altar stationed at Shiloh. The reproduction of the altar negates both the specificity and the eastern tribes' need for the central altar in Shiloh.⁹ The altar simulates the Shiloh altar as well as the twelve boundary stones at Gilgal and thus establishes both a competing cultic site (violating the interdiction of Deuteronomy 12:13–14) and a competing claim of memory. Memories that overtly threaten collective unity and resist cooption, according to Halbwachs, are slated for collective erasure. Since memory is a mode of authorization, alternate recollections disrupt extant authority by conferring it elsewhere.¹⁰ If, as the two and a half tribes imply, home lies both to the west and to the east of the river; if both banks are legitimate sites of memory; if Israel can encounter God on either shore, then the system of the land—built on distinction—collapses.

Analysis of the altar and the protest against it reveals the synthetic nature of memory. Israel's memory, like its land claims, is neither unified nor singular. The negotiations concerning representation and simulation, normative and deviant memory, homeland and diaspora proceed as such:

> When they arrived at Gelilot of the Jordan [that is in the land of Canaan], the people of Reuben, Gad, and half Manasseh¹¹ built an altar there beside the Jordan, a highly visible altar. When the People of Israel heard about it, they said, "Look, the people of Reuben, Gad, and half Manasseh have built an altar across from the land of Canaan at Gelilot of the Jordan, on the other side of the People of Israel!" The entire community of Israel gathered at Shiloh in order to go to war against them. But first the People of Israel sent the priest, Phineas the son of Eleazar, as a representative to Reuben, Gad, and half Manasseh in the land of Gilead. Ten tribal leaders accompanied him, one representative from the ancestral house of each tribal order; every one of them was head of his ancestral house among the divisions of Israel. They arrived in the land of Gilead and spoke to the people of Reuben, Gad, and half Manasseh. "All the community of Yahweh asks why you have devised this act of treason against the God of Israel and turned away from Yahweh

---

9. Sigmund Mowinckel takes the mention of Shiloh as a cultic center in Joshua as evidence of a postexilic, late Priestly hand that he labels P. Where the Deuteronomist maintains Shechem as the cultic focus, P, likely out of anti-Samaritan bias, moves it to Shiloh, where a center is recorded in 1 Samuel. Mowinckel, *Tetrateuch-Pentateuch-Hexateuch. Die Berichte über die Landnahme in den drei altisraelitischen Geschichtswerken* (Berlin: A. Töpelmann, 1964).

10. Halbwachs, *On Collective Memory*, 182–83.

11. Three different phrases describe the half-tribe in Joshua 22. Vv. 9, 10, 11, 21 (חצי שבט המנשה); vv. 13, 15 (חצי שבט מנשה) ; and vv. 30, 31 (בני מנשה). See A. Graeme Auld, *Joshua, Moses, and the Land: Tetrateuch-Pentateuch-Hexateuch in a Generation Since 1938* (Edinburgh: T. & T. Clark, 1980), 58.

by building an altar that celebrates your latest rebellion against God? Isn't the sin of Peor that brought the plague upon the community of Yahweh and from which we are not yet purified enough for us? Now you turn away from God! You rebel against God today, and tomorrow He will be enraged with the whole community of Israel. If the land of your holdings is impure, then cross over to the land of Yahweh's holding where the tabernacle of God resides and stake a claim among us, but don't rebel against God and against us by building yourselves an altar other than the altar of Yahweh our God. When Achan son of Zerah committed his treason by violating the ban, was God not angry with all of Israel? Was he the only one who died because of his transgression?" (Josh 22:10–20)

The altar initially appears to the west bank tribes as an affront to the very sanctity and immutability of the land. The suspicion is magnified by the belief that Yahweh cannot be served beyond the borders and therefore the altar must be devoted to some other god or at best to some radically different conception of God.[12] The current interlocutors, perhaps inspired by the prominence of the altar, speak explicitly of the two and a half tribes' "latest rebellion" as a pretext for civil war. They stick to what is by now the official line by framing the rebellion as the inevitable result of the incorrect practice of memory. The altar's location and cultic nature recall the Moabite orgy at Peor, and its violation of taboo reminds the leadership of Achan's transgression at the battle of Jericho. Feeling as imperiled by the sizable altar as by the orgy and the subterfuge, the west bank tribes present themselves as the self-suffering nationals who will pay for the sins of their diasporic counterparts.[13] They emphasize their sense of disjunction between the west and east bank by introducing a new dichotomy: pure and impure. Since the land beyond the Jordan is forever impure, the homeland party reasons, only immigration can repair the fissure and purify the east bank tribes.

Ever since the two and a half tribes articulated their desire for a Transjordanian home, they have been accused of subversion. The story of the east bank

---

12. Based on the poems in Numbers 23–24 that "use the divine name El more times than Yahweh for the God of Israel," Mark Smith argues that "this usage suggests an alternative Israelite view in the region on the eastern side of the Jordan River (Transjordan)." Smith, *The Memoirs of God: History, Memory, and the Experience of the Divine in Ancient Israel* (Minneapolis: Fortress Press, 2004), 20.

13. Elie Assis reproduces this tone in his contemporary analysis: "This striking suggestion reflects their understanding that the two groups are parts of one nation, and that the Cisjordanians are willing to pay the price and give up part of their land for the sake of mass-migration from the Transjordan." Elie Assis, "The Position and Function of Jos 22 in the Book of Joshua," *Zeitschrift für die Alttestamentliche Wissenschaft* 116 (2004): 538.

memorial extends the themes of Moses's and Joshua's confrontations with the tribes and suggests that every generation must contend anew with the problem of diaspora. Moses transmits the terms of the initial contract to "Eleazar the Priest, Joshua son of Nun, and the heads of the ancestral tribes" (Num 32:28). Joshua 22 describes the ways in which each of these figures interprets the contract: Joshua first presents his version (22:1–8) and then Phineas the son of Eleazar the Priest and the tribal heads have their say (22:13–20). The emphasis on future generations throughout the story promises the ongoing conundrum of insider and outsider Israel. As the two and a half tribes place themselves where Lot's descendants belong, they accrue associations with the problematic kin groups best left out of national figurations. Their environs distance them from Israel, and their position on the other side of the river sets them up to be opponents. The key to war or peace between the banks seems to hang on language, and indeed it is the clever rhetoric of the two and a half tribes that postpones confrontation.

> The people of Reuben, Gad, and half Manasseh answered the heads of the tribal divisions of Israel. "God, O God, Yahweh, God, O God, Yahweh knows and Israel knows that if we rebelled or committed treason against Yahweh, then you should not spare us today. If we built an altar in order to turn away from Yahweh or to offer burnt offerings, grain offerings or peace offerings, let Yahweh be the Judge. Did we not do this only out of the anxiety that in the future your children would say to our children, 'what have you to do with Yahweh the God of Israel? God set the Jordan as a boundary between us, children of Reuben and Gad, you have no portion in God!' Thus your children would prevent our children from following Yahweh. So we said to ourselves, 'let's build an altar, not for offerings or sacrifices, but as a witness between us and between the generations after us that we can worship God through our burnt offerings, sacrifices and peace offerings and so that your children will not say to our children in the future 'you have no portion in God.' We figured that if you say such things to us or to our descendants in the future, we will say, 'look at the replica altar to Yahweh that our ancestors made, not for offerings or sacrifices, but as a witness between us.' God forbid that we rebel against Yahweh or turn away from Him by building an altar for burnt offerings, grain offerings or sacrifices other than the altar of Yahweh our God that stands before His tabernacle." (Josh 22:21–29)

The two and a half tribes exhibit the same rhetorical skill in this situation as they did in the dialogues with Moses and Joshua—again they talk themselves out of trouble and present their commitment to the east bank as pragmatic and nonthreatening. While the Israelite representatives are ultimately convinced, the combination of rhetorical aplomb and "God forbid" causes the reader to wonder if they are telling the truth. Did the east bank tribes initially intend their altar to be a mere replica, or do they dissemble? They neither address nor refute the charge that their land is impure, but instead defer to the homeland party by describing their culture as a practice of reverent reference. By invoking the image of children, the east bank tribes connect their altar with the normative Israelite memory formulated as a legacy for future generations. We are behaving just like you, the tribes of Reuben, Gad, and half Manasseh explain; after crossing the Jordan we built a memorial out of concern for our children. Thus the east bank tribes neutralize the accusation of rebellion and align themselves with the national concern for continuity.

As an imagined conversation with future children accompanied the establishment of the west bank memorial, so the tribes of Reuben, Gad, and half Manasseh append a motivating dialogue. The story within the story comments upon the very events transpiring. The east bank tribes put the sentiments of Phineas and the Israelite representatives into the mouths of their descendants: "What have you to do with Yahweh the God of Israel? God set the Jordan as a boundary between us, children of Reuben and Gad, you have no portion in God" (Josh 22:24–25). Before they are even born, these future children of Reuben and Gad have a mixed identity and a dual idea of home. Their future west bank counterparts will have it wrong if they exclude them on the principle that the Jordan marks the limit of God's land. What the future counterparts and the current interlocutors should do, in their view, is accept the east bank tribes and allow them to forge and cultivate a connection to their virtual homeland. The altar's visibility is sanctioned, in their telling, by its function as a witness. Its scale is not intended to accommodate numerous sacrifices, but as a measure to prevent the east bank tribes from becoming invisible, forgotten, and ultimately barred from Israel. Although accused of mimicking transgressors, the east bankers claim to mirror the pious action of building an altar to serve memory. It is only the desire of the west bank tribes to forget about their distanced brothers that necessitates the grandiose memorial.

While ignoring the plea for immigration, the east bankers vow to cross the Jordan perpetually in order to visit the central shrine. They will continually come "home" if permitted a replica on their own turf. While conceding the fact that tribes of Israel dwell beyond the boundaries of the land, the writers forbid the construction of functioning altars anywhere east of the Jordan.[14]

---

14. I do not see how Joshua 22 can be taken as sanctioning altars outside the borders of the land. See J. G. Vink, "The Date and Origin of the Priestly Code in the Old Testament," *Old Testament Studies* 15 (1969):1–144; see p. 77.

Phineas the Priest, the leaders of the community, and the heads of the tribal divisions of Israel that were with him listened to what the people of Reuben, Gad, and Manasseh had to say and it made sense to them. Phineas the son of Eleazar the Priest said to the people of Reuben, Gad, and Manasseh, "Today we know that Yahweh is in our midst. You have not committed this treason against God and have, in fact, saved all Israel from the hand of God." Phineas the son of Eleazar the Priest and the leaders left the people of Reuben and Gad in the land of Gilead, returned to the land of Canaan, and reported the events to the People of Israel. The People of Israel received the report gladly and blessed God; they stopped speaking of going to war in order to destroy the land where the people of Reuben and Gad had settled. The people of Reuben and Gad called the altar Witness because "it is a witness between us that Yahweh is our God." (Josh 22:30–34)

The image of a memorial built to remind children of their God persuades Phineas the Priest and the tribal leaders that the east bank tribes are not working at cross-purposes from the west bank tribes. The opposition of the two and a half tribes is a feature of geography and not, for the time being, of ideology. At this juncture, negotiations avert civil war. In a conciliatory gesture, Phineas states that the lengths to which the east bank tribes will go to insure their inclusion in Israel attests to the fact "that God is in our midst." Phineas thus reverses his earlier accusation that the altar imperiled Israel (Josh 22:31). The clarifying dialogue has "saved all Israel from the hand of God" that would punish infighting among the tribes at such an early stage of settlement. Gilead is left to its ambiguity as the leadership returns to God's country of Canaan.

The story of the east bank altar concludes with a split scene. West of the Jordan, the altar is deemed a legitimate mnemonic device and God blesses the Israelites for no longer harboring resentments that might stir them to destroy the habitation of the two and a half tribes. The tribes of Reuben and Gad (half Manasseh is absent here, but restored in the Septuagint), now left at peace, name the altar "Witness," because "it is a Witness between us that Yahweh is our God." Through their justification, the altar finds its function and Gilead gains a second border through another witness to a negotiation.[15]

---

15. Remember that according to biblical etymology, the region gains its name when Jacob establishes his rock mound "as a witness between me and you," the "you" in question being Jacob's uncle/father-in-law Laban (Gen 31:44). The rock mound is to serve as a frontier that neither Jacob nor Laban can cross in the name of war, and so Gilead comes to serve as a frontier between Israel and Aram. If that act of witnessing establishes the eastern limits of Gilead, then that of witnessing the agreement between the east and west bank tribes reinforces the Jordan as its western boundary. The mediation

The split scene suggests that the hard-won unity does not involve unification. The two groups think differently about the same things—including the border between them—, yet they both remain a part of Israel.

The Half-Tribe

What is a half-tribe? What does it mean to have a tribe split between two banks of a river? In terms of the modern Middle East, the scenario becomes imaginable through the position of the Druze, a distinct religious group that accepts the sovereignty of whatever country claims Druze communities within its borders.[16] This philosophy means, for example, that since 1967 some Druze families have been split between Israel and Syria. Because there are only ceasefire lines, no formal treaty, between Syria and Israel, those Druze who hold Israeli passports are forbidden from visiting Syria, and the Syrian Druze cannot exit Syria in order to enter Israel. Since the categories of Golan Heights Druze and Syrian Druze have already emerged by default, we can begin to imagine a western half of Manasseh associated with the tribe of Ephraim and an eastern half linked with the tribes of Reuben and Gad. The very notion of a split tribe animates the tension between regional assimilation and the maintenance of tribal identity. In the case of the Druze, contact is partially maintained by a festival convened in the Valley of Shouts at which counterparts from either side of the ceasefire line gather at opposite ends of a valley to see one another, shout out family news, and commemorate collective history.[17] Imagining a similar function for the festival commemorating Jephthah's daughter can further serve to elucidate the biblical material. Perhaps the festival for Jephtah's daughter—problematic as it is for biblical writers—served to reunite Northern women from the two banks of the Jordan and generate common memories. As I have mentioned, the gendered nature of these common memories render them anti-national. Insofar as the sustained connection between the Druze on either side of the border entails Syrian leanings by some Druze of the Golan Heights, the extra or anti-national bent of group identity that straddles a border comes into relief.

David Jobling sees the half-tribe as a synecdoche of a split Israel. "The tribe of Manasseh, in its bipartition, recapitulates and *mediates* the bipartition of Israel; Israel's being split in two is softened if there is a point at which these two

---

of witnessing defines Gilead in two directions; it is both a unique and an intermediate space between Israel and Aram.

16. This position is troubled somewhat by the Israeli occupation of the Golan Heights as many Druze remain invested in a Syrian identity.

17. Martin Asser, "Golan Druze celebrate across barbed wire," http://news.bbc.co.uk/2/hi/middle_east/7353494.stm.

are still one."¹⁸ The two halves of a tribe split by the Jordan then dramatize the insider-outsider situation of Israel. If a tribe can be split and still maintain integrity, then Israel can support communities beyond its borders while remaining a discrete and distinct people. On this count, the half-tribe enables Israel to transcend the very borders that define it. However, the half-tribe is such an inconsistent presence that the softening of bipartition seems at best partial. What does half Manasseh's fading in and out communicate about Israel?

Half Manasseh is composite with Gilead and Machir, eponymous ancestors more vivid than Manasseh, firstborn son of Joseph. We first hear of Machir at the end of Numbers 32, after the tribes of Reuben and Gad have negotiated with Moses for the east bank.

> The sons of Machir son of Manasseh went to Gilead, captured it, and dispossessed the Amorites who inhabited it, so Moses gave Gilead to Machir the son of Manasseh and he settled there. Yair the son of Manasseh went and captured their villages then named them the villages of Yair. Novach went and captured Kenat and its environs and called them Novach in his name. (Num 32:39–42)¹⁹

Machir and the other sons of Manasseh stake a claim on Gilead through raids rather than rhetoric. Like the People of Israel, they dispossess Amorites from Gilead; yet the sons of Manasseh act on their own behalf rather than on behalf of Israel. Martin Noth already recognized the anti-national quality of such regional land claims. "Only in Num. 32 do we find survivals of an old tradition of the occupation which deals specifically with particular tribes."²⁰ Invested in the history behind oral traditions, Noth could only imagine traditions in historical terms; the Machir tradition, in his opinion, had to be a survival from a pre-state period.²¹ This could well be the case, but according to Noth's larger

---

18. Jobling, "The Jordan a Boundary," 116.

19. Deuteronomic texts contain variations on the theme. In Deuteronomy 3:13, 15, Moses confers part (or all!; see 3:15) of Gilead and the northern region of Bashan to the half-tribe of Manasseh. However, in Deuteronomy 3:14, Yair son of Manasseh conquers the Bashan and marks it with his name. "The Book of Joshua gives a third account of the conquest and distribution of East Jordan, as in Deut. 2:24–3:17 in retrospect. The description of the conquest and the fixing of the frontier of the Amorite kingdoms in 12:1–6 reveal the strong resemblances to the Deuteronomic prologue, while the section on the division in Josh. 13:8–32 contains far more details of the topography of the area and information about the towns in a fashion reminiscent of Num. 32. The accounts of the distribution in Josh. 13:8–32 and Num. 32 differ on essential points." Ottosson, *Gilead*, 129.

20. Noth, *A History of Pentateuchal Traditions*, trans. Bernhard Anderson (Chico, CA: Scholars Press, 1981), 42.

21. Noth dates the Machir tradition as later than that of Reuben and Gad. Martin Noth, *Numbers: A Commentary*, trans. J. D. Martin (Philadelphia: Westminster, 1968), 240–41.

principle that the incorporation of local heroes brought clans into the federation of Israel, we might also conclude that alliance with the families of Gilead required mention of their founders. In other words, the fact that this tradition marks a different founding myth does not make it decidedly "old."

However, it seems true that Machir and Gilead are the earlier appellations and half Manasseh the "artificial construction."[22] In the Song of Deborah, one of the oldest pieces in the Hebrew Bible, the Prophetess Deborah praises "Machir" for sending its princes to war and chastises "Gilead" for waiting out the battle "on the other side of the Jordan" (Judg 5:14, 17). Deborah's praise along with its founding myth suggests that Machir may have been a mercenary tribe.[23] Based on Machir's differentiation from Gilead, scholars speculate that Machir was initially a Canaanite group that migrated east. However, Joshua 17 narrates how Machir's descendants, the daughters of Zelophehad, migrated west after marrying their tribesmen (Josh 17:6). Ultimately, the texts defy historical reconstruction. The best one can say is that the half-tribe of Manasseh seems to be an administrative category intended to manage the peoples of Gilead as members of Israel. Because the people of Gilead have identities beyond the administrative category, friction surrounds their record.

The friction is to some degree resolved by a genealogy in which Machir is son of Manasseh and father of Gilead (Num 26:29; 1 Chr 7:14); however, the later version of this genealogy problematizes Gilead in a familiar way.[24] "The sons of Manasseh: Asriel, whom his Aramean concubine bore; she bore Machir the father of Gilead. And Machir took wives for Huppim and for Shuppim. The name of his sister was Maacah. And the name of the second was Zelophehad and Zelophehad had daughters" (1 Chr 7:14–15). According to this genealogy, the founding father Machir is an Israelite-Aramean half-breed. As if their blood mirrors their territory, Transjordanians are portrayed as always at least half-foreign. When they are not mixed-bloods (and some-

---

22. M. Patrick Graham, "Machir," *ABD*, 459.

23. "Machir's name, which meant 'sold,' may be taken, along with the remark in Josh 17:1 that the family included many fighters, to indicate that the tribe sold its military services to the Canaanites, probably during the Amarna age." M. Patrick Graham, "Machir," *ABD*, 459. I am thinking slightly differently that Machir "sold" its military services to Israel.

24. The complicated genealogy of Manasseh includes a version in which "Machir is the father of Gilead, i.e. a resident there, though certain districts of Mount Ephraim are also associated with him" (Judg 5:14; Josh 17:1–2). In Joshua 17:1–2, Hepher, for example, is Machir's brother, "but in the following verse a simpler family lineage is expressed: 'Zelophehad the son of Hepher, son of Gilead, son of Machir, son of Manasseh'" (Josh 17:3). 1 Chronicles 7:14–17 is perhaps the most forthcoming about its position on Transjordan: "according to this arrangement the western clans have descended from Manasseh's first wife while those in Transjordan are the progeny of his Aramean concubine." Jobling, "The Jordan a Boundary," 131. The Chronicles passage provides a wonderful example of the tendency to mark territory as problematic through association with a marginal or troubling character.

times when they are), then they are women, a gender position that we have learned to read as another device of marking Transjordanians as somewhat alien and potentially subversive. All the same, these strange breeds belong to Israel.

Half Manasseh is sometimes directly linked to the tribes of Reuben and Gad—hence the two *and a half* tribes—, sometimes omitted from tribal lists, and sometimes mentioned indirectly through association with Machir. Returning to the question of what half Manasseh's fading in and out communicates about Israel, we can detect a strategy of absorbing "new" groups into Israel while maintaining the myth of Israel's distinction. The tribal system proves flexible enough to accommodate the entrance and exit of groups without structural compromise. The overlap of Machir and half Manasseh then can serve as an example of "the flexibility of myth," the way in which myth reflects the changing social structure while professing to be immemorial.[25]

With an intentional avoidance of timelines, I am arguing two points here. The first is that the designation "half Manasseh" offered a means of domesticating Gileadite clans in the twelve-tribe system. It is not a question of a tribe splitting into two, but rather the extension of a tribal name in order to incorporate allies. Gileadite clans were brought into Israel under the rubric of half Manasseh, and the "half" in their name communicates both that they are partial and that they are mixed. My second point is that mixed ethnicity operates both as a border and as a frontier. As we saw in chapter three, the land of Gilead serves as a border between Israel and Aram. Because an Aramean cannot be a member of Israel, the region of Gilead represents an ethnic limit. However, a genealogical tradition credits an Aramean-Israelite with founding the eastern tribe of Manasseh, thereby recognizing his descendants as a marginal but included category. The mixed-blood opens up a frontier at once geographic and ethnic. In its function as a border, Gilead—reinforced by the Jordan River—keeps Arameans out of Israel, but in its function as a frontier it brings Aramean-Israelites in.

"The Land of the Amorites Who Lived Beyond the Jordan" (Josh 24:8)

The balance between the Israelite and the alien characteristics of the east bank is under constant negotiation in biblical texts. The half-tribe of Manasseh represents one example of such negotiation and the settlement history of the land claimed by the tribes of Reuben and Gad another. Suspicion that Reuben and Gad's territory is integrated into that of Moab and Ammon

---

25. See Theodore Van Baaren, "The Flexibility of Myth," in *Sacred Narrative: Readings in the Theory of Myth*, ed. Alan Dundes (Berkeley: University of California Press, 1984), 217–24.

directly challenges the various myths claiming that the Jordan River sets Israel apart from other nations.[26] For this reason, the proximity is managed through the claim that Israel conquered its Transjordanian holdings from Amorite foes, not from Moab, Ammon, and Edom.[27] The proscription of Moabite, Ammonite, and Edomite land on the principle that God granted circumscribed territory to these relatives of Israel just as He did for His beloved nation reinforces the claim. These alleged Amorite opponents, I suggest, are a fiction intended to break the contiguity of Reuben and Gad with Ammon and Moab and perhaps to obscure the indigenous origins of the east bank tribes. Sihon and Og, the two larger-than-life Amorite kings, create a legendary space between Israel and Moab. The Amorite tradition attempts to distance Israel from Moab and to draw clear lines in a region or at a time when such imagined lines did not operate.[28]

Assessing Transjordanian land claims requires some backtracking. Although we have already crossed the Jordan with Joshua and witnessed the resolution of the dispute between the east bank and the west bank tribes, let us return to the wilderness where a landless Israel faces its future neighbors. Israel turns to its brother nation Edom only to be denied passage and greeted, as was Jacob by Esau, by battalions of armed men (Num 20:14, 20). Like Jacob, Israel turns in the other direction. In the Negev, Israel seizes part of the holy land in a battle to regain Israelite captives (Num 21:1–3). From the Negev, Israel enters Moab, and here the text begins to dissemble with the insistence that Israel reaches only the edge, not the heart, of Moab. Israel first camps "just east of Moab in the wilderness" (21:11) and then crosses the Arnon River to camp "in the wilderness alongside the Amorite border" (21:13). The insertion of Amorites into Moabite space becomes conspicuous as the text elaborates on the "Amorite border": "the Arnon, you see, is the border of Moab, the border between Moab and the Amorites" (21:13).

---

26. Martin Noth claims that Israelite tribes, and Gad in particular, participated in cultic gatherings at Baal Peor, the reviled site of miscegenation in Numbers 25:1–3. "As a *boundary sanctuary* to which the 'peoples' gathered, the sanctuary of Baal Peor enjoyed widespread respect and esteem, not only among the narrower circles of the tribe of Gad in the north or on the other hand the Moabites who inhabited the southern region, but also among the nomadic Midianites of the eastern wilderness and the Israelite tribes living opposite in the country of West Jordan." Noth, *A History of Pentateuchal Traditions*, 74–75.

27. The Amorites "appear as a sub-group of the pre-Israelite population in the stereotyped census in Gen. 10:16; 15:21; Ex. 3:8; 13:5; 23:23; 33:2; 34:11, but as a specially emphasized racial group only in Gen. 15:16 and 48:22. In Gen. 15:16 the substantive is a generic name for the pre-Israelite population whose misdeeds culminated on the return of Abraham's descendants from Egypt." Ottosson, *Gilead*, 71,

28. For the historical debates on settlement in Transjordan, see Robert G. Boling, *The Early Biblical Community in Transjordan* (Sheffield: Almond Press, 1988).

The story continues with Israel requesting passage, as it did from Edom, from Sihon king of the Amorites. When Sihon in turn attacks Israel, he forfeits the chance that Israel might turn from him as they did from Edom. Israel emerges victorious from the battle and wins the territory between the Arnon and the Jabbok Rivers, the Jabbok here cited as the border with Ammon (21:24). Israel comes to possess land between Ammon and Moab without raising arms against either one. By protesting too much, the explanation of how this came to be the case betrays signs of a cover-up.[29] "Now Hesbon [the central city that Israel occupies] was the city of Sihon king of the Amorites, who had fought against the first king of Moab and taken all of his land up to the Arnon" (Num 21:26). The narrator seems to address incredulity on the part of the audience that the lands in question were seized from Sihon by admitting to a Moabite presence, albeit in a remote era.[30] Even a cautionary tone that it is in the audience's best interest to accept this explanation is audible: "Yes, the Moabites once were there," the narrator seems to say, "but it is best to celebrate how Israel took the land from Sihon king of the Amorites."

Sihon king of the Amorites and his counterpart Og king of Bashan assume mythic proportions in Deuteronomy's narration of the events. Deuteronomy describes a "fraternal process of distribution" in which "the God of Israel is described as distributor of land to Edom in 2:5, Moab 2:9 and Ammon 2:19."[31] And although there is a slip of admission that Israel crossed over the border right into Moab (Deut 2:18), the only land seized allegedly belongs to Sihon and Og.[32] A prehistory in which giants roamed Transjordan further heightens the stakes of Israel's defeat of Og, last of the giants. As part of its passage into national adulthood, Israel slays a giant—a task memorialized through the monument of Og's enormous bed frame in Rabbat Ammon (Deut 3:11).[33] The narrator in Deuteronomy substitutes an oedipal struggle for the theme

---

29. That the claim is backed up assiduously in the genres of poetry (Num 21:27–30; Ps 135:11–12, 136:19–22), historical reporting (Num 21:21–26; Deut 2:24–36, 4:46–49; Josh 24:8; Judg 11:18–22; Neh 9:22), and narrative mapping (Josh 12:2–5, 13:8–12, 21, 27) adds to my suspicion.

30. Geographic rhetoric like the phrase "from Arnon to the Jabbok" also participates in the minimization of Ammonite presence. See John Bartlett, "Sihon and Og, Kings of the Amorites," *Vetus Testamentum* 20 (1970): 257–77.

31. This myth may well be a variant of the daughters of Lot story in Genesis 13. Ottosson, *Gilead*, 92.

32. See Maxwell Miller, "The Israelite Journey Through (Around) Moab and Moabite Toponymy," *Journal of Biblical Literature* 108 (1989): 578.

33. On Israel's wilderness journey as a process of national development, see Ilana Pardes, *The Biography of Ancient Israel* (Berkeley: University of California Press, 2000). For the defeat of giants as an image of overcoming parents, see Bruno Bettelheim, *The Uses of Enchantment: The Meaning and Importance of Fairy Tales* (New York: Vintage Books, 1977), 27.

of contest between brothers. This substitution grants Israel hereditary rights to land already promised to Moab and Ammon. Although adding another historical claimant to contested land seems only to complicate matters, it is a move common in ethno-territorial struggles as a way to discredit and distance the immediate opponents.

The Frontier

Edom, Moab, Ammon, Aram, Gilead, Gad, Reuben, and Manasseh are best thought of as regional players in a borderland who alternately battle and blend with one another.[34] The overlap among these peoples is readable in terms of coexistent claims and the synonymous substitution of the terms as ethnic markers. The Amorites, as I have shown, are part of a frontier legend about Israel bringing down giants in the wild east. The size of Og's bed and its carnivalesque display in Rabbat Ammon are key indicators of a tall tale.[35] Since claims to the frontier often rest on defeating "bad guys," the Amorite tales provide the ideal enemy and play a central role in supporting Israelite claims to Transjordan. In the absence of these true villains and the pioneer fathers Abraham and Lot, the difference between Israel and its neighbors begins to erode.

The erosion of difference, however, is not in the interest of either Israel or Moab. The Mesha Inscription, the longest extant text in Moabite, cites "the men of Gad" as the catalytic opponent. The very presence of this tribe mobilizes disparate groups under the banner of Moab. In line 17 of the Mesha Inscription, the formerly Israelite town of Nebo is placed under a restric-

---

34. For example, the lists of cities said to be built by Gad and Reuben (Num 32:34–36, 37–38; Josh 13:15–28) are commonly attributed to the Moabites; this is the case with Dibon (Isa 12:2; Jer 48:18, 22) and Aroer (Judg 11:26; Jer 48:19). In Joshua 13:25 Aroer belongs to "the land of the Ammonites" at the same time that it is granted by Moses to the tribe of Gad (13:24). Judges 11:33 reports Jephthah's defeat of the Aroer Ammonites. "While Num 32.34 attributes Aroer to Gad, I Chr 5:8 makes it a possession of a descendant of Reuben." Burton MacDonald, *"East of the Jordan": Territories and Sites of the Hebrew Scriptures* (Boston: American Schools of Oriental Research), 2000, 133, 134, 146. The building of some of these very towns falls among the claims of Mesha the King of Moab, as inscribed in the Mesha Inscription. Tracing his own roots to Dibon, Mesha boasts that "I myself built Aroer and I myself made the highway through the Arnon" (26). These projects occur in the wake of Israelite defeat after all "the men of Gad" (10) are killed and all Israelites in Nebo eradicated (14–17). As the biblical writers recognize the antiquity of Moabite claims in the battle reports with Sihon, so the Mesha Inscription confesses "now the men of Gad (had) dwelt in the land of 'Atarot from of old and the King of Israel built for (them) 'Atarot" (10). An important difference between biblical texts and the Mesha Inscription arises from the fact that Mesha prides himself on replacing Israel where the Bible inserts buffer foes between Israel and Moab.

35. Dina Stein, "Believing Is Seeing: A Reading of Bab Bathra 73a–75b," *Jerusalem Studies in Hebrew Literature* 17 (1999): 22 (Hebrew).

tive ban, an alternate attestation of the biblical *herem*. In a stunning book on Moab, the archeologist Bruce Routledge observes that such a ban prevents not only the ransom of captive persons and property, but also exchange in general. This case "emphasizes the oppositional (and hence equivalent) nature of Moab and Israel by denying the possibility either of incorporating subunits associated with Israel (e.g., Men of Gad) into Moab or of mutual recognition via exchange (as in the case of tributary relations)."[36] Through the refusal to accommodate Israel, Moab establishes itself as an equal rival and "a workable, and independent, national identity."[37]

Because distinctions are vital yet not particularly apparent, a manner of frontier justice operates in Transjordan. This is evident in the way in which Phineas the priest punishes the participants in the orgy at Baal Peor,[38] in Ehud's stealth killing of Eglon king of Moab, and in Mesha king of Moab's sacrifice of his first-born son in order to repel a victorious Israelite army (2 Kings 3:27).[39] In these scenes, the importance of distinction sanctions acts of extreme and perverse violence. The shibboleth incident, one of the most memorable accounts of frontier justice, occurs during a civil war and shows the absurdity of insisting on difference at a frontier.

The Jordan River separates the two and a half tribes from mainland Israel while, at the same time, distinguishing Israel from Moab and Ammon. This positions the east bank tribes as essential intermediaries between Israel and their cousins, Moab/Ammon. The equation that comes into balanced tension in the book of Joshua spirals into rampant violence in the book of Judges. It is not surprising that the book of Judges portrays how things fall apart in Transjordan, since the book chronicles the dissolution of all domestic and national institutions. When the perennial problems arise in Transjordan, language serves to exacerbate rather than resolve them.

The (mis)speaker whose language wreaks the havoc is Jephthah of Gilead. As if born into the very line of Gileadite mixing, Jephthah is said to be "the son of a foreign woman with whom Gilead had Jephthah" (Judg 11:1). Even in Gilead, such maternity involves shame, particularly because the epony-

---

36. Bruce Routledge, *Moab in the Iron Age: Hegemony, Polity, Archaeology* (Philadelphia: University of Pennsylvania Press, 2004), 150.

37. Routledge, *Moab in the Iron Age*, 150.

38. If we take seriously Noth's suggestion that Baal Peor served as a central cultic site for all of Transjordan, then Phineas's actions there would represent a kind of fundamentalist reaction to regional syncretism.

39. Frontier justice is also evident when Israelite tribes block Midianite access (Judg 7:23–25) and when Nebuchanezzar dispatches Transjordanian bands during his initial campaigns against Judah (2 Kings 24:2).

mous Gilead also has sons with a woman deemed more legitimate. These sons banish Jephthah from the household and deny him inheritance rights with the accusation, "You are the son of an Other woman" (Judg 11:2). As an opening scene, the banishment of Jephthah provides a justifying trauma for the man who will sacrifice his daughter and further sets the stage for a civil war. Jephthah flees to the Gileadite region of Tov where he gathers a "low-life posse" around him and earns a reputation as a warrior (11:3). When the Ammonites attack Gilead, Jephthah's brothers come to realize the value of a half-breed intermediary. They send a delegation asking Jephthah to return as their commander. Jephthah wonders how the very people who scorned and dispossessed him could suddenly turn to him for help, but is answered by the goodwill of the distressed suddenly prepared to have him rule over all of Gilead. His brothers assure him that they harbor none of their previous resentments, but the sense of Jephthah as a border figure whose status depends on the Ammonite disposition toward Israel persists.[40]

The status of Gileadite territory also remains in dispute. In his first act as ruler, Jephthah inquires as to why the king of Ammon attacked. The Ammonite king's answer—a retelling of the story of the conquest—upsets the legitimacy of Israelite claims:

> (I attacked) because when Israel came from Egypt, it took my land from the Arnon to the Jabbok and all the way to the Jordan. Now then restore it peaceably. (Judg 11:13)[41]

Bluntly pressing his claims, the king of Ammon says outright what the Amorite tales have worked hard to obscure. The king evokes precedence as determinative of the right to territory: if the land *at first* belonged to Ammon, then it properly belongs to Ammon thereafter.

In response, Jephthah rehearses all the implicit arguments of the Amorite myth. He begins with a straightforward denial, "Israel did not conquer lands from Moab or Ammon" (11:15), but then details in great length how Israel avoided Edom, Moab, and Ammon, ultimately fighting the Amorites alone. Jephthah next counters that God dispossessed the Amorites for Israel

---

40. Steven Weitzman observes that "ambivalent inhabitants" with dual affiliations ultimately contribute to "border formation." Weitzman, "The Samson Story as Border Fiction," *Biblical Interpretation* 10 (2002): 174.

41. According to Miller, this text as well as Deuteronomy 2 is invested in setting the border of Moab at the Arnon and establishing Israel's claim to the north of this river. Maxwell Miller, "The Israelite Journey Through (Around) Moab and Moabite Toponymy," 582.

and that therefore divine will is a determinative stronger than precedence. Jephthah openly admits that Israel was not the first in Transjordan and prefers to speak instead of national destiny. Based on the premise that the god Chemosh has granted Ammon its patrimony, the king should respect what Yahweh has assigned to Israel. That, correctly speaking, Chemosh is the god of Moab and not Ammon leads Jobling to observe "that while addressing Ammon, [Jephthah] cannot get *Moab* off his mind."[42]

In his final point, Jephthah asks why the Ammonites have done nothing to recover Gilead during the "three hundred years that Israel dwelled in Heshbon and its satellite cities, Aroer and its satellites and in all of the cities alongside the Arnon" (11:26). Here the claim to land is based on continued presence rather than on mythic battles or divine sanction. One must admit that three hundred years of residence constitutes a sound rationale for a territorial claim. By speaking in terms of the Ammonite *recovery* of Gilead, Jephthah tacitly admits that Ammon (and/or Moab) was the prior tenant. It may be too much to say that Jephthah gives up the Amorite myth, but never before and never again will an Israelite speaker verify the Ammonite claim.

Their contending narratives lead to war. Jephthah calls upon God to judge between Ammon and Israel and receives "the spirit of God" that animates war heroes. A tendency toward excess leads Jephthah to pledge a sacrifice to God "of whatever comes out of the doors of my home to greet me when I return in peace from war with Ammon" (Judg 11:31). He wins the war and subdues the Ammonites only to come home and slaughter his only child, the inevitable first and only one to exit from the doors of his house. As Mieke Bal has shown, Jephthah transfers the contested front to the home by punishing his daughter's breach of the household's confines. Jephthah, in Bal's estimation, "is a verbal killer, a killer 'by the mouth' instead of 'by the mouth of the sword.'"[43]

The border with Ammon and the border of the household having been enforced by word and sword alike, the second dimension of the Jordan border now requires Jephthah's attention. The men of Ephraim, the west bank counterparts of Gilead, demand to know why Jephthah failed to call on their allegiance. Like the king of Ammon, the Ephraimites put forth a succinct claim: "Why did you go to war with Ammon without calling us to accompany you? We will burn you down in your house!" (Judg 12:1). Jephthah then goes to great lengths to explain himself and express outrage at the accusation.

---

42. This reads to Jobling as "evidence of a repressed fear that the land Israel took from Sihon belonged properly to *Moab*." Jobling, "The Jordan a Boundary," 129. On the Moab theme in Jephthah's speech, see also Judg 11:25.

43. Bal, *Death and Dissymmetry*, 65.

"I was a man at war—me, together with my people—with Ammon. I called upon you and yet you did not deliver me from them. When I saw that you were no deliverers, I took my life into my own hands and went to war with Ammon. God granted me victory, so why have you come to me today in order to fight against me?" (Judg 12:3)

For Jephthah every battle is personal; he sees each pretext as an affront and perhaps this is why he cannot but bring the battle home. Unable to defuse the situation with language, Jephthah gathers his Gileadite army in order to fight Ephraim. Throughout their interaction, allusions to other Jordan stories highlight the breakdown of every tenuous alliance staged at the river

In the report of Ephraim's defeat at the hands of Gilead, the narrator includes the hurtful taunts fired at the east bankers: "you people of Gilead are just fugitives from Ephraim'" (12:4). Like the king of Ammon, the Ephraimites accuse the people of Gilead of being both illegitimate and out of place. This echoes the initial expulsion of Jephthah from his father's house as well as the underlying suspicions about the two and a half tribes. Jephthah's subsequent actions are then best explained as aimed at proving the distinct and sovereign nature of Gilead. The soldiers of Gilead fortify the fords of the Jordan and stop the literal "fugitives of Ephraim" at the border (12:5). In what Derrida calls "this barred right of passage at the border of the Jordan," the future violence of passports, border guards, and watchtowers is presaged.[44] In the present of the text, Transjordanians for the first time assert rather than accept the difference ascribed to them while parodying the artifice of Jordan distinctions.

Fugitives of Ephraim: Let me cross.
Gileadites: Are you from Ephraim?
Ephraimite: No.
Gileadites: Say "shibboleth."
Ephraimite: Sibboleth.[45]

This test and its failure, we are told, are repeated forty-two thousand times (Judg 12:6). The Jordan brings into relief a minute but irreducible variant of pronunciation—a consonant that determines who can cross and live. The absurdity of difference hanging on a consonant reflects back the emptiness of

---

44. A "shibboleth" is, in Derrida's estimation, "any insignificant, arbitrary mark... once it becomes discriminative and decisive, that is, divisive." Jacques Derrida, "Shibboleth," in *Midrash and Literature*, ed. Geoffrey H. Hartman and Sanford Budick (New Haven: Yale University Press, 1986), 322.

45. I have translated the relevant verses (Judges 12:5–6) into a dramatic dialogue.

territory as a signifier. The border has created the arbitrary differences that it claims to protect. As the difference becomes drawn in blood, the reader is left with the sense that Gilead and Ephraim are not really different at all. The pronunciation of "shibboleth" is a "mark of an alliance inverted and turned against oneself"; the tribal structure has abetted civil war.[46] Marking an Other side makes things clear, but it sows an inevitable opposition, which comes to feel natural. The shibboleth incident shows how lines in the sand—or in the water—deepen as bodies pile up around them. By narrating the violence at the borders established by Joshua, the book of Judges complicates the very form of the nation.

Always the Other Side
There are yet more Jordan crossings to consider. Because the next chapter jumps forward in time to include the New Testament, I present an overview of how the east bank remained the other side under a line of empires as well as during the period of the autonomous state of Judea. Throughout the rise and fall of the Israelite tribal federation and the defeats and restorations of the Judean state, Transjordan continued to be a distinct territory that supported Jewish communities. At certain junctures the Jordan constituted a definitive boundary and at others it was mostly a vital water source claimed within a larger tract of land.[47] As we have seen, Israel's "actual" and represented borders always existed in a state of flux determined by migration and settlement patterns, the waxing and waning of imperial powers, and shifts in the balance of power among Israel and its immediate neighbors. That said, the Jordan persisted as a remarkably resilient border.

As the biblical texts, in their manner, suggest, the ninth and eighth centuries B.C.E. saw the development of Israel, Aram, Edom, Moab, and Ammon with the contests among them constantly impacting the character of Transjordan. The shifts in regional power made Transjordan particularly vulnerable to the forces of the Assyrian Empire. In 722 B.C.E. the Northern Kingdom of Israel fell to Assyria, which subsumed both of its banks. During the Assyrian reign, the Jordan delineated the district of Megiddo from those of Damascus, Karnaim, and Gilead to the east as well as the district

---

46. Derrida, "Shibboleth," 346.

47. "The actual political boundaries for the area east of the Jordan shifted considerably during the various monarchic and colonial governments, but generally the Jordan River as a natural feature in the landscape caused it to function as a rather constant border line on the one side." Douglas A. Knight, "Joshua 22 and the Ideology of Space," in *'Imagining' Biblical Worlds: Studies in Spatial, Social and Historical Constructs in Honor of James W. Flanagan*, ed. David M. Gunn and Paula M. McNutt (Sheffield: Sheffield Academic Press, 2002), 61.

of Samaria from the southern portion of Gilead. Judah, which maintained a precarious independence in this period, was separated from Moab by the Salt Sea.[48]

Although it weathered the Assyrian storm, the Kingdom of Judah was conquered by the Babylonian Empire in 586 B.C.E. and subsequently defined as a province with restricted boundaries, including the Jordan to the east.[49] Inheriting the lands of the Babylonian Empire, the Persians called the stretch west of the Euphrates "Abar-Nahara"—the province "on the other side" of the river. Despite the fact that the structuring river was the Euphrates and not the Jordan, the region was subject to internal subdivisions outlined by the Jordan. For example, the Jordan separated the district of Galilee from those of Karnaim, Hauran, and Gilead to the east, Samaria from south Gilead and the Northern region of the Tobiads and, finally, the restored district of Yehud from the Tobiads just north of the Salt Sea in the environs of Jericho. "The status of Ammon was peculiar: although largely inhabited by Ammonites, it had a Jewish governor from the Tobiad family (Neh. 2:19, 4:1, 6:1)."[50]

Biblical sources malign the Transjordanian Tobiads in familiar ways. Nehemiah stresses the "uncertain genealogy" of the ostensibly Israelite Tobiad families that dwell east of the Jordan by referring to Tobiah as "the Ammonite servant" (Neh 2:10, 19)[51] and rejecting his gifts intended for the restored Temple (Neh 13:4–9).[52] The Tobiad position on the east bank involves contact and even the supervision of Ammonities. Such contact, in turn, leads to a charge of impure background.[53] The irony of Nehemiah's charge is that the Tobiad

---

48. See Anson F. Rainey, *The Sacred Bridge: Carta's Atlas of the Biblical World*, ed. Anson F. Rainey and R. Steven Notley (Jerusalem: Carta, 2006), 236. For Babylonian conceptions of the region and Israelite/Judean responses, see chapter one.

49. The province as defined by Nebuchadnezzar "extended from Bethel in the north to Beth-Zur in the south and from the Jordan in the east to Emmaus in the west." M. Avi-Yonah, "Historical Geography of Palestine," in *The Jewish People in the First Century: Historical Geography, Political History, Social, Cultural, and Religious Life and Institutions*, ed. S. Safrai and M. Stern (Philadelphia: Fortress Press, 1974), 79.

50. Avi-Yonah, "Historical Geography of Palestine," 80.

51. Ziona Grosmark understands the epithet not as an insult, but as an indicator of Tobiah's official administrative capacity with the Persian imperial system. See Grosmark, "Transjordan in the Hellenistic, Roman, and Byzantine Periods," in *Jordan and Its Sites: The Early Geography and History of Jordan*, ed. Gavriel Barkai and Eli Shiller (Jerusalem: Ariel Publishers, 1995), 93 (Hebrew).

52. See Tamara Eskenazi, "Tobiads," *ABD*, 584–85.

53. The episode recalls Jephtah, who, when dispossessed for having a foreign mother, flees to "the land of Tob," possibly the origin of the Tobiad family name. It also brings to mind the near civil war when the two and a half tribes build an altar. Nehemiah's vow to Tobiah that "you will have no share, claim, or record in Jerusalem" ולכם אין-חלק וצדקה וזכרון בירושלם (Neh 2:20) bespeaks the fears of the two

family seems to have remained in place throughout the exiles and imperial reconfigurations of the region, while Nehemiah is a recently returned Babylonian exile granted political influence through the Persians.[54] This attempt to exclude east bankers from a sanctified Israelite genealogy corroborates James Kugel's suggestion that the notion of Israel as a discrete and holy "seed" arose during fifth-century Persian rule, when the identity of Israel could no longer be supported by a pure, bordered land.[55] "If the People of Israel were *not* necessarily coterminous with the people living in Israel's territory, then the issue of borders needed to be handled in some nongeographic fashion, and this is precisely what Ezra-Nehemiah, *Jubilees*, and other early texts seek to do."[56] In the centuries just before the Common Era, a restricted genealogical line assumed force similar to that of geographic borders.[57] At the same time, earlier biblical texts bespeak a larger trend of classifying groups at the peripheries as socially marginal.

Josephus embellishes the themes of Tobiad wealth and rhetorical cunning in his Tobiad Romance (*Antiquities* 12.154–222), in which the Tobiad hero Hyrcanus settles in Transjordan, much like Jephthah, due to a dispute with his half-brothers. However, biblical evidence along with the Zenon papyri and a passage in the book of 2 Maccabees (3.10–11) militate against Josephus's proposal by suggesting Tobiad longevity east of the Jordan.[58] Perhaps Josephus chronicles in narrative form the political shift in which the Ptolemies created a series of regional city-states (Gaulanitis, Auranities, Trachonitis, and Moabitis) and detached Ammon from Tobiad rule in order to establish

---

and a half tribes that their west bank counterparts would deny their children religious inclusion with the claim that "you have no share in the Lord" אין־לכם חלק בה' (Josh 22:27).

54. Grosmark notes the Tobiad presence before the period of Nehemiah and cites the possibility that the Tobiads are descendants of the "Jewish" inhabitants of Gilead. Grosmark, "Transjordan," 93–94. The Zenun papyri, in Grosmark's opinion, provide sound evidence of Tobiad longevity.

55. Kugel, "The Holiness of Israel and the Land in Second Temple Times," 28. Kugel also considers the possibility that "the very notion of the land's holiness" may have resulted from the construal of Israel as a holy people. This would mark a shift from the Priestly view that the land is holy because it belongs to God to a view held by Ezra-Nehemiah and the Jubilees writers that the land's holiness reflects that of the People.

56. Kugel, "The Holiness of Israel and the Land in Second Temple Times," 28.

57. Transjordan generates high anxiety because it never hosted a single dominant ethnicity, but rather was a place of contact and mixing for various groups. See Uriel Rappoport, "The Jewish Community in Transjordan in the Second Temple Period," *Jordan and ts Sites: The Early Geography and History of Jordan*, ed. Gavriel Barkai and Eli Shiller (Jerusalem: Ariel Publishers, 1995), 97 (Hebrew).

58. See Erich S. Gruen, *Heritage and Hellenism: The Reinvention of Jewish Tradition* (Berkeley: University of California Press, 1998), 103–4.

it as a Greek city.⁵⁹ During this shift, the Tobiads retained autonomy just east of the Jordan and served as a geographic buffer between Judea and Ammon. However, in the ensuing era of Seleucid rule, "the Tobiad domain, now called Peraea, 'the land beyond' [the Jordan] was attached to Samaria," and the Tobiad dynasty came to an end."⁶⁰ Josephus codes this event as Hyrcanus's suicide upon learning of the new Seleucid dominance.

During the Hasmonean rebellion against Seleucid rule, Judah Maccabee led campaigns in Transjordan in order to liberate Jewish communities there (1 Macc 5:3–13, 24–55). After a successful campaign, Judah crossed the Jordan from east to west and marked his victorious return with sacrifices offered at the rededicated Temple (1 Macc 5:54). The Judean state established by his Hasmonean heirs resembles, in certain ways, the nation described in the book of Joshua (figure 5).⁶¹ The Jordan again functioned as something of a border, yet a sizable region beyond the river fell under Hasmonean jurisdiction and wars with Transjordanian leaders were common.⁶² These territorial gains show that the restoration of Israel's ancient boundaries was a key aspect of national revival (or, as some would argue, that the ancient boundaries are created in the name of justifying the Hasmonean state). However, the restored borders did not stand for long. Pompey reduced Hasmonean hegemony and "separated the areas inhabited by the Jews into two, cut Judea off from access to the sea and encircled it with a belt of Greek cities."⁶³ The Jewish regions of Judea and Galilee became noncontiguous, broken by Samaria and Greek cities. All these regions, along with Transjordan, eventually fell under the control of Herod the Great. When Herod's kingdom was divided among his

---

59. Avi-Yonah, "Historical Geography," 81; see also 84. The figure of Hyrcanus in Josephus's story would then represent the new Greek imprint on Jewish Transjordan.

60. Avi-Yonah, "Historical Geography," 84.

61. "At the time of his death Alexander Jannaeus ruled the whole of the coastal plain from the Kishon to the Brook of Egypt, the whole of the mountains west of the Jordan from Dan to Beer-sheba, almost the whole area west of the Jordan from Paneas down to Zoar and the river Zared." Rainey, *The Sacred Bridge*, 333.

62. "John [Hyrcanus] first conquered territory across the Jordan River that he apparently failed to retain, probably because the east bank of the Jordan was claimed also by the increasingly powerful Nabatean kingdom . . . Alexander Yannai (reigned 103–76) greatly extended the Hasmonean conquests, concentrating on the Greek cities of coastal Palestine and the mainly Greek cities east of the Jordan River . . . He was not invariably successful in his campaigns, especially in Transjordan, where he came up against the Nabateans." Schwartz, *Imperialism and Jewish Society*, 37–38. The Arnon River functioned to separate Jewish Transjordan from the Nabateans, who lived south of the river.

63. Michael Avi-Yonah, *The Holy Land: A Historical Geography from the Persian to the Arab Conquest (536 B.C. to A.D. 640)* (Jerusalem: Tsafrir, 2002), 79.

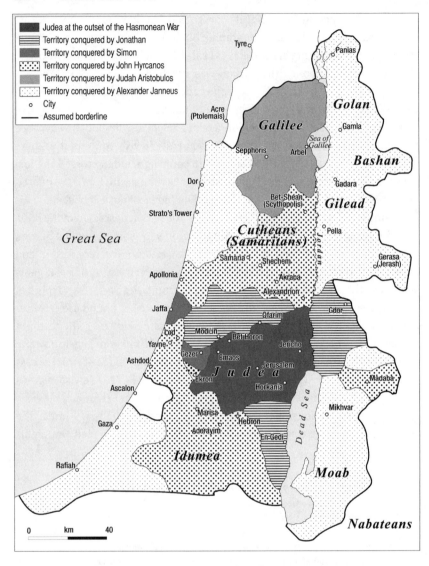

5. Judea under the Hasmonean kings (167–37 B.C.E.). Soffer Mapping.

sons, again the Jordan operated as a district border separating the territory of Archelaus west of the Jordan from that of Herod Antipas to the east (Perea) as well as Herod Antipas's Galilean holdings from Philip's territory northeast of the Sea of Galilee (figure 6).[64]

---

64. "The elder son, Archelaus, was assigned Judaea with the bulk of the Jewish population; but this was counter-balanced by Idumaea and Samaria (including the cities of Caesarea and Sebaste). He was honored by the title of ethnarch, whereas the two other heirs received only the title of tetrarch.

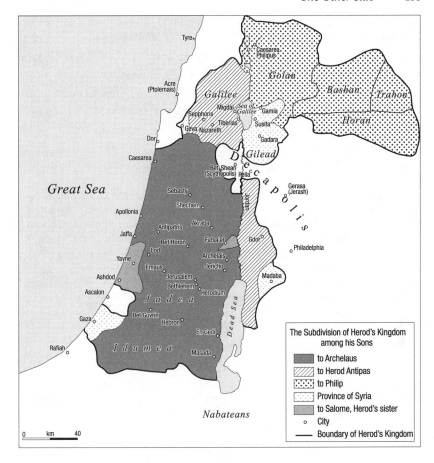

6. The land and its administrative districts in the time of Jesus. Soffer Mapping.

Under the rule of Herod and his sons there was a conscious effort to break up the Jewish character of the land that had so inspired Hasmonean battles and victories. Splitting the territory was a strategy of eviscerating the potency of the land as a symbol. Through administrative "redistricting," the Romans and the heirs of Herod sought to impede the logistical planning and the passions of Jewish nationalism. Based on Josephus' description of Judea stretching from the Jordan in the east to Joppa in the west (*War* 3:51), it seems that the biblical paradigm trumped administrative districting and that Jewish

---

The second son, Herod Antipas, ruled over purely Jewish Galilee and Peraea; but his two areas were widely separated. The third son, Philip, received the lands east of the Jordan: the district round Paneas (a town which he embellished and which became known as Caesarea Philippi), the Batanaea, Trachonitis, Gaulanitis and Auranitis. His territory was compact, and its population evenly balanced between Jews and gentiles." Avi-Yonah, "Historical Geography," 93.

nationalism remained intact.[65] A Priestly Jew like John the Baptist as well as a Galilean Jew like Jesus would have had a sensibility in which the Jordan constituted both a symbolic and a national boundary.[66] At the same time, John the Baptist's activities "beyond the river" points to Perea as a site of Jewish communities as well as an important route from Galilee to Judea on which Jewish pilgrims traveled in order to avoid Samaria.[67] In this capacity, the eastern side of the Jordan again played an important, intermediate role.

---

65. The Jews who dwelled east of the Jordan also remained nationally oriented. See Avi-Yonah, "Historical Geography," 103.

66. "The border of Galilee . . . In its last stretch near the Sea of Galilee the border ran along the Jordan, including Sennabris, but excluding Philoteria-Beth Yerah which then was situated east of the river. The remainder of the eastern border of Galilee followed the shores of the Sea of Galilee to the entry of the Jordan and ran along the Jordan to a point slightly north of Chorazin." Avi-Yonah, "Historical Geography," 95–96. At the same time, if we understand Jesus the Galilean Jew as a Northerner, then it is possible that he comes from a tradition in which geographic boundaries are not so deep or so significant. From this perspective, Jesus's clash with Jerusalem authorities represents just another itieration of a dispute between Northern and Southern positions.

67. Avi-Yonah, "Historical Geography," 96. Jewish communities continued in Transjordan in late antiquity. On the communities and synagogues of Livias and Naaran, see Sivan, *Palestine in Late Antiquity*, 58–69.

Chapter Six

CROSSING OVER
Prophetic Succession at the Jordan

Saintliness leads to the gift of the holy spirit, and the holy spirit leads to the resurrection of the dead. And the resurrection of the dead shall come about through Elijah of blessed memory.
—Mishnah Sotah 9:15

Much of power is placement. After all, a monarch reigns largely because he sits in a palace. Heredity plays an important role, but a king with the right blood and the wrong location holds little sway. In the case of leaders not as strongly associated with fixed place, their site of initiation can be the key to authority. In this chapter I show the eastern shore of the Jordan to be a site of initiation, particularly for prophets. While some kings are initiated at the Jordan, the prophetic cases warrant more attention since the Jordan is the singular site of prophetic succession. Because only two prophets appoint successors, the episodes in which Joshua succeeds Moses and Elisha succeeds Elijah operate paradigmatically as type-scenes with recognizable parallels. Using the type-scene as a means of intertestamental comparison, I argue that the Gospel writers modeled the narrative of Jesus's baptism on the scenes of prophetic succession in order to figure Jesus as similar to Joshua and Elisha as well as to Moses and Elijah.

In these cases, the Jordan functions less as a border of homeland and more as a ritual boundary. The border again appears as a productive site because the prophets experience the "wrong" side prior to their elevation and are transformed by God in the midst of its crossing. The three-part ritual structure advanced by Arnold Van Gennep and Victor Turner illuminates the cases: the initiates separate from society, undergo a liminal period during which they are between categories, and then are reincorporated with a new status. In the biblical succession scenes, the structure literally maps onto the Jordan and its two banks. The east bank falls outside the societal structure, the liminal stage transpires at a border, and return involves a westward crossing.[1]

Notions of the rupture and potential reconciliation of distinct periods of time are expressed spatially through the setting of succession scenes at a border. Moses and Elijah both appoint their successors on the east side of the Jordan River, and the successors carry the spirit of their teachers into the new era by crossing the Jordan from east to west. The idea of a predecessor's spirit carried into a new age and there redeemed influences the accounts of Jesus's baptism in the Jordan River. Where Joshua and Elisha are initiated by crossing the Jordan from east to west, Jesus is initiated by immersing in the Jordan. In some unexpected ways, these movements indicate hierarchical order: Joshua and Elisha have a lower status than their predecessors, while Jesus enjoys a status above that of John the Baptist. Jesus's baptism also operates under quite different assumptions about the relative values of the future and the past. As I will show, all three initiations at the Jordan stage movements around this border as a means of outlining the limits of the past as well as how much of the past can be accommodated by the future.

Other than the obvious chronology in which the Gospels follow the Hebrew Bible, I am not concerned with issues of dependence. Most prior studies of the initiations of Joshua and Elisha concern themselves with which of the two scenes derives from the other. I remain unconvinced that derivation can be clearly illustrated in one direction or the other because reaching a conclusive timeline requires much speculation. Furthermore, such arguments do not necessarily bring us any closer to understanding the meaning or relevance of the texts. A more provocative line of inquiry concerns the connection between prophetic and monarchical initiations. The writers of the Deuteronomistic History (in which the successions of Joshua and Elisha appear) are invested in succession as an example of how covenant is transmitted from one generation to the next. Since kings always appoint successors and prophets rarely do, it is likely that the prophetic scenes underwrite the royal ones. Why talk about kings through prophets? By equating the two, divine approval transfers from prophets to the throne and the law delivered by prophets becomes the domain of kings.

---

1. On Transjordan as a site of exile in the Deuteronomistic Court History, see Jeremy Hutton, *The Transjordanian Palimpsest: The Overwritten Texts of Personal Exile and Transformation in the Deuteronomistic History* (Berlin: Walter de Gruyter, 2009), 372–73.

A similar conception of authority as an essence transmitted from a figure of one generation to another characterizes the medieval conception of kingship. In his study of the king's two bodies, the physical body and the body politic, Ernst Kantorowicz identifies the "holy spirit," passed from one medieval monarch to the next, with the body politic. This spirit, according to Kantorowicz, represents "the Immutable within Time" and indicates a belief in the immortality of the collective manifested in continuous office.[2] The spirit, descending on the regent from above, elevates the king to an intermediary station between heaven and earth and perfects the king's body by eliminating any defects. At the very moment that it attains such splendor, the royal body becomes a hybrid of the human and the divine. The dichotomous elements blend in the king's person only to be separated during the expiration of the king's physical body. Not until death is the difference between the ephemeral body and the immortal spirit brought into such relief. The immortal spirit has two options: To vacate earth and return to heaven or to abandon a withered body and enter a vital one. Political institutions like the monarchy are prone to be invested in the theory that upon his death the king's authority is transferred to the next in line while remaining ever-present in the throne itself. Although most biblical prophets don't rise to their station in this manner, it is from the two instances of prophetic succession in the Hebrew Bible that these theories of transferable, immortal authority derive.[3] As I will show through the examples of Moses and Elijah, the individual bearer of spirit stands vicariously as a kind of prototype of the immortal People Israel. The immortality conferred by Jesus extends to a self-selecting group redefined by a collective fate in the afterlife.

Succession

Continuous hereditary succession characterizes the offices of king and priest in the Hebrew Bible (Deut 18:5). While biblical narrative attests to the rupture of succession through tales of usurpation of the throne and of the revolving door of families who occupy the priesthood, these offices are theoretically understood as eternal and present in the eon that follows the Day of the Lord. Even the apocalyptic imagination conceives of authority located in purified but familiar forms of kingship and priesthood. The prophets, however, the

---

2. Ernst Kantorowicz, *The King's Two Bodies: A Study in Mediaeval Political Theology* (Princeton: Princeton University Press, 1957), 8.

3. Recognizing a royal motif in prophetic narrative is not particularly anomalous since "it is fair to say that the institution of prophecy appeared simultaneously with kingship in Israel and fell with kingship." Frank Moore Cross, *Canaanite Myth and Hebrew Epic: Essays in the History of the Religion of Israel* (Cambridge: Harvard University Press, 1997), 223.

very conveyers of visions of renewed forms of authority, do not imagine their station as enduring. In contrast to the kingship and the priesthood, the office of prophecy is not portrayed as continuous or successive. No prophet ever passes the mantle to his son; the prophets seem to emerge from nowhere called at random with no consistent set of characteristics, and each prophet structures a unique mode of relating to the authority of kings and priests.

The concept of prophetic succession is treated only in Deuteronomy 18, where prophets are contrasted with other sorts of diviners as a legitimate form of the same. The prophet, spoken of theoretically without a definite article, is established by God, resembles Moses, and should be listened to at all costs (18:15). The origin of prophecy is traced to Sinai, where the People requested an intermediary to buffer the fiery voice of God with its annihilating potential. As Robert Wilson points out, prophecy is here "redefined as a way of continuing the work of Moses throughout history," and Moses is above all a figure of mediation who binds the People to God through Law and perpetually binds God to the People through Exodus memory.[4] A conduit of divine speech, the prophet transmits inspired language to an audience stipulated by God (Deut 18:18). The criteria of language help the audience to determine whether the prophet is true or false: If what he says come to pass, then his speech originates with God; if what he says fails to occur, then he has no divine authority and can be disregarded (Deut 18:22). The office of prophecy is itself a memorial to Moses, and recalling Moses becomes the species of memory that makes a place for prophetic authority in the Israelite context.

The legacy of Moses is so important that he is only one of two prophets in the entire Hebrew Bible who appoints a successor.[5] This account as well as the other, that of Elijah designating Elisha as his successor, is ultimately structured by the Deuteronomistic writers as a treatise on continuous authority. These narratives of prophetic succession domesticate the future by characterizing authority as a bridge between distinct periods of time made possible by an immortal, transferable link to the Divine. With Jesus's baptism, the movement at the Jordan that enacts transformation changes from crossing the river to immersing in the river. As the action shifts, so does the nature of the predecessor-successor relationship. While the careers of Joshua

---

4. R. R. Wilson, *Prophecy and Society in Ancient Israel* (Philadelphia: Fortress Press, 1980), 12.

5. Because Joshua is not named as a prophet, Robert Carroll speaks of Elisha's initiation as the only biblical occasion of prophetic succession. He questions "whether this specimen of succession is to be understood as a definitive act, characteristic of prophetism in general, or simply an isolated case with no more significance to it." R. P. Carroll, "The Elijah-Elisha Sagas: Some Remarks on Prophetic Succession in Ancient Israel," *Vetus Testamentum* 19 (1969): 403. I present Joshua's initiation as an instance of prophetic succession not because Joshua becomes a prophet, but because he is appointed as successor to a prophet.

and Elisha attest to the fact that the prophetic disciple can neither equal nor surpass the teacher, Jesus's career definitively transcends John's. Furthermore, while Joshua and Elisha gain their authority by being initiated at the Jordan, Jesus's baptism lends authority to John the Baptist. I read these differences as indicative of the synoptic interpretive practice of employing motifs from the Hebrew Bible while contextualizing them in a manner that promotes novel concepts of community, time, and salvation.

The three successors, Joshua יהושע, Elisha אלישע, and Jesus ישוע, all have names based on the Hebrew root ישע, with the meaning "he redeems."[6] Beyond the charge of performing acts of redemption for their followers, the three share the burden of redeeming the memory of their forerunners through the actualization of their aspirations. Joshua, Moses's successor, never receives the title of prophet. As commentators point out, perhaps as head strategist and acting general of the Israelite conquest he has too much blood on his hands for such an ethereal title, or perhaps Moses's main spin doctors among the Deuteronomists deny Joshua the title in order to maintain a clear hierarchy.[7] Since Joshua is so close to Moses, receives his spirit and ultimately achieves his dream of leading the People into the land, the writers prevent dangerous suppositions by marking Joshua from his first appearance as Moses's "apprentice" (Exod 24:13, 33:11; Num 11:28; Josh 1:1). Moses's authority may live on in Joshua, but Moses's memory reigns supreme.

Elisha, Elijah's successor, does receive the title of prophet as well as Elijah's epithet "Israel's horseman and chariot" (2 Kings 2:12, 13:14). More debate surrounds the status of Elisha vis-à-vis Elijah; Elisha requests a double portion of Elijah's spirit, performs twice the miracles, and fulfills all the political dictates outlined by God to Elijah during the prophet's crisis of faith atop Sinai. Although the apocryphal writer of Ben Sira lauds Elisha as twice the prophet of

---

6. Bob Becking surveys alternate ways of understanding Elisha's name. Becking, "Elisha: 'Sha' Is My God'"? *Zeitschrift für die Alttestamentliche Wissenschaft* 106 (1994): 113–16. "The parallel between Elisha and Joshua is not limited to the plot, but extends to their names, which combine a theophoric element with the language of redemption or delivery." Yael Shemesh, "'I Am Sure He Is a Holy Man of God' (2 Kings 4:9): The Unique Figure of Elisha," in *And God Said: 'You Are Fired': The Narrative of Elijah and Elisha*, ed. Mishael Caspi and John T. Greene (North Richland Hills, TX: Bibal Press, 2007), 31.

7. Postbiblical commentators cast a more favorable light on Joshua: "Ben Sira (46:1) refers to him as 'a minister of Moses in the prophetical office'. According to Josephus, Moses 'appointed Joshua to succeed him both in his prophetical functions and as a commander-in-chief, whensoever the need should arise'" (*Antiquities of the Jews*, 4, 7, 2). Shemesh, "I Am Sure He Is a Holy Man of God," 46. Rabbinic commentators reproduce the Deuteronomistic impulse by saying little about Joshua apart from the fact that he wrote the last lines of the book of Deuteronomy, thereby chronicling Moses's death (BT Men. 30a) was like the moon reflecting the bright light of Moses, the sun (Sifre 23); and married Rahab, sole survivor of the battle of Jericho (BT Meg. 14b).

Elijah, Elijah remains forever relevant through his ascent to immortality and status as forerunner to the Day of Yahweh (Mal 3:23). In the end, Moses and Elijah tower as the famed immortal prophets unsurpassed in spiritual ranks, while Joshua and Elisha, who actually fulfill the mandates of their predecessors, are familiar only to avid Bible readers. This distinction between teachers and disciples falls in line with a larger program in the Hebrew Bible of privileging the past as the site of truth. It is precisely this program that is at issue when Jesus seeks out John the Baptist on the shores of the Jordan. The Baptist, who as the last of the prophets is the very manifestation of the Israelite past, is explicitly marked as inferior through several devices. In fact, the Gospel writers so prefer the new to the old that the predecessor is beheaded just as the successor is getting warmed up.

The fate of John the Baptist reveals a revisionist orientation to time on the part of the Gospel writers, whose narrative goal is the reversal of the traditional reverence for the past in favor of a future-oriented plot. John the Baptist then, the last of the prophets, crops up at the opening of Jesus's story only to attest to his own diminution, perform a vestigial prophetic-Priestly ritual, and step aside for the miracles of Jesus Christ. In addition to the inverted status of master and disciple, Jesus does not cross the Jordan River into a new era but instead goes down into the water thereby tilting the temporal axis of the biblical plot from horizontal to vertical. No longer is the horizontal plane of space the site of redemption; instead the Kingdom of Heaven, at the zenith of the vertical axis of earth and heaven, functions as such a location. As the Baptist fades from the narrative, the relevance of the past dissipates and the present into which Jesus enters following his baptism becomes at worst an illusion and at best a rehearsal for a heavenly future. During this shift the Jordan remains a border, but one that separates heaven from earth, the future from the present, rather than one that delimits the territory of Israel.

Moses and Elijah appoint their successors on the east bank of the Jordan River prior to their disappearance from the earth. The river in both cases signifies a limit. For Moses, it signals the end of the wilderness wandering, the edge of the land and the term limit of his prophetic career. For Elijah, crossing to the east bank is a return of sorts to a location of his early prophetic wanderings as well as to Gilead, his place of origin; it is likewise the indicator of his limited days on earth. The geographic boundary conveys that prophecy, like all good things, must come to an end. The death of the prophet, synonymous in these cases with the end of his prophecy, threatens the notion of the eternal covenant by terminating the mediator of the covenant. In other words, the prophet's absence raises questions of how Israel will be able to hear the divine voice as well as who will buffer Israel from divine punishment. The end of the prophet's life causes anxiety concerning the perpetuation of covenant, the continuity of Israel, and its relationship to God. Such anxiety breeds theories of prophetic immortality.

Biblical texts claim that the disappearance of Moses and Elijah does not necessary mean that they died. The claim is made explicitly in the case of Elijah, who does not perish, but rather ascends to heaven in a chariot of fire. Although death visits Moses, his soundness of body and mind in his final hour, his lack of a grave, and his end on the peaks east of the Jordan (where Elijah ascended) undermine a definitive sense of the prophet's demise. Elijah sails into eternity on a whirlwind and Moses vanishes while his eyes are still moist. The immortality of these prophets suggests not only their larger-than-life stature, but also that in another time they may return from beyond the Jordan. The possibility of such a return leads to acts of commemoration intended not only to pay homage to the past, but also to ensure that the import of prophetic reappearance be understood in the future. Memory thus becomes the second mode through which the lives of these prophets are sustained. The idea of immortality conferred by memory is developed in the book of Malachi, where the return of Moses and Elijah is the sign of the future's threshold. Indeed, the successors serve the prophets as apprentices and resemble them as disciples, yet gain authority only upon the transfer of spirit from master to student.

In the Jewish biblical canon (Tanakh), the Torah (the Five Books of Moses) ends with the death of Moses and Nevi'im (Prophets) ends with the command to remember Moses and anticipate the return of Elijah:

Remember the Torah of Moses my servant in which I instructed him on Horeb—laws and rules for all Israel. Behold I will send you Elijah the Prophet before the arrival of the great and awesome day of God. He will turn the hearts of the fathers back to the sons and the hearts of the sons back to the fathers, so that, when I come, I do not strike the land with destructive holiness. (Mal 3:22–23)

As Prophets, the second section of the Hebrew Bible, comes to a close with the book of Malachi, Moses and Elijah gain a preeminence that transcends time. The Torah is "the Torah of Moses," a title that attests to the fact that Moses's legacy is imbedded in the book. Remembering the Torah of Moses is simultaneously an act of recalling Moses. Living according to the "laws and rules" of the Torah is a means of preserving Moses's legacy in present actions. This present is delimited by Elijah's return. Upon his return, time will undergo a great upheaval during which the present will fold into the future. As Elijah departed the world at a geographic border, so he will return to mediate between generations at the border between present and future. It becomes clear in the book of Malachi that mediation between human and Divine is likewise necessary in the future. The destructive side of God, remembered and experienced, is the primary visitor to the People in the era of the apocalypse. Then more than ever, the prophets will shield the People from the consuming presence of the Holy. The assurance of prophetic return is salvific in the sense that Elijah will protect the People during the trials of the end while allowing them to undergo experiences that lead to a redeemed survival. Like Moses, Elijah shuttles the People to the space of the future. Moses and Elijah are kept alive by memory, literary representation, and the promise of return, at the same time that the political agenda of prophecy is sustained through succession.

Type-Scene

Along with the setting east of the Jordan, the scenes of prophetic succession share five thematic elements. These recurring elements combine to form a type-scene attested in three narrative forms. The type-scene has been popularized as a mode of reading across biblical texts in order to uncover similarities and references shared among narratives as well as to highlight nuanced thematic difference and consider the conventions of biblical literature. The application of a type-scene tends to be restricted to a discrete corpus of literature so it is applied to either the Hebrew Bible or the New Testament. Here I expand the range of the type-scene by reading across testaments and showing how certain plots were not only transmitted among writers of the Hebrew Bible, but also consciously employed by the writers of the Gospels. One implication of this is that by using type-scenes and other thematic forms, Gospel writers presented their work as a continuation of the Hebrew Scriptures and ensured that their work would be familiar to its first audiences. Furthermore, familiar forms enabled a sense of "at-homeness," which in turn allowed for the introduction of novel ideas about time, redemption, and the nature of spirit. That is, recognizable plots conveyed new takes without too much disjuncture.

As Robert Alter has observed, "the biblical type-scene occurs not in the rituals of daily existence but at the critical junctures in the lives of the heroes, from conception and birth to betrothal to deathbed."[8] The successions are deathbed scenes when the symbolic sons of the prophets stand before them to receive their legacy. That the departing prophets do not die in the strict sense of the term is the twist in the plot that asserts the miraculous dimension of the whole tale. The juncture is critical not only for the departing prophet, but also for the People of Israel as the role of divine intermediary threatens to be vacant. The danger of such a scenario is great, for Israel is warned that without reminders of its covenant, its very existence is threatened. The ten-

---

8. Robert Alter, *The Art of Biblical Narrative* (New York: Basic Books, 1983), 51.

sion thus mounts when the prophet approaches the limit indicated by the Jordan River.

The death of Moses stands for the end of a generation; the Exodus will give way to conquest and settlement once he departs and the People cross the Jordan. At the same time, his death raises the question of how Israel can survive the encounter with other peoples and the outbreak of divine rage without Moses. As Elijah nears the Jordan, God's injunction to appoint kings of Israel and Aram (1 Kings 19:15–16) remains unfulfilled. His impending departure raises the concern of who besides Elijah can appoint these kings and continue the war against institutionalized idolatry in Israel. As the final prophet, John the Baptist is the manifestation of the end of an era and his acts of baptizing are preparations for a new temporality. As a liminal figure in a precarious state, John the Baptist points to the instability of his shifting times. In each case, the fate of the nation hangs in the balance of prophetic succession.

The succession scene concerns the impending "death" of the master and the transfer of spirit and its attendant authority to the disciple. With 2 Kings 2 as the paradigm, I identify five elements in the succession type-scene: (1) the impending death of the master; (2) the transfer of spirit to the disciple; (3) the presence of witnesses; (4) the death of the master; and (5) the disciple's crossing of the Jordan River. When the disciples cross the river, they mediate the spatial rupture and carry the legacy of their predecessor into a new period of time. In this way, prophetic succession involves not only a change in leadership, but also the juncture between an old and a new era. The disciple's river crossing is the spatial corollary to carrying on the mission of a previous prophet.

Moses and Joshua

The book of Deuteronomy presents a Moses haunted by the implications of having reached the eastern shore of the Jordan. In this book-long speech, the river forms a key signifier of the differences between here and there (2:29, 9:1, 11:31, 12:10). The law with its potential blessings and curses becomes operative west of the Jordan, and the foods of God's promise are consumed across the river, where Israelite political offices become manifest and Joshua replaces Moses as leader. Moses's articulation of the dichotomy between east and west, not-land and land, becomes fraught with pathos insofar as the difference hinges on the presence and absence of the prophet himself. As Deuteronomy's Moses recaps the wilderness journey and elaborates on the law, he grapples with his own demise. This aspect lends the book a tragic tenor while enabling the writer(s) to hold forth on issues of mortality, continuity, and political succession. As a point of geographic rupture, the Jordan serves as a proving ground and site of transformation. Disjuncture occurs because certain figures are barred from crossing; continuity results from the renewal of national institutions that existed in inchoate form in the wilderness.

Moses bequeaths law and leadership to Joshua like a father bestowing inheritance on a son.

> Be strong and bold, because you will go with this people into the land that God promised their ancestors that he would give to them, and you will settle it for them. And God Himself will go before you. He will be with you; He will not fail you or abandon you. Do not be afraid and do not be discouraged. (Deut 31:7–8)

The Deuteronomistic exhortations to be strong and bold, unafraid and undiscouraged form part of a personal address with which Moses hopes to confer success on Joshua. In Moses's initiatory vision on Horeb, God promised "to be with" Moses (Ex 3:12) as Moses here promises Joshua. Unlike Moses (Deut 1:37), Joshua will enter the land and enact the ancestral promise (1:38, 31:7). In contrast to the sons of Aaron, who inherit the priesthood and whose sons will inherit the office from them, Moses appoints a nonhereditary successor. The appointment is not surprising since Joshua has long been the apprentice משרת to Moses (Ex 24:13, 33:11; Num 11:28; Josh 1:1)[9] and Moses's sons have not been heard from since their grandfather, Jethro, restored them to their father (Ex 18:2–7). As Moses, the prophet of unsurpassed standing (Deut 34:10), appoints a nonhereditary successor, the principle that prophecy has no lineage is established.

The cases of Joshua and Elisha attest to the fact that the prophetic spirit, in unique cases, can be transmitted from one person to another. The spirit that both Joshua (Deut 34) and Elisha (2 Kings 2:9) receive is the spirit of their masters, not a spirit directly bestowed by the Divine. Succession and the transmission of spirit occur in these two cases in order to ensure the continuity of the specific imperatives of Moses and Elijah, not to establish a prophetic lineage. The spirit allows Joshua and Elisha to realize the goals of Moses and Elijah, but not to replicate their visions. Let us now attend to the steps through which the transfer of spirit occurs.

Impending Death of the Master

The impending death of Moses frames the book of Deuteronomy. As Moses recounts the collective history, he passes through the stages of blame (Deut 1:37–38), denial (Deut 3:23–26), and finally acceptance (Deut 31:1). At

---

9. Elisha is likewise designated as serving as an apprentice משרת to Elijah in 1 Kings 19:21.

the same time, Moses behaves as a responsible leader concerned about the future of his people who both accepts and promotes Joshua as his successor (Deut 31:3, 7–8, 23).[10] The official succession begins when Moses recognizes the Jordan as the border between his life and death.

> Today I am one hundred and twenty years old and I can no longer come and go, besides God has said to me, 'you will not cross this Jordan.' Yahweh your God will be the one to cross ahead of you. He will clear the nations from your path and you will dispossess them. Joshua is the one who will cross ahead of you as Yahweh has spoken. (Deut 31:2–3)

Moses concedes that his ripe age prevents him from leading the battles anticipated in the land while concomitantly contesting the justice of his end.[11] The prophet cites God a final, weary time as if cognizant of the indelibility of divine writ. Repetition of the verb "cross" contrasts Moses's station, "you will not cross this Jordan" לא תעבר, with the immortal God, "the one to cross ahead of you" הוא עבר, and the continuous presence of Joshua, who also "will cross ahead of you" יהושע הוא עבר. In the following paragraph (31:7–8, examined above), Moses transmits the partnership with God to Joshua. With Moses left behind, the two will cross the Jordan together.

The public transfer of authority from Moses to Joshua illustrates the doubly embodied nature of the law: It exists as a scroll written by priests and referred to by the political leader (Deut 17:18),[12] and it lives in the successors to Moses (Deut 18:15). Moses then first writes "this Torah" as an authorizing legacy for the "priests, People of Levi" and "elders of Israel" (31:9) and then publicly installs Joshua. The divine word that Moses spent a career articulating takes on the flesh of written Torah, and the spirit with which he led Israel settles in his apprentice. The chiastic envelope of instating Joshua (31:7//31:23) and conferring the written Torah on the priests (31:9//31:24–25)[13] consecrates the two offices as consecutive with the great leader and lawgiver.

Because Moses announces his approaching death but then continues speaking, God halts him, "the time of your death has arrived," and dispatches Moses to "call Joshua and position him in the tent of meeting so that I may

---

10. Where Joshua is authorized by the spirit of Moses, the priests derive authority by being the custodians of Moses's teaching (Deut 31:9).

11. The phrase "come and go" refers to entering and exiting the battlefield.

12. Deuteronomy 17:18 speaks specifically to the legal check on kings. That is, if the king does not observe the law, then his reign and his son's succession become imperiled (17:20).

13. The Kohanim and elders of Israel of verse 31:9 are absent in 31:25, where only the Levites are featured.

instruct *him*" (Deut 31:14).[14] Here we see the impending death of the master and the transfer of authority with "all Israel" witnessing (31:7). Joshua is initiated through a theophany of cloud at the tent of meeting: "The Lord appeared in the tent, in a pillar of cloud standing at the entrance of the tent" (Deut 31:15). That Joshua is assuming Moses's role becomes clear in his position inside the Tent of Meeting, a space reserved for Moses's audiences with God. He shares the sacred space representative of the wandering with the prophet of the Exodus, absorbs his spirit, and carries it into the land.

The death of Moses, like his prophetic initiation and reception of the law, transpires on a mountain. His brother Aaron similarly joined the community of the dead on a mountaintop (Num 20:28).[15] Moses's final mountain is geographically over-determined: called both Mount Nebo and the Mountain of Abarim עברים (Crossings) (Num 27:12), it is a peak above the plains of Moab and across from Jericho. Positioned on the border between the Moabite setting of Israel's recent past and the Jericho of its near future, Moses beholds the land. As compensation and consolation, the view encompasses the places where Moses's feet will not tread (Deut 32:52, 34:4), the land promised to the ancestors and denied to Moses (Deut 34:4). Moses is shown an anomalous map both more specific and less conventional than others.

> All the land from Gilead to Dan and all Naphtali, Ephraim, and Manasseh, and all the land of Judah until the bordering sea. The Negev and the river plain from the valley of Jericho city of date palms until Zoar. (Deut 34:1b–3)

To begin, Moses's vision does not conform to the mythic structure of most maps of the land, which span from an eastern river (the Jordan or Euphrates) to the sea.[16] His vantage point encompasses both the eastern (Gilead, Manasseh) and the western (Dan, Naphtali, Ephraim, Judah, and the Negev) sides of the Jordan. Rather than mentioning the patrimony of all twelve tribes, this map refers to five tribal regions (Dan, Naphtali, Ephraim, Manasseh, Judah), three geographic ones (Gilead, Negev, Jordan River Plain הככר), and two cities

---

14. Paradoxically, God does not directly address Joshua here. The next record of God's speech is addressed to Moses (Deut 31:16).

15. As Moses dies at the border between Moab and Israel, so Aaron dies at the border of Edom (Num 20:23). Graves are here seen as a manner of boundary markers. For the structural similarities between the death scenes of the brothers, see Elie Assis, "Divine versus Human Leadership: An Examination of Joshua's Succession," in *Saints and Role Models in Judaism and Christianity*, ed. Marcel Poorthuis and Joshua Schwartz (Leiden: Brill, 2004), 27–28.

16. The Samaritan Pentateuch harmonizes this map with others by having Moses's eyes fall "from the River of Egypt until the great river" in place of "from Gilead to Dan."

of the Jordan Valley (Jericho, Zoar). In other words, the map crosses geographic signifiers and omits the inhabiting nations. Of course, we note that the map includes both banks of the Jordan.

With a view of what is to come, Moses dies as he lived "at the command of God" (Deut 34:5) and is buried by God at a site both ignominious and mysterious. The eternally unknown gravesite is somewhere in Moab across from Beit Peor, the site of Israel's apostasy (Deut 31:6; Num 25). On one hand, the site stresses the proximity of past transgression to the threshold of the land, thus offering a sense of continuity; on the other, it seems unjust that Moses should have to face Israel's infidelity in perpetuity.[17] However, the unmarked grave conspires with other details to suggest that Moses in fact transcended mortality.[18] When Moses expires at one hundred and twenty, his eyes have not lost their moisture and his vigor is not depleted (Deut 34:7). Along with the "king's" two bodies (the ephemeral physical and immortal body politic), it seems as if Moses has two spirits, one that passes to Joshua and another that is translated into the immortal realm. This issue stands at the center of the Elijah-Elisha succession where a "double portion" of Elijah's spirit is at stake. As indicated by the prescribed period of mourning (34:8) and the elevation of Joshua (34:9), Moses departs from Israel as he ascends Mount Nebo, but the terminal destination of the unequaled prophet (34:10) remains uncertain. Moses is further seen to have two bodies in the sense that he lives on in both the body politic and in the corpus of Torah. With his death, the corpus of the Five Books of Moses draws to a close while promising to be sustained through recitation, reading, and reference (Deut 31:11–13). Moses's multiple literary inheritors become evident not only by the continuation of the story

---

17. See George Coats,"Legendary Motifs in the Moses Death Reports," *CBQ* 39 (1977): 38–39.

18. In Jewish tradition, Moses's grave is said to remain unknown in order to prevent religious attachment to a place outside the borders of the land of Israel. In Palestinian Muslim tradition, it is believed that after Moses's death, Allah lifted up his bones, carried them across the Jordan, and buried them on a hill in the West Bank. The grave of Moses is enshrined in the Nebi Musa Mosque, whose environs serve as an exalted burial place; it is also a site at which to pray for the health of sick relatives. The annual Nebi Musa festival is a central rite of Palestinian culture. See Roger Friedland and Richard D. Hecht, "The Pilgrimage to Nebi Musa and the Origins of Palestinian Nationalism," *Pilgrims and Travelers to the Holy Land*, ed. Bryan F. Le Beau and Menachem Mor (Omaha: Creighton University Press, 1996), 89–118. Moses has another gravesite to which Muslims and Christians make pilgrimage beneath Mount Nebo by the Spring of Moses. A local legend has it that Moses drank his last water from this spring before crawling into his grave. Because he was not ready to die, Moses disguised himself as a Bedouin and traveled through the wadis along the Jordan. An exasperated God dispatched the Angel of Death to bring an end to the prophet's life. The Angel of Death disguised himself as an old man. Confronting a fatigued Moses, the angel consoled him with water to drink and a cool cave in which to rest. After Moses drank the water and entered the cave, the Angel of Death covered the mouth of the cave with a large stone and put Moses to rest.

in the book of Joshua, but also in the centrality of Torah to all subsequent Hebrew texts.

Transfer of Spirit to the Disciple

Upon Moses's death, Joshua son of Nun becomes "filled with the spirit of wisdom" (Deut 34:9). The spirit, transferred to Joshua when Moses laid his hands on him in the wilderness, seems to have lain dormant until Moses's death. By reading Deuteronomy 34:9 together with Numbers 27:20, a sequence emerges in which Moses transfers some authority to Joshua through the laying on of hands (Num 27:18), but the spirit is not imparted to Joshua until after the death of Moses.

The transfer is chronicled in Numbers 27:12–23, where Moses transmits some (but not all) of his authority (הוד) to Joshua through the laying on of hands. The scene begins on Mount Avarim, where God speaks to Moses of his death. The prophet's demise is here presented as punishment for his failure to properly sanctify God with the waters of Meribah. Without quarrel, Moses accepts his impending death, expressing concern only about the continuation of the community. The health of the collective, as Moses sees it, depends upon a single leader who can lead them to war. Thus, Moses initiates the idea of a certain type of successor so that the People are not left "like sheep that have no shepherd" (Num 27:17).[19] Succession in this case is a human rather than a divine imperative. There are different ways to look at Moses's request. Moses appears ever the selfless leader concerned only with the well-being of his people in the face of his own death. At the same time, Moses's legacy depends upon Israel's successful entrance and settlement of Canaan. He wants to know that his absence will not entail the failure of his mission. So Moses requests the realization of his vision at the hands of another. The perpetuation of his efforts by a leader that resembles Moses, I argue, represents the striving for a kind of immortality. Moses's choice of words further suggests that he has a particular successor in mind. Without naming names, Moses alludes to a military man with a share of charisma.[20]

---

19. Noth believes that Numbers 27:15–23 is the original pre-Priestly, pre-Deuteronomic tradition about Joshua. Although he acknowledges that this passage is incorporated in a Priestly narrative, Noth asserts its antiquity on the basis of its concern with the conquest of the land, not a usual concern of the P writer. According to Noth, P cares about institutions and not the land. Martin Noth, *Überlieferungsgeschichte des Pentateuch* (Stuttgart: W. Kohlhammer Verlag, 1948), 193.

20. Moses mentions spirit in connection with God, "the God of spirit that abides in all flesh" אלהי הרוחת לכל־בשר (Num 27:16), not his successor. But God responds to Moses by designating Joshua son of Nun, "a man endowed with spirit" איש אשר רוח בו (Num 27:18), as the recipient of Moses's charismatic authority הוד. Prior to his succession, Joshua seems to possess an innate quality necessary for the office. The repetition of "spirit" רוח (once in connection with God and once with Joshua) implies

The transfer of authority intends for the entire community to heed Joshua (27:20).

Presence of Witnesses I
The "entire community" that is expected to follow Joshua witnesses the transfer of authority (Num 27:22). Most prominent among the community of witnesses stands Eleazar the Priest, who participates in the installation ritual and will lead alongside Joshua after Moses's death (Num 27:19, 21). Once Moses is gone, Eleazar will mediate between God and Joshua, consulting the oracular Urim and divining the auspicious times for initiating and concluding battles (27:21). As Moses stands Joshua before Eleazar, so Joshua is to stand before the Priest in regular consultation. Joshua's power is here presented as shared and checked by the Kohanim. The emphasis on Eleazar's sanction may be a mode of justification for this anomalous succession in which Joshua, a nonhereditary heir, receives authority from Moses, who is both prophet and priest, while becoming neither prophet nor priest. Not only is Joshua not Moses's son, but by succeeding Moses he also disrupts the transmission of Moses's Priestly legacy to another family member. Joshua's office is portrayed as requiring the priesthood and, in this sense, may serve as a template for kingship.

Death of the Master
As Moses draws closer to his death, Joshua assumes greater status. When Moses's fate is decreed, Joshua is designated his replacement through the laying on of hands (Num 27:18, 23). After Moses dies, the spirit visibly animates Joshua (Deut 34:9), yet God does not address him until Moses's death becomes an event of the past (Josh 1:1–2). On the occasion of the Jordan crossing, God finally elevates Joshua in the eyes of Israel to the status of Moses (Josh 4:14). Dispersed among the books of Numbers, Deuteronomy, and Joshua, the steps of Joshua's succession are staggered. Joshua takes Moses's place as he performs the very actions denied to Moses. God first addresses Joshua with the words Moses so desired to hear, "get up and cross this Jordan" (Josh 1:2). While Joshua will never warrant the title of prophet, his footsteps stake Israel's claim (Josh 1:3) and his partitions put the tribes in place.

God's motivational pre-crossing speech to Joshua (Josh 1:1–9) echoes the public initiation by Moses (Deut 31:7–8) with God shifting from third to first person. In addition to a general exhortation to "be strong and bold" (Josh 1:9), God's speech introduces the idea of being strong and very bold in upholding the Torah transmitted by Moses (Josh 1:7). This, combined with the injunc-

---

that Joshua's spirit comes from God and that the aura of a leader הוד will be conferred once Moses lays his hands upon him (Num 27:18, 20).

tion to "never let this scroll [book] of Torah cease from your lips" (Josh 1:8), suggests that effective leadership relies on both spirit and word mediated by Moses. In the absence of Moses, only adherence to the words of Torah can ensure correct action and successful endeavors (Josh 1:8).[21] The central difference between initiation at the hands of Moses in Deuteronomy 31 and initiation by the word of God in Joshua 1 is that Deuteronomy expressly designates Torah as the domain of the priests (31:9).[22] The priests may have a kind of implicit presence here if we read Joshua through the figure of the king in Deuteronomy 17:18–20, which "is indeed obvious ... [as] the closest parallel,"[23] whose copy of the Torah is provided by the levitical priests.[24] In any case, Joshua, who must preserve the words of Moses's Torah on his lips, is more like a king who refers to a written corpus than a prophet into whose mouth God directly places the word (Deut 18:18).[25]

Disciple's Jordan River Crossing and Presence of Witnesses II

Not until the final step of succession, the Jordan River crossing, does God publicly elevate Joshua before all Israel and enable recognition that God accompanies Joshua as He did Moses (Josh 3:7, 4:14). Joshua undergoes the transformation into Israel's leader during the crossing so that his term begins west of the Jordan. Insofar as Joshua's position is necessitated by the impending wars to conquer the land, the Jordan distinguishes between the time when Israel required a lawgiver and the time when it needs a general. God's assurance, "no one will stand up to you as long as you live" (Josh 1:5), hints at the character of Joshua's office, but his role as military leader becomes starkly apparent during the event of the crossing. In the account narrated in Joshua 4:10–19, an infantry comprising the east bank tribes is followed by forty thousand warriors marching toward war in Jericho (4:13). It is in this consciously military context that the authority transferred to Joshua actually takes root. Since Joshua directs the priests throughout the crossing (Josh 3:8, 4:10, 4:16), a hierarchy emerges that resembles the scenario of Moses and Aaron more than the one imagined in Numbers 27:12–23.

---

21. Parallels are found in Ps 1:3, 1 Kings 2:3, and 1 Chr 22:13. J. Roy Porter, "The Succession of Joshua," in *Proclamation and Presence: Old Testament Essays in Honour of Gwynne Henton Davis*, ed. John I. Durham and J. R. Porter (Macon, GA: Mercer University Press, 1983), 116.

22. For the summary of arguments concerning authorship and original form of Josh 1:7–9, see Porter, "The Succession of Joshua," 109.

23. Porter, "The Succession of Joshua," 112.

24. The image of a copy of the Torah appears only in Deut 17:18 and Josh 8:32, where Joshua imprints a copy of Moses's Torah on the rocks of Mount Ebal: See Richard D. Nelson, "Josiah in the Book of Joshua," *Journal of Biblical Literature* 100, no. 4 (1981): 533.

25. Porter sees this as evidence that the Deuteronomist describes Joshua in royal terms and that Joshua, like a king, received an actual scroll when he succeeded Moses.

As Joshua crosses the Jordan, the effects of succession are publicly witnessed by all Israel so that the People will obey Joshua as they did Moses (Josh 4:14). Although the text itself strikes a solemn tone, some irony arises if one recalls Israel's behavior toward Moses. Nevertheless the Jordan crossing ushers in a new beginning for Joshua and inaugurates a process in which "Joshua is gradually introduced to the office of Moses, the servant of Yahweh, until he himself becomes a 'servant' (Josh. 24:29)."[26] Through the act of crossing the river, Joshua becomes the authoritative leader who directs battles, communicates with God, and exhorts the People to obey the tenets of the Torah.[27] This marks the fifth step of the succession, the disciple's crossing of the Jordan.

Working inductively, a few general points about prophetic succession can be made. While a prophet like Moses or Elijah can be initiated privately, their

---

26. J. Alberto Soggin, *Joshua* (Philadelphia: Westminster, 1972), 57.

27. Scholars have long harbored suspicions of monarchical investment in Joshua's succession. Based on the root צוה used in Numbers 27:19 and 23 and translated as "commission" in the JPS Tanakh, J. Roy Porter concludes "that the installation formula in them has its background in the royal practice and administration of the Judaean monarchy." Porter stresses the authorial hand of the Deuteronomic writer. J. Roy Porter, "The Succession of Joshua," in *Proclamation and Presence: Old Testament Essays in Honour of Gwynne Henton Davies*, ed. John I. Durham and J. R. Porter (Macon, GA: Mercer University Press, 1983), 108. In addition, Porter sees the repeated call for Joshua to abide by the Torah as fitting "perfectly with the use of the installation formula at the accession of a king." By reading the Joshua passages through the scenes of David bequeathing the throne to Solomon (1 Kings 2.1–4, 1 Chr 22–23.1, 28–29), he identifies "the royal Davidic covenant, where the security of the succession and the continuation of the royal house forever is one of the major concerns," as the institution that motivates the representation of prophetic succession. Ibid., 122. Mark Smith and William Schniedewind see Joshua as a paradigmatic figure drawn to anticipate a later good king of Judah, Hezekiah. Smith, *The Memoirs of God*, 61, and William M. Schniedewind, *How the Bible Became a Book* (Cambridge: Cambridge University Press, 2004), 80. The perceived parallel leads Smith to speculate that the conquest is in fact "a late-monarchic expression of hope for the reestablishment of the kingdom back to what was thought to be its earlier boundaries." Richard Nelson reconciles the royal trappings of Joshua's installation and subsequent career as redolent of the Deuteronomically favored king, Josiah. Since Josiah becomes the standard against which his predecessors and successors are measured, his qualities are retrojected onto the hero of the conquest; this in turn paints Josiah's efforts as a national revival. Marvin Sweeney makes more modest claims about the resemblance: "Given the Josianic associations of the language in these chapters . . . this would suggest that Joshua 1 and 23 are the product of redaction that is designed to establish the analogy between Joshua and Josiah. Likewise, Joshua 8:30–35 also appears to be the product of such redaction." Sweeney, *King Josiah of Judah: The Lost Messiah of Israel* (New York: Oxford University Press, 2001), 135. The image of the ideal king thus finds a reflex in Joshua. Like Josiah (2 Kings 22:2) and the theoretical king of Deuteronomy 17:20, Joshua does not deviate from the course of law (Josh 1:7, 23:6). In both cases, a book or scroll of Torah facilitates the navigation of this course (Josh 1:8; 2 Kings 22:8, 11). Fidelity to the book has ritual as well as military implications, as evinced in the successful ways that the counterparts reinstate the commemoration of Passover (Josh 5:10–12; 2 Kings 23:21–23) and annex "promised" land (Josh 1:3; 2 Kings 23:19). Joshua's succession, which blends elements of royal and charismatic installation, does double duty to dramatize the Deuteronomic theory of both the king (Deut 17:14–20) and the prophet (Deut 18:15–22). In this sense, the fact that Joshua is named neither king nor prophet safeguards its applicability.

successors, whose authority relies on appointment, require public witness. Public witness affirms both their resemblance to a predecessor and their legitimacy. However, resemblance is not equivalence. Joshua may actualize Moses's vision, but Moses remains distinct. Although the land is preferred to the wilderness and the new generation superior to the old, Joshua pales in comparison to his predecessor. There is even some suggestion that Joshua is only as praiseworthy as he is faithful to Moses's example.

The loss of a prophet entails rupture. The writers confront the fact that political stability is often only as long as a life by setting the scenes of prophetic demise on the eastern side of the Jordan, where the temporal rupture figures as a spatial rift. By crossing the river, the disciples bridge the rupture and contribute to a political myth of continuity. The ritual process as defined by Victor Turner becomes most evident in succession scenes in which the east side of the Jordan indicates separation, the Jordan functions as the liminal space, and the west side represents reintegration and institution. Separation from everyday activities occurs in Joshua's case when Moses lays his hands upon him (Num 27:23) as well as when he is removed to the Tent of Meeting (Deut 31:14). Yet separation from the sphere of fixed institutions is signaled most strongly by the fact that the prophets bequeath leadership to their disciples on the east side of the Jordan, outside the domain of binding law. The ritual crossing of the Jordan detaches the successors from the confines of the disciple's identity and dissolves its characteristics. For its duration, the successors are neither disciples nor leaders, but rather fall between the two states as they navigate the border of the land. West of the Jordan, the successors take their place as leaders who command a new set of titles and face a novel reality. Entering the space of the land initiates a new era with visible connections to a recent past. The ritual crossing enacts the succession and grants it credibility.

Elijah and Elisha

The scene of Elijah and Elisha in 2 Kings 2:1–18 sets the paradigm for prophetic succession at the Jordan. The elements of impending death, transfer of spirit, the presence of witnesses, the death of the master, and the disciple's river crossing are all present in the text of Elijah's ascension and Elisha's succession. The ritual phases of separation, liminality, and reaggregation stage Elisha's transformation from apprentice to man of God while ushering him into Elijah's place as prophet in Israel. Scholars have remarked on the disparities between the careers of Elijah and Elisha; Elijah struggles against Baal worship and syncretism while leading a solitary, tortured life, and Elisha performs miracles to make the quotidian bearable while working closely with a band of prophets and the kings of Israel and Aram alike. The respective genres of the prophetic traditions reflect the differences insofar as Elijah's narrative sustains continuous themes with quasi-historical reportage in contrast to the Elisha legends, a series of quixotic miracle tales that read as an assemblage of intermittent traditions.[28]

These differences in character and genre point toward a theory of biblical succession in which prophets commanding a systematic vision designate successors only in the name of its implementation. So Joshua traverses the Jordan, conquers the land, and establishes a geographic framework for the law, and Elisha advises kings and determines the outcome of battles. Variant genres like the Torah vs. the book of Joshua and prophetic biography vs. a compendium of legends safeguard the crucial difference between prophets and their successors.[29] Longer narratives that chronicle the careers of the vaunted prophets have a biographical cast where the records of the successors are episodic and focused on action rather than interiority.

As Elisha shifts from disciple to master, Elijah undergoes a transformation from mortal to immortal. Elijah's removal to the geographical-ritual limen indicates a departure from society in several senses. He leaves the company of other prophets, the monarchical society of Israel, and the realm of humanity. When Elisha returns from the east bank alone, he reintegrates into society and assumes a stature similar to Elijah's. In contrast to Moses and Joshua, for whom the Jordan is an unknown terminal point, the Jordan is familiar to the two Northern prophets. Elijah underwent a manner of initiation at the Jordan, where he drank from a stream and was fed by ravens during a drought (1 Kings 17:2–6). Miraculous events at the Jordan thus frame Elijah's prophetic biography and to some degree his life.[30] Both Elijah and Elisha come from Gilead, on the east bank of the Jordan (1 Kings 17:1, 19:16). Elisha establishes a prophetic school by the Jordan and helps his disciples in its construction by miraculously rescuing a sinking ax from the river (2 Kings 6:1–7). In the case of Moses, the Jordan is a constant reminder of death, but for Elijah, it is a place of sustenance and vitality toward which he goes willingly at the end

---

28. See Alexander Rofé, *The Prophetical Stories: The Narratives about the Prophets in the Hebrew Bible, Their Literary Types, and History* (Jerusalem: Magnes Press, 1988), 55–74; Yael Shemesh, "'I Am Sure He Is a Holy Man of God' (2 Kgs 4:9): The Unique Figure of Elisha," in *And God Said, 'You Are Fired': Elijah and Elisha*, ed. Mishael M. Caspi and John T. Greene (North Richland Hills, TX: Bibal Press, 2007), 20–30. Joseph Blenkinsopp calls Elisha "a more 'primitive' figure, embodying the more destructive forces that could be concentrated in that kind of personality." Blenkinsopp, *A History of Prophecy in Israel* (Louisville: KY: Westminster John Knox Press, 1996), 63.

29. Resemblances are drawn in shared titles such as "the chariots of Israel and its horsemen" (2 Kings 2:12, 13:14) and "prophet" for Elijah and Elisha and "servant of God" 'עבד ה for Moses (Ex 14:31, Num 12:7–8, Deut 34:5, Josh 1:1, 2, 7, 13, 15, 8:31, 33, 11:12, 15, 12:6, 13:8, 14:7; 18:7; 22:2, 4, 5, 2 Kings 18:12, 21:8) and Joshua (Josh 24:29; Judg 2:8).

30. Robert L. Cohn sees this as a feature of the chiastic symmetry of Elijah's biography. see his *2 Kings* (Collegeville, MN: Liturgical Press, 2000), 10.

of his time on earth. In conjunction with Elijah and Elisha, the Jordan functions as a locus of miracle and revelation.

Impending Death of the Master

Elijah's impending transport becomes evident in the first, scene-setting verse, "when God was about to take Elijah up to heaven in a whirlwind" (2 Kings 2:1). In preparing for his stormy departure, Elijah makes no requests that his flock not be like sheep without a shepherd. And, although Elisha walks with Elijah in the opening scene, Elijah states his preference for being alone. Elijah releases him three times and three times Elisha insists on remaining.[31] Elisha follows his master to all of the key sites around the Jordan: Gilgal of the twelve-stone memorial,[32] Beit-El, where Jacob saw the staircase that leads to heaven,[33] Jericho, where Israel won its first legitimate battle, and finally to the Jordan where Moses met his end.

At Beit-El and Jericho, Elisha encounters other prophets, who taunt him about Elijah's imminent removal. At both cities, the prophets speak as if God's translation of Elijah were intended only to punish or devastate Elisha (2 Kings 2:3, 5). Competition may lie at the root of the exchange, since the prophets question the degree to which Elisha knows what is to transpire. Knowledge of what is to come is, after all, central to the definition of a prophet, and so we can read the prophets' question as an inquiry or even an expressed doubt about Elisha's ability to succeed Elijah. Curtly defending his knowledge, Elisha urges the prophets to be quiet during the solemn occasion. The scene's repetition reveals the tenuous status of Elisha's authority among his cohort and makes clear why they must witness his succession.

The Presence of Witnesses I

Elijah tries to separate from Elisha a third time, but a persistent Elisha follows him to the Jordan shore. Fifty prophets stand at a distance as master

---

31. In a similar fashion, Naomi before crossing the Jordan tries to persuade Ruth to leave her three times (Ruth 1:7–18).

32. In the books of Amos and Hosea, Gilgal is excoriated as a site of Israel's wickedness (Amos 4:4, 5:5; Hosea 4:15, 9:15, 12:12).

33. Ottosson remarks that Elijah and Elisha commute "between cult sites of great importance to the North." The fact that such visits are "reported without any polemical features from Jerusalem's aspect, is striking." Ottoson, *Gilead*, 230. Beit El was the national sanctuary of the Northern Kingdom of Israel, although one that some prophets maligned (Amos 3:14, 4:4, 5:5–6; Hosea 5:8, 10:5, 15). In ancient Israel the account of Jacob's vision likely justified and sanctified the Beth El temple. James Montgomery notices the problem of Elijah's "going down" from Gilgal to Bethel since "this site lies lower than Bethel—so went down to Bethel is inaccurate." James A Montgomery, *International Critical Commentary: A Critical and Exegetical Commentary on the Books of Kings*, ed. Henry Snyder Gehman (New York: C. Scribner's Sons, 1951), 151.

and disciple cross eastward. Like Joshua distinguished among the People of Israel (literally "children of Israel") at his succession, Elisha is separated from the prophetic band (literally "children of the prophets"). The biblical scholar Ze'ev Weisman compares the investiture of Elisha with the commissioning of the seventy elders in Numbers 11—another occasion where spirit is transferred—and sees the presence of eyewitnesses as evidence of "the conferring of authority by a public bestowal of the spirit." He reads the spirit in question as "akin to an external supra-individual entity that causes a radical shift in the recipient's status."[34] In other words, the transfer becomes effective because an audience recognizes and endorses it. In the absence of witnesses, one can imagine that the reception of spirit would be devoid of social potency. Elisha's prophetic witnesses are anomalous since they are present during his succession, but unable to fully see what transpires or fully comprehend its implications. Perhaps for this reason Elisha's authority as a prophet in Israel depends upon the consistent performance of miracles.

Transfer of Spirit to the Disciple I
Elijah and Elisha traverse the Jordan in the only instance where miracle attends an eastward crossing. Elijah parts the Jordan by removing his cloak, rolling it up, and striking the water. At the same time that the crossing brings Elijah and Elisha to the proper site for initiation, it references the national crossings of the Jordan and the Red Sea. An obvious intertextual link is חרבה, the "dry land" on which the prophets walk (2 Kings 2:8), as did the People across the Red Sea (Ex 14:21) and the Jordan River (Josh 3:17).[35] The parallels between these water crossings suggest that more is at stake than simply the transmission of Elijah's powers. Although the death of a prophet may not seem as momentous as national liberation or conquest, it is perhaps more expressly political.

In a vision on Mount Sinai, God instructed Elijah to transform the political landscape by appointing Hazael as king of Aram, Jehu as king of Israel, and Elisha as his successor (1 Kings 19:15–16). This coup, God promises, will give the Northern Kingdom of Israel a new start while undermining the apostates in Israel's midst (1 Kings 19:18). However, by the time Elijah is summoned to the Jordan, he has appointed Elisha alone. Between Elijah's reluctance to cross the Jordan with Elisha and the doubts expressed by the other prophets, one

---

34. Ze'ev Weisman, "The Personal Spirit as Imparting Authority," *Zeitschrift für die Alttestamentliche Wissenschaft* 93 (1981): 226–27.

35. "The striking (*nkh* Hiphil) of the water with his mantle (v. 8) is reminiscent both of Moses's magical rod at the Sea (Exod 14:16, 21) and his water-producing strike (*nkh* Hiphil) at the rock (Exod 17:6)." Robert L. Cohn, *2 Kings* (Collegeville, MN: Liturgical Press, 2000), 14.

wonders if Elijah has the ability to fulfill any part of his vision. Since the political scenario in no way resembles what God dictated to Elijah, Elisha's succession represents the only available means of fulfilling God's mandate. Or, said differently, the succession scene is set at the Jordan River in order to heighten the importance of the political transition from the dynasty of Ahab to that of Jehu. Only after Elisha installs Hazael as king of Aram (2 Kings 8:7–15) and anoints Jehu as king over Israel in Ramoth-Gilead east of the Jordan (2 Kings 9:1–14) does the future revealed to Elijah becomes Israel's present.

East of the Jordan Elijah articulates no expectations for his successor, but instead asks him, "What can I do for you before I am taken from you?" (2 Kings 2:9). Elisha initiates the transfer of spirit by requesting that "a double portion" of Elijah's spirit pass on to him. The language of "a double portion" is familiar from Deuteronomy 21:17, where it is legislated that a firstborn son, even if he is the son of an unloved wife, must receive a double portion of inheritance.[36] What does this correspondence mean? Does Elisha feel like Elijah's son who is entitled to the rights of the firstborn? In this case, Elisha may be asking for twice as much as the other prophets, a request potentially motivated by his resentment. Elijah explicitly promises Elisha nothing, but leaves the spiritual inheritance dependant on a test. Since Elisha has "asked for something difficult," his vision must be proven worthy if he is to receive it. Elijah says, "If you see me taken from you, you will have it and if you don't, you won't" (2 Kings 2:10). The vision commences as a flaming chariot divides the two and carries Elijah to heaven in the eye of a storm.

"Death" of the Master

Departing the world at long last, Elijah ascends to immortality. The Jordan River, where a chariot of fire drawn by horses of fire touches down, functions as a portal to heaven as well as a site of contact between human and Divine. It is further a junction between life and death where death can be transcended.[37] Elijah's ascent adds a vertical dimension to the horizontal border. In all of the instances surveyed thus far the Jordan's primary importance lay in the fact that it demarcated distinct spatial zones—east and west bank, Moab/Ammon and Israel, illegitimate and legitimate landscapes of memory, exile and home. As we have often seen, a temporal distinction attends the spatial divide while

---

36. Alternately, some have understood Deuteronomy 21:17 to refer to the "two-thirds" of a portion that the firstborn son receives when there is another heir.

37. "It is interesting to note that the place where Elijah ascended to the heavens—across the Jordan opposite Jericho—is also the place that Moses ascended before his death (Deut. 34:1). To this, let us add that Elisha, who died a natural death, was evidently buried in the very same area, as we see from the story of the revival of the dead man who came into contact with Elisha's bones (II Kgs 13:20)." Yair Zakovitch, *The Concept of the Miracle in the Bible* (Tel Aviv: Mod Books, 1990), 101.

still being overlaid on a horizontal plane. As the chariot lifts Elijah, the Jordan appears as the site where heavenly figures descend and mortals can rise to heaven. The Jordan now seems to divide space into four coordinates: the land to the west, the more ambiguous space to the east, heaven above, and earth below. As we will see in the Gospels, a different temporal order is associated with the vertical dimension. And, as we will see in later Jewish and Christian texts, the Jordan figures as a site where past, present, and future collide. For our purposes, we see the vertical dimension stressed over the horizontal for the first time.

Transfer of Spirit II

With Elijah removed, the focus turns to Elisha. The answer to whether or not Elisha receives a double portion, never addressed directly, can be surmised only by textual clues that contain contradictions and evade a definitive conclusion. The terms of the test rested on whether Elisha could observe Elijah's translation. The act of seeing comports with the larger theme of witnessing, a key component of the succession scenes. Immediately after Elijah departs in the storm, the text reports, "Elisha saw," which seems to indicate that Elisha passed his test. When he sees Elijah no more, he rips his clothing and makes mourning part of his initiation. Along with the theme of seeing, allusions to hereditary transmission inform the scene. Elisha asks to be recognized as first-born among the prophets (2 Kings 2:9) and cries out "father, father" to the ascending Elijah (2 Kings 2:12).[38] He then turns to claim his inheritance—Elijah's discarded mantle. By taking the cloak imbued with Elijah's spirit, Elisha assumes Elijah's position.[39]

The same cloak has before designated Elisha as successor. On his return from Sinai, Elijah sought out his promised disciple and found him plowing with twelve yoke of oxen.[40] Without introduction, Elijah cast his mantle upon

---

38. The whole address reads "Father, father, Israel's chariots and horsemen," a cry of loss also uttered by Joash the king of Israel upon Elisha's death (2 Kings 13:14). As Eviatar Zerubavel notes, "spiritual pedigrees" developed through acts like prophetic succession are modeled after bloodlines. Zerubavel, *Time Maps: Collective Memory and the Social Shape of the Past* (Chicago: University of Chicago Press, 2003), 56.

39. In Pseudo-Philo's *Liber antiquitatum biblicarum*, "the succession of the prophet forms a closer parallel: As Elisha took the mantle of Elijah, Joshua takes Moses' garment." Erkki Koskenniemi, *The Old Testament Miracle-Workers in Early Judaism* (Tübingen, Germany: Mohr Siebeck, 2005), 204.

40. Elisha plows with twelve oxen, a number symbolic within the succession pattern of the ability to lead twelve tribes of Israel. "The only sensible explanation is that the reference to twelve, however unrealistic, is there to summon Moses and Joshua traditions (Moses: Exod 24:4, Dt. 1:23; Joshua: Josh 3:12; 4:2, 3, 4, 8, 9, 20). The depiction of Elisha as Joshua to Elijah's Moses is thus enhanced through a connection to Elijah that is also a reference to Moses and Joshua. Elijah and Elisha are depicted as presiding over the twelve tribes, invoking their predecessors Moses and Joshua, even though the

Elisha (1 Kings 19:19). From that moment on, Elisha was in some sense adopted by Elijah, whom he served וישרתהו (1 Kings 19:21) as Joshua served Moses (Ex 24:13). The assumption of clothing as a means of succession resembles both Kohanim (Num 20:26) and kings; the "mantle" that plays so important a part in conferring his position to Elisha (1 Kings 19:19b; 2 Kings 2:13) is properly a robe of state, commonly worn by kings.[41] The royal parallel seems particularly operative since Elijah is first told to "anoint" Elisha as his prophetic successor (1 Kings 19:16).[42] Since no scene of anointing is narrated, one can assume that the casting of the cloak is its functional equivalent.[43] With reference to the idea of the king's two bodies, it can be said that Elijah gains immortality in two forms. His mission continues as his disciple steps into his cloak and his physical body, while absent from the earthly plane, is translated into a heavenly form. Elijah, like Moses, becomes doubly immortal.

Disciple's River Crossing and Presence of Witnesses II

Does Elisha inherit a double spirit?[44] With the cloak in hand, Elisha returns to the eastern edge of the Jordan and imitates Elijah by striking the water with it. The repetition of the verb "to strike" builds a sequence in which Elisha's first attempt to part the waters is unsuccessful (the Lucian version of the LXX and the Vulgate insert "the water was not divided"). By next calling out "where, O where is Yahweh the God of Elijah," Elisha solicits divine aid and brings himself into a relation with his master's God. This is similar to Joshua, who was first addressed by God on the eastern shore prior to his Jordan crossing.[45]

---

unity of the twelve was no longer a historical reality in their time." Marsha White, *The Elijah Legends and Jehu's Coup* (Atlanta: Scholar's Press, 1997), 8–9. Moses built a twelve-stone altar at Sinai (Exod 24:4), Joshua set up twelve stones at Gilgal (Josh 4:3), and Elijah built an altar of twelve stones during his showdown with the prophets of Ba'al (1 Kings 18:31). In each case, the stones represent the tribes. Elisha, however, is associated with twelve oxen rather than twelve stones. He conducts some sort of ritual slaughter and feeds "the People" without building an altar of twelve stones (1 Kings 19:21). The correspondence of number and substitution of oxen for stones may foreshadow Elisha's focus on feeding people and his neglect of cultic matters.

41. See J. A. Montgomery, *A Critical and Exegetical Commentary on The Books of Kings* (New York: Scribner's, 1951), 316.

42. Porter, "The Succession of Joshua," 120–21.

43. Otto Eissfeldt understands the anointing to be figurative. Eissfeldt, *Könige*, 4th ed. (Tübingen: J. C. B. Mohr, 1922), 329.

44. The author of Ecclesiasticus is sure that he did. "When Elijah was enveloped in the whirlwind, Elisha was filled with his spirit. He performed twice as many signs, and marvels with every utterance of his mouth. Never in his lifetime did he tremble before any ruler, nor could anyone intimidate him at all. Nothing was too hard for him, and when he was dead, his body prophesied. In his life he did wonders, and in death his deeds were marvelous" (Sirach 48:12–14).

45. Like Joshua, Elisha's first miracle after crossing the Jordan River involves Jericho (2 Kings 2:19–22).

After invoking God, Elisha strikes the Jordan a second time. The waters separate for him and he returns alone to the assembly of prophets.

As Elisha "saw" Elijah depart, so the prophets "see" Elisha upon his return and announce, "the spirit of Elijah has rested on Elisha" (2 Kings 2:15). The acknowledgment echoes the end of Deuteronomy when, after Moses's death, "Joshua son of Nun was filled with the spirit of wisdom because Moses had laid his hands upon him" (Deut 34:9). The effect on the two audiences, the prophets and the People Israel, offers another parallel. The People submit to Joshua's authority and "listen to him" (Deut 34:9), and the prophets initially defer to Elisha.[46] Upon recognizing the endowment of spirit, the prophets bow to Elisha (2 Kings 2:15), an obeisance paid to Elijah only once, by Obadiah (1 Kings 18:7).

Even after a confirmation that Elisha received Elijah's spirit, the question of the double portion persists. Although the prophets recognize Elijah's spirit in Elisha, they perceive only a limited quantity, since they insist on dispatching a search party to find Elijah (2 Kings 2:16). It is not surprising that they did not see Elijah ride to heaven, since he stressed the difficulty of such a vision to Elisha, but their sense that Elijah is still at large makes it seem as if Elisha commands considerably less spirit than did Elijah. The prophets also mention "the spirit of God" that may have conducted Elijah to a distant mountain or valley (2 Kings 2:16). This signals that the prophets sense the operation of divine spirit apart from Elisha. When the search for Elijah proves fruitless, the prophets come to no additional conclusions.

Bible scholars tend to be particularly confident that Elisha received Elijah's spirit. According to Hermann Gunkel, Elisha's ability to miraculously ford the Jordan proves that "Elijah's power *has* passed to him" and "that a marvel could be transferred by tradition from one man of God to another." Alexander Rofé concurs that Elisha's ability to emulate Elijah's Jordan crossing indicates "that Elisha is the true heir of Elijah, inheriting not only his mantle but also his spirit." Robert Carroll is certain that Elisha's request was entirely granted and that "the reception of the double share identified Elisha as the first-born among the prophets."[47] Although the text leaves the question of the double portion somewhat open, it seems that Elisha receives enough spirit to

---

46. "We see how Elisha had to validate his position as successor to Elijah vis-à-vis "the sons of the prophets" by deeds of power similar to those of his master. His first act, therefore, was to divide the waters of the Jordan with the help of Elijah's mantle." Robert Wilson, *Prophecy and Society in Ancient Israel* (Philadelphia: Fortress Press, 1980), 37.

47. Hermann Gunkel, "Elisha: The Successor of Elijah (2 Kings II.1–18)," *Expository Times* 41 (1929–30): 185; Alexander Rofé, *The Prophetical Stories* (Jerusalem: Magnes Press, 1988), 46; R. P. Carroll, "The Elijah-Elisha Sagas: Some Remarks on Prophetic Succession in Ancient Israel," *Vetus Testamentum* 19 (1969): 405.

perform miracles and to execute the political tasks assigned Elijah, but not enough to surpass his master. As shown, Elisha cannot part the Jordan before invoking God, and his spirit is not powerful enough to convince his witnesses that he is the only remaining vestige of Elijah.

Wesley Bergen follows a midrashic tradition (Midrash Hagadol I. XIX) and notes that Elisha performs double the miracles of Elijah.[48] However, Bergen characterizes Elisha as a degenerate Elijah who never assumes the historical import of his predecessor. Elisha's initiation, in his view, is necessary because his "authority rests [only] upon his being the legitimate successor of another prophet" rather than on any innate qualities.[49] I do not adopt Bergen's resolutely negative portrait of Elisha; nevertheless, his analysis facilitates certain observations. To begin with, Elisha obtains some of Elijah's spirit in order to enact his prophecies (1 Kings 19:15–17). Like Joshua, Elisha's task is not to receive his own prophecies, but rather to bring his predecessor's vision into being. Because Elijah's zeal stood in the way of his achieving the tasks assigned him by God, a successor must be designated to fulfill them.[50] The second point facilitated by Bergen's analysis is that Elisha is Israel's last prophet of deeds. Although subsequent prophets heal and perform pneumatic signs, their primary medium is language. Within the sequence of the biblical canon, Elisha marks the end of a particular sort of prophetic vocation and is the transitional figure between the former and the latter prophets. Joshua is a similar bridge between the first, unmatched prophet and the era of judges during which prophecy seems to subside and authority depends upon military charisma.

Moses and Elijah have long been recognized as analogs. The two prophets are both visionary, ardent ideologues who conceive of a utopia, but do not achieve it. Their stories begin at a life-giving river during a time of danger. Both engage in spectacular competitions with representatives of foreign religions, speak with God on Mount Sinai, and end their careers east of the Jordan. While the masters reach heaven, the disciples remain on earth. Where Moses dreams, Joshua enters; where Elijah envisions a reformed kingdom of Israel, Elisha brings it about. In the two scenes of prophetic succession, the disciples inherit unfulfilled tasks together with the spirit of their masters. The masters even seem to have more at stake in the succession than their

---

48. After Elisha crosses the Jordan, he immediately begins his miracle working (2 Kings 2:20–22); he also draws upon the miraculous powers of the Jordan at other points in his prophetic career (2 Kings 5:10, 6:1–7).

49. Wesley Bergen, *Elisha and the End of Prophetism* (Sheffield: Sheffield Academic Press, 1999), 42–47.

50. God designates Elisha as Elijah's companion and successor after Elijah's breakdown on Mount Sinai (1 Kings 19:19).

disciples. Both Joshua and Elisha carry out the missions of their predecessors, but do not equal their accomplishments. The scope of their power is forever limited by their status as heirs. Moses and Elijah are leaders credited with transformative speech whereas Joshua and Elisha actualize the words. As the disciples cross the Jordan into their careers, the territory west of the river appears as the only place where prophecy can find its fulfillment.

John the Baptist and Jesus

John the Baptist's every feature evokes the time of prophets. His leather belt and hairy cloak recall Elijah (2 Kings 1:8),[51] his predictions draw from established tropes of biblical prophecy (Isa 40:3), and he foretells of the impending new age from the prophetic locus of the Jordan River. In Mark (1:2), John the Baptist is announced as a forerunner ("Behold, I am sending my messenger ahead of you"), in much the same tone as the book of Malachi (3:24) predicts the arrival of Elijah ("Behold I will send Elijah the Prophet to you before the arrival of the mighty, frightening day of the Lord").[52] Such features not only classify the Baptist among the prophets, but they also figure him along with his cohort as belonging to the era of the old, one that anticipated redemption without providing the key to salvation.[53] Time then takes on a new cast. In

---

51. "The second detail concerning John's life in the wilderness, namely, the camel's hair garment and the leather girdle, tradition has also seized upon, making them the garment of Elijah and the girdle of Elisha (Syriac Life of John) . . . The LXX version of 2 Kings 1:8 uses virtually the same words as Mark 1:6, when it describes Elijah as one who had 'a leather girdle girded about his loins.'" Carl H. Kraeling, *John the Baptist* (New York: Charles Scribner's Sons, 1951), 14. "The perceptive reader cannot miss Mark's point: John is the prophet of the end-time, the eschatological messenger of Malachi; yes, he is Elijah who is to 'come first to restore *all* things' (Mark 9:11)." Walter Wink, *John the Baptist in the Gospel Tradition* (Cambridge: Cambridge University Press, 1968), 3.

52. See the parallels in Matthew 11:10 and Luke 1:76, 7:27. For the symmetry between the relation of Elijah and Elisha and that of John the Baptist and Jesus, see Joan E. Taylor, *The Immerser: John the Baptist within Second Temple Judasim* (Grand Rapids, MI: Eerdmans Publishing Company, 1997), 281, and Aharon Wiener, *The Prophet Elijah in the Development of Judaism* (London: Routledge & Kegan Paul, 1978), 141. Thomas Brodie tries to unravel the tight knot of association between Elijah/Elisha and John the Baptist/Jesus: "Elijah's name occurs [in the New Testament] twenty-nine times, all but two in the Gospels . . . On some occasions Elijah is associated with John the Baptist—suggesting that John is the returned Elijah (Luke 1:17; Mark 9:10–12; Matt 17:10–12; 11:14), yet in John's Gospel the Baptist denies that he is Elijah (John 1:21–25). On other occasions Elijah is associated not with John the Baptist but with Jesus. Elijah is a model, along with Elisha, for Jesus's ministry (Luke 4:25–26), and he is a companion, along with Moses, of Jesus's transfiguration (Mark 9:3, 4; Matt 17:3, 4; Luke 9:30, 33)." Thomas L. Brodie, *The Crucial Bridge: The Elijah-Elisha Narrative as an Interpretive Synthesis of Genesis-Kings and a Literary Model for the Gospels* (Collegeville, MN: Liturgical Press, 1999), 80.

53. Further support for this timeline is evident in Matthew 11:13, where the use of the past-tense "prophesied" "makes it unlikely that Matthew wanted to incorporate John into the new era." Ernst Bammel, "The Baptist in Early Christian Tradition," *NTS* 18 (1971–72): 101.

distinction from the scenes of prophetic succession at the Jordan, the past becomes degraded in the face of the future. John the Baptist appears already in his diminishment and recedes after coming into contact with the agent of the altered future called the Kingdom of Heaven.[54] As a necessary premise of Gospel narrative, this switch of temporal valuation, where the future trumps the past, occurs at the Gospels' very outset. So Mark, the earliest Gospel, begins with John the Baptist in order to enact the new temporal hierarchy. His baptism transpires at the Jordan River so that a familiar scene promotes a revised schema of authority.

The stylistic adaptations and dialogic interactions with the Jordan successions are manifold. As the transfer of power from Moses to Joshua is structured as the transition from the corpus of Torah to that of Prophets, so authority is transferred from the prophets—manifest in the figure of John the Baptist—to Jesus Christ through the literary transition from an "Old" Testament to a "New" Testament.[55] Ritualized movement at the Jordan, in both cases, depicts the crossing from one corpus to another. The shift from John the Baptist/the prophets to Jesus mirrors that from Moses to Joshua. In both cases, the predecessor and successor differ, but in the case of Jesus, status becomes detached from chronology. Within the sphere of the New Testament, the transition occurs at the beginning of the Gospel of Mark, which as the oldest initiates the Gospel genre, as well as in Matthew and Luke, where Jesus's baptism occurs after he has been established as resembling yet surpassing the Hebrew patriarchs and Moses. The baptism of Jesus further privileges the future over the past by reversing the nature of the master-disciple relationship. As we will see, the baptism scene directly references prophetic succession at the Jordan and then employs the familiar pattern in order to introduce novel concepts of authority, redemption, and time.

Like Moses and Elijah, John's voice "cries out in the wilderness" and delimits the present by heralding the future and, like these prophets, John does not live to see the redemption that he predicts, but initiates a successor at the Jordan River.[56] Like Joshua and Elisha, Jesus undergoes an initiation at the

---

54. "John the Baptist stands at the dividing line between the period of anticipation and the period in which the Kingdom is present but in conflict." Carl H. Kraeling, *John the Baptist* (New York: Charles Scribner's Sons, 1951), 156.

55. This certainly occurs in both cases at the point of editing and canonization. In the case of the Gospel of Mark, the transition occurs earlier with the genre decision of a Gospel and the use of Koine Greek. Since the canonical Gospels begin with Matthew and not Mark, Jesus's placement in biblical genealogy precedes the shift from prophets to messiah.

56. "Johannes ist nach des Intentionen des Markus der Mann 'zwischen den Zeiten.'" J. Ernst, *Johannes der Täufer: Interpretation—Geschichte—Wirkungsgeschichte* (Berlin: Walter de Gruyter, 1989), 37.

Jordan and carries the vision of his predecessor forward in time.[57] The Jordan as baptismal site recalls Israel's momentous transitions and functions as a boundary between heaven and earth. Most importantly, it is the only locus of prophetic initiation with, as we have seen, ramifications for royal initiations as well. Without embroiling myself in centuries of Christological debates, I want to argue that Jesus's baptism operates as an initiation.[58] In contrast to the prophetic succession scenes, the spirit that descends upon Jesus is not that of his predecessor, but rather the Spirit that stems directly from a Divine source. Reception of this Spirit begins the plot of Jesus's miraculous career insofar as it drives Jesus through a concentrated wilderness test lasting a day for each of Israel's forty years in the wilderness, inaugurates his public exposure, and marks an era when this Spirit operates within human society.[59] That John the Baptist effects the initiation signals continuity between prophetic powers and those of Jesus as well as a radical break from an era marked as the old. We will revisit the five steps of the succession scene while investigating their applicability to the baptism of Jesus in the Jordan River.

Impending Death of the Master

The Baptist's inevitable demise multiply impacts the baptism scene. In the most important sense, John the Baptist is already on his way from relevance to obsolescence insofar as his influence is reduced in the wake of Jesus's appearance. He articulates the erosion of his mission with the promise that "the one who is more powerful than I is coming after me" (Mk 1:7), a statement that reverses the traditional predominance of predecessor above successor. With John's description of his role in the drama of redemption, prophetic stature

---

57. "Just as the first stage of Joshua's leadership took place in Transjordan, Jesus' appearance began with his bathing in the Jordan." Elchanan Reiner, "From Joshua to Jesus: The Transformation of a Biblical Story to a Local Myth: A Chapter in the Religious Life of the Galilean Jew," in *Sharing the Sacred: Religious Contacts and Conflicts in the Holy Land First–Fifteenth CE*, ed. Arieh Kofsky and Guy G. Strouma (Jerusalem: Yad Ben Zvi, 1998), 249.

58. "Jesus received the Holy Spirit at his baptism, when God told him that he was his Beloved Son (1:10–11). So for Mark, Jesus' baptism by John was both the moment when he was first identified as God's Son and when he received the Spirit that initiated his new role." Norman R. Petersen, "Elijah, the Son of God, and Jesus: Some Issues in the Anthropology of Characterization in Mark," in *For a Later Generation: The Transformation of Tradition in Israel, Early Judaism, and Early Christianity*, ed. Randall A. Argall, Beverly A. Bow, and Rodney A. Werlin (Harrisburg, PA: Trinity Press International, 2000), 235.

59. Looking through a social-scientific lens, Richard DeMaris understands the baptism and subsequent events in terms of spirit-possession. See DeMaris, "The Baptism of Jesus: A Ritual-Critical Approach," in *The Social Setting of Jesus and the Gospels*, ed. Wolfgang Stegemann, Bruce J. Malina, and Gerd Theissen (Minneapolis: Fortress Press, 2002): 137–57; and Richard E. DeMaris, "Possession, Good and Bad: Ritual, Effects, and Side-Effects: The Baptism of Jesus and Mark 1.9–11 from a Cross-Cultural Perspective," *Journal for the Study of the New Testament* 23 (2000): 3–30.

becomes detached from temporal order. This marks the most overt reversal of the previous succession stories in which the masters, Moses and Elijah, proved superior in status and more intimate with God than their successors. John, who predates and prophesies about Jesus, may initiate him; but Jesus, who comes second in time, comes first in heaven.

In the Gospel of Matthew Jesus qualifies John's stature in the new temporality: "Truly I tell you, among those born of women no one has arisen greater than John the Baptist; yet the least in the Kingdom of Heaven is greater than he" (Mt 11:11). The assessment becomes all the more provocative as the Baptist, the last of the prophets ("for all the prophets and the law prophesied until John came"), is explicitly identified as Elijah: "if you are willing to accept it, he is Elijah who is to come" (Mt 11:13–14). Elijah reincarnate, a figure at the axis of time, has no standing in the Kingdom of Heaven. John the Baptist's footing in the past disqualifies him from redemption in the future outlined by Jesus. Such separation of stature from order supports the concept that, in this new era, the last will come first and the first last.

John's elaboration of the new hierarchy, "I am not worthy to stoop down and untie the thong of his sandals" (Mk 1:7), not only further downgrades the Baptist, but it also safeguards Jesus's superiority even as John sends Jesus down into the waters of the Jordan. Scholars often take John's subordination as a device to discount the efficacy of his baptism and thus to resolve the paradox of why the Son of God would require "a baptism of repentance for the forgiveness of sins" (Mk 1:4).[60] By contextualizing Jesus's baptism among

---

60. This problem is navigated differently by each of the Gospels. In Matthew, John hesitates to baptize Jesus and insists that Jesus should play the role of baptizer and John the baptized. Jesus says, "Let it be so now; for it is proper for us in this way to fulfill all righteousness" (Mt 3:15), indicating that they remain at the end of a period marked as the past (in the "now") when John's ritual is still operative. Luke reports Jesus's baptism in the past, implying but not naming John, "when Jesus also had been baptized" (Lk 3:21) and ascribes more agency to Jesus by showing him engaged in prayer as the heavens open and the Spirit descends. Carl Kraeling opines that Matthew and Luke make of John "a Moses who had glimpsed but not entered the Promised Land" (Lk 16:16; Mt 11:11; Lk 7:28). Kraeling, *John the Baptist*, 179. In the Gospel of John, it is nowhere stated that John the Baptist baptizes Jesus in the Jordan River. The purpose of this baptism in water seems to be the determination of who is the "one who baptizes with the Holy Spirit" (Jn 1:33). In other words, baptism in the Gospel of John does not offer repentance for the forgiveness of sin, but rather locates the messianic figure. When the Pharisaic emissaries inquire as to why John baptizes, he responds, "I baptize with water. Among you stands one whom you do not know" (1:26). In other company, John elaborates on the purpose of his baptism: "I myself did not know him; but I came baptizing with water for this reason, that he might be revealed to Israel" (1:31). John's baptism in this case functions as a kind of messianic try-out in which the waters identify the redeemer to Israel. Without the ritual of baptism, neither the Baptist nor Israel can locate the Redeemer. In fact, Jesus gains two of his disciples through the Baptist's identification, "here is the Lamb of God" (Jn 1:36–37). God instructs the Baptist, "He on whom you see the Spirit descend and remain is the one who baptizes with the Holy Spirit" (1:33). Ernst Bammel identifies a

succession texts, it appears that the Baptist's emphasis on his inferiority is the primary means of establishing a hierarchy of disciple above master and future above past that stands in contrast to the hierarchy outlined in the other succession scenes. John's subordination is a technique of revision that bespeaks a larger narrative program. As the Baptist goes on to equate his merely anticipatory status with his baptism with water (Mk 1:8), the role of the prophets is similarly diminished to precursor and their recommended purifications likewise become valuable to the degree that they point toward John's baptism and, most importantly, toward the baptism in Holy Spirit.[61] In this case, it is John the Baptist's inferiority instead of his death that enables his successor to assume and transcend his position.

Still, the actual death of John the Baptist also plays a role in the narrative. However difficult it is to arrive at a precise sequence of events, it seems that John is arrested as the Spirit first moves Jesus. Parallel to how the deaths of Moses and Elijah allow their disciples to translate their master's visions into social reality, John's arrest enables Jesus to become the proclaimer of his own good news (Mk 1:14).[62] The arrest of John also seems to mark the definitive end of the old time, since Jesus states, "The time is fulfilled, and the Kingdom of God has come near" (1:15). The time seems to be fulfilled because John is arrested and will subsequently die, and the Kingdom of God has come near in the very body of Jesus infused with the Holy Spirit since his baptism. After John's arrest, Jesus begins to disseminate the Baptist's prophecy, "repent and believe in the good news" (Mk 1:15) and thus to fulfill the mission of his "master" (1:4). John's imprisonment precedes Jesus's public ministry. The Lukan version emphasizes the Baptist's obsolescence by telling how Herod locked John in prison prior to Jesus's baptism (Lk 3:20).

Whatever the time frame of John the Baptist's arrest and beheading and its

---

layer in the Gospel of John in which Jesus appears "as the true disciple and successor of John, whose ministry is given special distinction by miracles, in a way like that of Elisha, in comparison with his former master." Bammel, "The Baptist in Early Christian Tradition," 112. E. P. Sanders reads the emphasis on John the Baptist's subordination as potential evidence that "the opposite was the case, that Jesus was a follower of the Baptist." Sanders, *Jesus and Judaism* (Philadelphia: Fortress Press, 1985), 91. I see the device of John's subordination as evidence of the revision of the prophetic initiation scenes.

61. "Far from attesting the 'Jewishness' of Jesus and his early disciples, the reference to Jesus' baptism by John in Mark serves, in the end, only to make John, along with other forms of early Judaism, but a foreshadow of 'the more powerful one' to come." Leif E. Vaage, "Bird-watching at the Baptism of Jesus: Early Christian Mythmaking in Mark 1:9–11," in *Reimagining Christian Origins: A Colloquium Honoring Burton L. Mack*, ed. Elizabeth Castelli and Hal Taussig (Valley Forge, PA: Trinity Press International, 1996), 290.

62. According to Mark, Matthew, and Luke, Jesus's activity of teaching and healing began only after John was arrested (Mk 1:14, Mt 4:12–17; Lk 3:18–23)." Adela Yarbro Collins, "The Origin of Christian Baptism," *Studia Liturgica* 19 (1989): 35.

relationship to the events of Jesus's ministry, the death report in Mark recalls the relationship of Elijah and Elisha.[63] As Herod becomes aware of Jesus, the details of the Baptist's death are revealed (Mk 6:14). Herod knows of "Jesus's name" (in a wonderful representation of oral transmission) because of the theories in circulation concerning his identity. Each of the three circulating theories relates to aspects of succession. Jesus is understood as either a resurrected John the Baptist, Elijah, or "a prophet like one of the prophets of old" (6:15). In other words, witnesses classify Jesus's wonders according to a familiar typology. Even more interesting is that the public seems to read Jesus's baptism as a kind of succession in which he becomes John the Baptist, Elijah redivivus, or a prophet in the long line of prophets.

Herod, ever the self-centered interpreter, concludes that Jesus is a resurrected John the Baptist perhaps intent on a vengeful haunting. As a means of analyzing the guilt that drives Herod's interpretation, a back story comes to the fore. By violating the Law and marrying Herodias, his brother's wife, Herod earned the rebuke of John the Baptist (Mk 6:17–18; Lev 18:16). In a later scene of royal decadence, Herod asks his stepdaughter to dance before his guests and promises her any gift she desires in return. After a consultation with her mother, the girl asks for John the Baptist's head on a platter (Mk 6:17–28). That the murderous instinct toward John stems from Herodias, the mother, and that Herod finds himself ensnared in a web of deviant desire recalls Jezebel's drive to eliminate Elijah (1 Kings 19:1–2) and Ahab's necessary capitulation since he was in so deep with the Phoenician queen (1 Kings 18:18–19). Where flight saves Elijah, John the Baptist becomes trapped and imperiled by the court.

After his disciples bury the remainder of the Baptist's body, the focus returns to Jesus, who, at least in the eyes of the public, continues to advance the imperatives of John. When the reader next encounters Jesus, he ferries his disciples to a deserted place in order for them to retreat from the populace in the name of rest and revivification (Mk 6:30–32). Yet when they arrive at a refuge, a vast multitude surrounds them for which Jesus feels compassion because they are, true to Mosaic fears, "like sheep without a shepherd" (Mk 6:34; Num 27:17). Jesus engages the crowd until the disciples voice their concern that the people be released in order to acquire food. When he commands the disciples to feed the masses, Jesus faces their incredulity that a multitude can be satiated with five loaves of bread and two fish. With the aid of a blessing, Jesus multiplies the bread and fish (6:41–42) and sustains the crowd. The

---

63. Josephus's report that John the Baptist was imprisoned at Machaerus, a fortress in Perea east of the Jordan and the Dead Sea, adds to the parallels between the Baptist and Moses and Elijah (*Antiquities* 18:109–13).

scale of the miracle is described in terms of five thousand diners and twelve baskets to carry the leftovers (6:42–44).

The act exceeds a similar miracle in the career of Elisha. Feeding the people is a role passed from Moses to Joshua (Josh 5:11–12) and from Elijah (1 Kings 17:14–16) to Elisha (2 Kings 4:1–7, 38–41, 42–44). Among the wonders Elisha performs is the transformation of twenty barley loaves and unprocessed grain into a replenishing harvest that sustains one hundred people during a famine (2 Kings 4:42–44). The prophet's ability to create food stuns an incredulous assistant and recalls the manna in the wilderness.[64] The multiplication of bread and fish resembles Elisha's miracle more than any other act of divinely assisted sustenance. But similar to the nourishment provided by Joshua west of the Jordan (Josh 5:11–12), the loaves and fish also symbolize that Jesus has led his people into an anticipated yet unfamiliar space. The effect of Jesus behaving like Elisha following the report of John the Baptist's death advances the parallel between John the Baptist-Jesus and Elijah-Elisha.[65] The diminishment of the Baptist plays a more vital role than his impending death and further articulates Mark's program of reversal: the Jesus who is spatially subordinate to John during baptism is the elevated Son of God. The additional reversal that I am arguing for is the superior predecessor paradigm of the Hebrew Bible. This reversal impacts the plot of the Jordan initiation although the central elements remain identifiable. John's impending death may not motivate Jesus's baptism, but it still plays a role in the development of Jesus's career and his recognition by the public. After the Baptist is reported dead, Jesus appears most like Elisha.

Transfer of Spirit to the Disciple

Whether answering John's call for repentance or entering on cue as the one "coming after me," Jesus comes from Nazareth to the Jordan in order to be baptized (Mk 1:7, 9). In addition to transposing the import of predecessor onto successor, Jesus's baptism also transforms the initiatory act performed

---

64. In Elisha's case, God commands, "eat and have some left over" אכל והותר (1 Kings 4:43), but in the case of the manna, Moses explains that none should be left over in the morning, ממנו עד־בקר איש אל־יותר (Ex 16:19).

65. Jesus is modeled on Elisha and John the Baptist on Elijah, but role-reversal is also common in the Gospels, where Jesus sometimes plays Elijah and John the Baptist Elisha. Other connections between Jesus and Elisha include Luke 4, where Jesus's infusion with the Holy Spirit (4:1, 14) enables him to heal (4:18 by way of Isa 61:1) outsiders in the manner in which Elisha healed Naaman (4:27). In the Lukan chronology, Jesus then heals the leper, not ostensibly an outsider, in 5:12–14. Bruce Chilton sees this as an explicit parallel between Elisha and Jesus. As Elisha could heal a leper because he possessed the spirit of Elijah, so Jesus could heal a leper because he was filled with the Spirit. Chilton, *Jesus' Baptism and Jesus' Healing: His Personal Practice of Spirituality* (Harrisburg, PA: Trinity Press International, 1998), 76.

at the Jordan. Where Joshua and Elisha cross the Jordan, Jesus submerges in its waters. The Jordan persists as the liminal space, yet transcending the limen depends upon immersion rather than crossing. This subtle difference in the succession scene bespeaks a larger revision. The axis of difference that makes the difference shifts from the horizontal axis of the land of Israel vis-à-vis other lands and other peoples to the vertical axis of heaven and earth. As Jesus goes down under the water, the Spirit descends to earth. The Jordan, in this sense, is a medial point on the axis.[66] The divide that Jesus must bridge is not the one separating the land of Israel from other lands or that separating home and exile, but rather the divide between heaven and earth.[67] Although the coordinates change, the Jordan continues to draw the divide. Jesus negotiates the gap between heaven and earth at the Jordan, Israel's primary political border. Yet during the course of Jesus's baptism, the Jordan signifies only the boundary between heaven and earth and seems to have little force as a geographic border. Redemption no longer transpires on a spatial plane, but rather on a temporal axis kept aloft by the perpetual tension of heaven and hell.

The image of the Jordan as the border between heaven and earth finds precedent in the story of Elijah's ascent (2 Kings 2:11), but amidst this comparison another reversal is in play: where Elijah ascended to heaven, the Spirit descends upon Jesus. As Jesus emerges from the Jordan, the heavens split and the Spirit falls upon him. Where the Jordan split for Joshua (Josh 3:13) and parted for Elisha (2 Kings 2:14), the "heavens are torn apart" for Jesus (Mk 1:10). It is the splitting of the heavens, not the water, that enables the Spirit to fall with birdlike precision to earth.[68] John the Baptist, who in some sense helps to convey the Spirit through the ritual of baptism, is in no sense a medium of the Spirit. Immersion at John's hands causes *the* Spirit to descend into Jesus, as emphasized by the pronoun εις (1:10b). Where the spirit of Moses entered Joshua and that of Elijah entered Elisha as they crossed the Jordan, Jesus receives a divine spirit with which John has had no contact. This difference comports with the separate categories of baptism with water and baptism with Holy Spirit, the first achieved by John and the second contingent upon Jesus. In the Gospels of Matthew and Luke (Mt 3:11; Lk 3:16) the Bap-

---

66. The tradition that while in the waters of the Jordan, Jesus reckoned with the forces of the underworld deepens the notion of the axis with a hellish counterpart beneath the Jordan, see K. McDonnell, "The Baptism of Jesus in the Jordan and the Descent into Hell." *Worship* 69 (1995): 98–109.

67. In an additional structural variation, earth represents exile and heaven home.

68. "All canonical accounts have God's spirit descending upon him 'like a dove,' although Matthew reads 'as if a dove,' and Luke says that it was 'in bodily form as a dove.'" Glenn A. Koch, "Jesus' Baptism and Temptation Accounts in Mark's Gospel," in *A Multiform Heritage: Studies on Early Judaism and Christianity in Honor of Robert A. Kraft*, ed. Benjamin G. Wright (Atlanta: Scholars Press, 1999), 41.

tist predicts that Jesus's baptisms will involve both "the Holy Spirit and fire," reminiscent of Elijah's spirit and the fiery vehicle in which he was conveyed to heaven.[69] Despite the fact that John does not possess the Spirit himself, he does recognize the redemptive powers of the Jordan and initiates both the messianic age and its primary representative at the Jordan.

The fact that John facilitates Jesus's reception of Spirit yet does not bestow his own spirit not only contributes to John's subordination, but also disrupts the continuity extended by the smooth transfer of authority. Prophetic succession, as we have seen, legitimates royal succession and mirrors—in a more populist light—the succession of Kohanim. Succession works perfectly as a metaphor for the Israelite/Judean/Jewish theory of redemption through subsequent generations. One is redeemed, according to the theory, by the children of the next generation, who will carry on as Israel. When the Gospels alter the model of succession, this notion of redemption is transformed. The sudden point is to end the physical exile here on earth and to enter a heavenly homeland. To have children, as the Apostle Paul said famously, is "to bear fruit for death" (Romans 7:5). Beyond thinking of Israel's covenant in an entirely new way, multiple programs of the Gospel coalesce in the alteration of the succession model. Held by an imperial representative, the throne is no longer an office in the service of God. Where several prophets opposed a corrupt king, Jesus faces the severance of kingship from domestic religious affairs. The proposed solution is to discount civic government in a kind of return to the time when God alone was king (1 Sam 8:7). The Kohanim (called Sadducees in the Gospels), who ultimately fulfill their narrative role by delivering the proper sacrifice to God, are—whether because of their involvement in politics or because of the anti-priestly bent of the Gospel writers—cast as mortal enemies of Jesus and his followers. Redemption is not to transpire in a human future, but in an eschatological era that brings an end to all familiar things. The Spirit received at baptism serves as a portal to this era. This spirit cannot be transmitted among people; it breaks out as a force either from above or below to drive people toward their salvation or destruction.

The theme of discipleship as refiguring the father-son relationship appears in the baptism scene as well. As the Spirit descends, a voice declares, "You are my Son, the Beloved; with you I am well pleased" (Mk 1:11). For the duration of the baptismal ritual, genealogy as well as the divide between heaven and earth collapse. God's proclamation at the Jordan that Jesus is His Son dis-

---

69. The punitive aspect of the fire predicted in Jesus's future baptisms (Mt 3:12b; Lk 3:9, 17b) further recalls the fire called down by Elijah to consume two sets of Ahaziah the King's messengers (1 Kings 1:10–14).

solves all other bonds of filiation and kinship.[70] As the spirit of their masters made Joshua and Elisha their "sons," so the heavenly Spirit is an integral component of the recognition of Jesus as God's son. Sidestepping the theological implications and assessing the baptism scene as a literary work, let me say that the reception of spirit and sonship are a necessary pair in Jordan initiation scenes. Succession replaces genealogical order with position in a hierarchy. Along with the Spirit, God must bequeath Jesus His Son in order for the initiation to entirely remove Jesus from his previous context and recast him in a new public role. Again the ritual process is at work and the Jordan functions as the site of the liminal where relationships are realigned.

Where genealogy is transcended in the prophetic succession scenes and filial language indicates the transmission of authority, the genealogical premise becomes dislodged in the baptism scenes. The reception and transmission of Spirit have the potential to explode genealogical position as well as temporal order. This Spirit, which can detach someone from his standing in the familial and social orders, initiates those who receive it into the new society of the Jesus movement and the new temporality of the Kingdom of Heaven. The ramifications of baptism for negating hereditary standing become clear in Matthew and Luke, when John the Baptist upbraids the Pharisees and Sadducees (the crowds in Luke), "do not presume to say to yourselves, 'we have Abraham as our ancestor,' for I tell you, God is able from these stones to raise up children to Abraham" (Mt 3:9; Lk 3:8). God's children can presumably emerge from anywhere, particularly, it is implied, from baptismal waters. In his lashing, John nullifies the merit conferred on Abraham's descendants through their ancestry and covenant. Tenets of the past have little place in temporality that lies on the other side of baptism.

It is therefore most important that Jesus be declared God's Son during his baptism so that his heredity—despite its importance in the Gospels of Matthew and Luke—is annulled at the outset. For Jesus's followers, baptism will similarly function as a new beginning. Jesus, who has disciples but lacks a clear successor,[71] presents the possibility of immortality through individual acts of baptism and emulation while representing two figures, "Jesus *Christus* and Jesus *christus*, the Anointed from Eternity and the one anointed in Jordan during his ministry on earth."[72] The spirits of Moses and Elijah made

---

70. This is made explicit when Jesus asks rhetorically, "'Who are my mother and my brothers?' And looking at those who sat around him, he said, 'Here are my mother and my brothers! Whoever does the will of God is my brother and sister and mother'" (Mk 3:33–35).

71. "Jesus' call of the first disciples (Mark 1:16–20) is modeled partly on Elijah's call of Elisha (1 Kgs 19:19–21)." K. Brower, "Elijah in the Markan Passion Narrative," *JNTS* 18 (1983): 91.

72. Kantorowicz, *The King's Two Bodies*, 52.

their successors into sons and thereby conferred authority; the Spirit of God elevates Jesus and grants him authority in his earthly incarnation. Because he is prince in God's kingdom, Jesus's baptism conveys royal associations similar to the prophetic initiations; his initiation is also his inauguration as regent of the heavenly kingdom.

The Presence of Witnesses

The story of Jesus's transfiguration revisits baptismal themes, while also employing elements of the succession type-scene. On the Mount of Transfiguration, where Jesus's whites sparkle brighter than any bleach could achieve (Mk. 9:3), God reiterates Jesus's beloved Sonship for the benefit of Peter, James, and John. The voice that emanates from a cloud recalls the cloud over the Tent of Meeting that initiated Joshua (Deut 31:15). The declaration elevates Jesus and, similar to Joshua's and Elisha's reception of the spirit, impresses obedience upon Jesus's followers. God declares, "This is my Son, the Beloved; listen to him" (Mk 9:7). In this sense, the transfiguration makes the transformation of Jesus at the Jordan known to an audience. The addition of "listen to him" emphasizes the public implications of Jesus's authority as the Son. At the same time, the transfiguration contributes to the revision of the succession scene by associating Jesus, the initiated, with Moses and Elijah, the initiators. As Elijah and Moses appear and converse with Jesus (Mk 9:4), Jesus achieves the status of an immortal.[73] Jesus transcends mortality, while also surpassing Joshua and Elisha, the figures with whom he is equated through name, place, and nature of initiation. The writer of Mark asserts that Jesus may have begun his ministry like Joshua and Elisha, but he is not limited by the vision of his predecessor. Instead, Jesus is the Jordan River initiate who becomes like Moses and Elijah and promotes his own vision. The transfiguration reverses the typology of Moses, Elijah, and John the Baptist by associating Jesus with Moses and Elijah.[74]

If Jesus has an audience at his transfiguration, does he also have an audience at his baptism? It seems that, like Joshua and Elisha, some of Jesus's future constituents surround him during his initiation. Because, from the moment that John the Baptist appears in the wilderness, "people from the whole Judean countryside and all the people of Jerusalem" (Mk 1:5) come out for

---

73. In Acts 1:6–11 Jesus ascends to heaven in much the same way as Elijah. The gathered witnesses not only see the translation, but are charged to bear witness to Jesus "in Jerusalem, in all Judea and Samaria, and to the ends of the earth" (Acts 1:8).

74. At the bottom of the Mount of Transfiguration, the association between Elijah and John the Baptist is restored when Jesus explains: "But I tell you that Elijah has come, and they did to him whatever they pleased, as it is written about him" (Mk 9:13).

baptism, it makes sense that at least some of the masses would be with John when Jesus comes for his baptism. The gathered masses are also likely to have heard John's prophecy concerning the superior baptizer to come. However, no one else seems to witness the parting of the heavens and the descent of the Spirit. The seclusion of Jesus's baptism in Mark is suggested by the address in the second person, "you are My Son," as opposed to the parallel in Matthew and at the transfiguration, "this is My Son" (Mt 3:17; Mk 9:7).[75] The scene of Jesus's baptism contributes to the larger theme in Mark of Jesus being surrounded by followers while being misunderstood by them. That no one else seems to see or hear the miraculous events surrounding Jesus's baptism points to the failure of the witnesses' perception.

Death of the Master, and Disciple's River Crossing

The absence of the final steps of succession furthers the disparity between baptism and the initiations of Joshua and Elisha. By the time Jesus emerges from the water, John the Baptist has faded from view and the era of Spirit baptism that he foresaw is closer at hand. Through John's initiatory ritual and his disappearance, Jesus enters a new temporality quite unlike anything before. In this sense baptism resembles Joshua's and Elisha's Jordan crossings. However, the murkiness surrounding when exactly John the Baptist dies prevents the construal of Jesus as his successor in anything more than time. That Jesus goes down into the river rather than crossing it introduces the reader to the premise that the horizontal plane where political contest transpires is but another arena of the struggle between God and Satan that transpires along a vertical axis. This becomes all the clearer when the Spirit that settles on Jesus at baptism drives him to the wilderness to contend with Satan. Jesus's path of descents and ascents—as paradigmatic as Joshua's departure from Egypt, presence at Sinai, and crossing of the Jordan—maps redemption onto topography. Jesus's itinerary represents the movement of the individual soul rather than a collective march.

Just as a new national Israelite character forms when the Children of Israel cross the Jordan with Joshua, a new Israel comprised of believers rather than descendants materializes through acts of baptism. Many sorts of baptisms occur throughout the New Testament, but John's baptism at the Jordan initi-

---

75. "Matthew's alteration of Mark's 'you are my beloved Son' to '*this is* my beloved Son' (3.17) makes the event more public than does Mark's account. Since John and Jesus have engaged in conversation, it may be that Matthew is suggesting that John, and possibly the crowds, also heard the voice." Robert L. Webb, *John the Baptizer and Prophet: A Socio-Historical Study* (Sheffield: Sheffield Academic Press, 1991), 57.

ates the very project of a new testament. Like the prophets before him, John calls the People to repent and enacts collective transformation at the Jordan. The new ritual suggests just how different the new collective is imagined to be. Although baptism will not long be performed exclusively at the Jordan, the first baptisms must occur there in order to correctly situate the collective transformation of Israel. The setting further points to the momentous, irreversible nature of the transformation. Jesus must be baptized in the Jordan in order to resemble his prophetic counterparts and to reformulate the very course of their missions.

The scene of Jesus's baptism in the Jordan is modeled on prophetic initiation and makes use of its regal allusions. Whether, in a historical sense, John's baptism influenced Christian baptism or the Christian rite required a story of Jesus's immersion as a charter, the scene both mirrors and inverts the prophetic rites of passage at the Jordan. Each reversal of the succession scenes has a specific thematic aim. The predecessor no longer predominates because the past is no longer the privileged site of revelation. Jesus is not a metaphorical son of his predecessor; he is the Son of God. Despite the equations of John the Baptist with Elijah, the Baptist meets an ignominious end without the suggestion that he achieves immortality. The public witnessing of John's interactions with Jesus confers status on John rather than on Jesus. Immersion replaces crossing as the initiatory movement because the plot of the Gospels turns on the vertical axis of heaven and earth rather than the horizontal axis of Israel and its surrounding or encroaching neighbors. For similar effect, the heavens rather than the river split. The Spirit does not originate in John or any other person, but descends directly from heaven as a declaration that the person of Jesus replaces the office of prophecy as the vehicle of mediation between God and humanity. The authors modify the type-scene is ways subtle and dramatic in order to present a landscape of redemption both familiar and surprising.

By engaging in a cross-testamental reading, I have crossed a Jordan of sorts and read succession texts from the two contiguous, distinct banks of the Old and the New Testaments. My goal was neither to insist that the Gospels co-opted themes of the Hebrew Bible nor to support a strict typological reading in which Moses and Elijah point toward John the Baptist and eventually Jesus, but to highlight the shared structure of the three scenes of initiation at the Jordan. As the parallel structure is figured into the specific contexts of the prophets' lives, distinct images of the predecessors, Moses, Elijah, and John the Baptist, emerge as do the portraits of the successors, Joshua, Elisha, and Jesus. Rather than collapsing all of the characters into one flat type, I emphasized the particular characteristics of each pair. The most cohesive element, in my reading, is the Jordan River. As a ritual site the Jordan promotes the transmission of authority along nonhereditary lines, the encounter between heaven and earth, and the translation of exceptional men from mortals to immortals. In the Deuteronomistic texts as well as in the Gospels, the Jordan is a threshold of redemption. The distinction between acts of crossing and immersing in the Jordan comes to define such redemption as either a Jewish return or a Christian rebirth.

Chapter Seven

DIPPING IN

Baptism and the State of the Body

I came to a high place of darkness and light,
the dividing line ran through the center of town.
I hitched up my pony to a post on the right,
went into the laundry and washed my clothes down.
  —"Isis," Bob Dylan

W. D. Davies has written of how Christianity "increasingly abandoned the geographic involvement of Judaism" and shattered its system of territorial borders.[1] Expressing a preexistent strand of Jewish universalism, Christianity finally got over the obsession with ethnic boundaries. Stripping this formulation of its theological triumphalism and examining its conceptual implications constitutes the next stage of analysis. The Jordan has already led to the New Testament. Now we go upstream a bit to revisit a Northern story from the Hebrew Bible and to trace its interpretation in Jewish and Christian exegesis. Davies' thesis is modified through the argument that territorial borders are not shattered in Christianity, but rather transposed onto the body. In fact, the same border system persists with a necessary exclusion of a reviled ethnic Other. The rub, for the Jews, is that they become the Christians' Moabites. Rather than a geographic bar-

---

1. W. D. Davies, *The Gospel and the Land: Early Christianity and Jewish Territorial Doctrine* (Berkeley: University of California Press, 1974) 336. He does note a dialectical development in which the land retained significance as the site of the crucifixion and resurrection at the same time that its landscapes became eschatological symbols.

rier established by Abraham, the Jordan becomes for Christians a symbol of the afterlife conferred at baptism that separates them in the here and now from the Jews. The problem is not quite with Jews qua Jews, but with the resistance perceived by Christian thinkers in the Jewish refusal to undergo baptism. Through the act of not becoming Christian, the Jews are seen to perform a sickly attachment to tradition and territory. In contrast, a baptized body secures a place on the other side of the heavenly Jordan.

At the same time that the patristic and rabbinic texts in question work to draw a dividing line between Christians and Jews, both are premised on Priestly ideas of purity as well as on the conceptions of identity expressed in the Priestly national myth. Although the Priestly national myth with its borders drawn on the body exerts influence on early Christian as well as on early Jewish ideas of collective purity, the process traced in this chapter does not bear out a direct line of inheritance. In fact, the Priestly ideas in question are mediated and modified in 2 Kings 5, a text of Northern provenance collected in the Deuteronomistic history. 2 Kings 5 interacts with Priestly ideas of purity, in particular the ritual of purification for the leper in Leviticus 13–14, as it depicts an Aramean general healed of leprosy in the Jordan River.

2 Kings 5

In order to see how Christian and Jewish exegetes realign biblical binaries of self and Other by projecting them onto the body, we turn to 2 Kings 5, a text concerned with the fact that the Jordan is an insufficient buffer between Israel and Aram. The protagonists include Elisha the prophet, a leprous Aramean general named Naaman, a deceptive disciple of Elisha's named Gehazi, and the anonymous yet knowing servants of Naaman. If the story of Jacob establishing a frontier east of the Jordan concedes to Israelite proximity to Aram while also expressing the desire to mark its distinction from Aram, then that of Elisha and Naaman admits to the penetrability of Israel and attributes magical properties to the Jordan exactly as its powers of delimitation are shown to be lacking.

The opening scenes portray the intermixture of Israel and Aram in terms of a series of mix-ups; no one occupies the correct position and nothing is as it should be. Naaman, introduced by a chain of descriptors uncharacteristic of the adjective-averse biblical prose, is "the general of the king of Aram's army, a man of standing before his lord with an elevated positioned gained because God had given Aram victory through him, and a valiant warrior with leprosy" (2 Kings 5:1). The first thing out of place in the description is that Yahweh, Israel's God, has sided with Aram and made Namaan victorious. The Deity's seemingly perverse alliance conforms to the paradigm through which biblical writers account for Israel's defeats. Israel is victorious when deserving of

God's intervention and defeated when its inattention to the terms of covenant cause God to withdraw. All the same, the delivery of Israel into Naaman's hands suggests a general state of disorder in Israel's relationship with God. Naaman, afflicted with leprosy, is himself in a state of disorder. Leprosy is cause for quarantine; according to Leviticus, the leper "is in a state of social death, cut off from human society and from the ritually available presence of God."[2] Where Naaman should be removed from society human and divine, he instead takes center stage. The Syrian general constitutes an exception: the Priestly rules, although operative, are not applied in this Northern, prophetic text and, even if they were operative, could not apply to a foreigner.[3]

The text then introduces an anonymous "young girl from the land of Israel" captured in an Aramean raid and now serving Naaman's wife. Because the girl is out of place, her local knowledge commands authority. She tells her mistress, "if only my lord could go before the prophet in Samaria, then he would cure him of his leprosy" (2 Kings 5:3). The sudden value of knowledge out of place becomes apparent when Naaman hurries with her advice to the king of Aram, who then dispatches Naaman with a letter and tribute to the king of Israel. Displacement makes the knowledge of a young girl pressed into servitude exotic, which, in turn, confers value. The folk register of her advice becomes evident through the fact that she knows the prophet's location, but not his name. This is in line with the populist tone of the story in which servants and a local prophet command power that mystifies generals and kings.

The girl's advice further points to a link between invasion and illness as well as between peaceful contact and healing. It seems that Naaman's leprosy is a punishment to fit his crime. Because he caused the borders of Israel to bleed, so to speak, by carrying off its members into Aramean captivity, an outbreak violates the boundary of his skin.[4] Leprosy, as Mary Douglas first made clear, is matter out of place. "The breach of the body's containing walls evidenced

---

2. Ron Hendel, "Analogy in Priestly Thought," *Journal of Ritual Studies* 18 (2004): 173.

3. "Gentiles also are simply profane (i.e., not consecrated to God). If Gentiles engage in immoral and idolatrous activity, as the Canaanites did, they are not only profane but also defiled with a moral impurity. If they do not engage in such activities but lead a moral and God-fearing life (as do *gerim* and even foreigners like Naaman the Aramean), they are (morally) pure, though still profane," Christine Hayes, *Gentile Impurities and Jewish Identities: Intermarriage and Conversion from the Bible to the Talmud* (New York: Oxford University Press), 43–44.

4. In the Hebrew Bible, leprosy is not the flesh-eroding Hansen's disease or what we think of as afflicting those quarantined in leper colonies, but an outbreak on the skin that can disappear. That the term "leprosy" is an inaccurate translation of צרעת is well-chronicled by Jacob Milgrom, *Leviticus 1–16: A New Translation with Introduction and Commentary*, the Anchor Bible (New York: Doubleday, 1991), 816–20. He translates it as "scale disease" and approximates that psoriasis is the closest contemporary condition. Still, "one can say with equal assurance that the identification of צרעת is uncertain." Ibid., 817.

by escape of vital fluids and the failure of its skin cover are vulnerable states which go counter to God's creative action when he set up separating boundaries in the beginning."[5] As Naaman breached national boundaries, so leprosy violates the integrity of his body. Healing and purification will result from a nonmilitary encounter with his Israelite rivals. Borders are thus depicted as necessary for collective health and safety at the same time that the crossing of such borders can have a salutary effect.

The captive in Naaman's household sets him on the path to restoration. When the girl's folk advice moves through the official channels, the result is comedy. The king of Aram acts swiftly on the advice "of the girl from the land of Israel" (5:4), where the king of Israel becomes unhinged when the leprous Naaman comes before him in search of a cure. Ripping his clothes, the anonymous king of Israel demands to know, "Am I God who deals death and gives life that this one sends me a man to cure of leprosy? You should all know and see for yourselves that he is seeking a pretext against me" (2 Kings 5:7). The king anxiously imagines that his failure to cure Namaan will result in further invasion and casualties. When word gets to Elisha the Prophet, the intended recipient, he measures the king's ineffectualness against the prophet's superior powers. "Why have you torn your clothes?," he challenges the king, "let him come to me and know that there is a prophet in Israel" (2 Kings 5:8). Against the current of official transmission, Naaman reaches Elisha as the servant girl had intended.

Naaman arrives like Pharaoh "with his horses and chariots," yet halts at the threshold of Elisha's home like a supplicant (2 Kings 5:9). Elisha does not go out to meet him, sending instead a messenger with the instruction, "Go and bathe in the Jordan seven times and your flesh will be restored and you will be purified" (5:10).[6] The servant girl, the king of Aram, and Naaman all use the phrase "to be cured of leprosy (אסף מצרעתו)," but Elisha employs the language of purity: "wash ... and you will be clean" (ורחצת ... וטהר).[7] Elisha directs

---

5. Mary Douglas, *Leviticus as Literature* (Oxford: Oxford University Press, 1999), 190.

6. When Naaman travels to Elisha, he comes no closer than the threshold of the prophet's house פתח-הבית as the leper is stationed by the priest at the threshold of the Tabernacle פתח אהל מועד (Lev 14:11). Unlike the priest of Leviticus who must examine the leper before pronouncing him unclean, Elisha does not go out to see Namaan. Like the leper undergoing purification in Leviticus 14, Naaman washes (רחץ) his flesh in water in order to become clean (טהר) (Lev 14:9; 2 Kings 5:10). In contrast to the priest who performs the steps of the cleansing ceremony (Lev 13:4, 31, 34; 14:7), Elisha instructs Naaman to wash himself in the Jordan in order to be clean (2 Kings 5:10, 13, 14). The seven lustrations (שבע פעמים) (2 Kings 5:10) echo the seven days and seven sprinkles of Leviticus (שבע פעמים) (Lev 14:7).

7. Zakovitch notes that the narrator "substitutes the verb רחץ (wash) of the prophet's words to the verb טבל (dip) in order to draw an additional association to the Priestly ritual surrounding the leper (Lev. 14:17)." Yair Zakovitch, *Every High Official Has a Higher One Set Over Him: A Literary Analysis of 2 Kings 5* (Tel Aviv: Am Oved Publishers, 1985), 69–70.

Naaman to retrace a number of his steps and immerse in a border delineating Israel and Aram. Because Naaman crossed the Jordan to attack Israel and again to seek healing, his immersion involves the bridging of a border and a symbolic reconciliation. This, it seems, is a source of his initial resistance. As Naaman rails against Elisha, he expresses disappointment in Elisha's lack of ritual performance and articulates a sense of national superiority.[8] In terms of ritual performance, Namaan expected a kind of shamanistic engagement: "I said to myself he will come out, stand, call out in the name of Yahweh his God, wave his hand over the source and cure the leprosy" (5:11).[9] Naaman thus disparages the so-called Israelite holy man and then moves on to belittle the Jordan. "Aren't the Avanah and the Parpar, the rivers of Damascus, better than all the little streams of Israel? Couldn't I bathe in them and be purified?" (2 Kings 5:12). The fulmination expresses two dimensions of nationalistic sentiment. The first is a sense of national superiority; since Aram is more impressive in terms of natural resources as well as military might, what can the small and weak Israel offer him? The second sentiment correlates purity and national integrity by imagining purity as a state of being that can be offered to a national only by the features of his homeland.[10]

Naaman's servants, no doubt accustomed to making concessions, placate him with an alternative way of understanding Elisha's recommendation. Addressing him as "father," they reason that since he would certainly undertake a difficult task recommended by the prophet, the smaller challenge of immersion should not be rejected. Their most effective rhetorical strategy is to omit reference to any specific body of water as they paraphrase Elisha: "All he told you to do was wash and become clean" (2 Kings 5:13). Stripped of a specific geographic or national referent and presented by his underlings, the instruction poses no threat.

---

8. "The writer focuses on the discomfiture of the haughty but ailing field marshal by offering a rare biblical look into the thoughts of this, until now, silent character . . . As a 'great man,' he had imagined a 'great thing.' All at once the blank figure of Naaman is shaded in, and we suspect that Elisha's instruction, in its simplicity, is designed to cure this arrogant Aramean of more than his leprosy." Robert L. Cohn, *2 Kings* (Collegeville, MN: Liturgical Press, 2000), 37.

9. Naaman's frustration highlights the fact that "bathing in the Hebrew Bible involves no incantations or prayers, never mentions water as a source of life and death, offers little discussion of the nature of the water, and separates bathing from healing." Jonathan David Lawrence, *Washing in Water: Trajectories of Ritual Bathing in the Hebrew Bible and Second Temple Literature* (Leiden: Brill, 2006), 7.

10. National self-portrayal in physical terms aims to lend the nation the qualities of being "organic," subject to stages of growth and decline and threatened by "alien 'bodies.'" Jonathan Boyarin, "Space, Time, and the Politics of Memory," in *Remapping Memory: The Politics of TimeSpace*, ed. Jonathan Boyarin (Minneapolis: University of Minnesota Press, 1994), 25. Competition, conquest, and exclusion in such a scheme are all justified as a manner of prophylaxis.

A convinced Naaman goes down to the Jordan, then immerses in the river seven times "according to the Man of God's word" (5:14). As Naaman is not only healed and purified but also restored to a youthful state, the little stream in the vanquished land of Israel reveals an unsuspected potency.[11] A sense of Israelite superiority is further conveyed when Naaaman returns to Elisha proclaiming that after his immersion he knows that God resides nowhere but in Israel. Now that Naaman returns to Elisha as something of a believer, he stands before him and beseeches Elisha to receive material tribute.[12] However, Elisha's stance remains relatively unchanged. He finally invokes the name of Yahweh his God when refusing Naaman's gift, but otherwise resists engagement. His scruples are admirable particularly in contrast to his conniving disciple, but one wonders what else is embedded in Elisha's distance.

Despite the fact that Elisha rejects his gifts, Naaman submits a request. "Can your servant not be given two mule-loads of earth because your servant will no longer make offering or sacrifice to gods other than Yahweh?" (2 Kings 5:17).[13] Naaman does not ask to remain on Israelite ground, but rather to take some of it with him. Along with his physical restoration, he has grasped the Israelite God's demand of fidelity as well as the distinction between the clean land where sacrifices can be made and unclean land where they are anathema. Since he cannot stay in Yahweh's domain, Naaman hopes to transport its soil and build his own altar east of the Jordan.

In order for Naaman to fashion an identity of a God-fearing Aramean, he requires more than a souvenir from Elisha. His next request is for Elisha to condone his necessary adaptations. "May God forgive your servant this one thing—when my lord comes to the temple of Rimmon to prostrate there and he leans on me and I bow in the temple of Rimmon—for my bowing in the temple of Rimmon may God forgive your servant this one thing" (5:18). Although chiastically balanced, the request is a clumsy attempt to secure God's approval for the worship of other gods. In this sense, Naaman cuts the figure of an uncertain initiate wanting acceptance while violating the fundamental tenet. At the same time, Naaman's grappling is genuine. Since he cannot cease to be Naaman the Aramean, an identity that involves bowing with his king to the national god, he attempts to incorporate some of his gratitude and

---

11. Naaman's healing also figures as a rebirth. His flesh becomes like that of "a young boy" נער קטן causing Naaman to resemble the young girl נערה קטנה who initiated the visit to Elisha.

12. As the leper "returns to the outside of his tent" וישב מחוץ לאהלו Lev 14:8) after bathing, so Naaman's flesh "returns to its youthful state" וישב בשרו 5:14) and he "returns" וישב אל־איש האלהים to finally gain audience with the Man of God (2 Kings 5:15).

13. The anticipated sacrifices for which Naaman requests the two mule-loads of earth further recall the sacrifices which follow the leper's ablutions (Lev 14:10, 12, 13, 21–25). Unlike the hypothetical leper, Naaman, who never offers a sacrifice, never "returns" to the camp of Israelite worshippers.

recognition of Israel's God into his postimmersion existence. Elisha releases Naaman, without a gift of earth, assistance or judgment, with the terse blessing, "go in peace" (2 Kings 5:19).

In the schema articulated here national identity, nearly impossible to sever from ethnic and religious identity, is imprinted on the body and cannot be effaced by geographic dislocation or purifying rituals. The body is thus like the nation insofar as both can be captured or assimilated while not undergoing a substantive change. That Elisha never takes up the request indicates, as many commentators have noticed, that no altar whether constructed by Israelites or Arameans can be built beyond the borders of the land. As the body cannot fully be assimilated by another nation, so Israel's holiness cannot be translated to another ground. Mary Douglas has noted that when Deuteronomistic sources use "a conception of the body it is the body politic."[14] The body, the body politic, and the land are all depicted as impervious to changes in material conditions. 2 Kings 5 presents a static picture in which nation and homeland persist in integral form despite political fluctuation.

Alexander Rofé reads the crux of Naaman's altar as expressing the dilemma of Israelite exiles about how to worship their God in an 'unclean' land.

> The comparison between the Jordan and the rivers of Aram in the cleansing of Naaman, whose purpose is to ascribe wondrous curative powers to the Land, signifies an idealized conception of the Land reflecting the longings of the descendants of exiles living far from their coveted homeland.

Elisha's "silence is in keeping with the deuteronomistic redactions of Kings, which forbids sacrifice to God anywhere outside Jerusalem, even in the Land of Israel, not to mention in foreign lands." Rofé's interpretation of Elisha's silence as a kind of official prohibition is more convincing than his claim of an "original form" of the story that sanctioned Naaman's altar as a means of permitting the Israelite exiles "to erect altars and sacrifice to God on foreign soil ... [with] earth imported from the Land of Israel."[15] But if "actual" Israelites were culpable, why go to the trouble of fashioning a story about an Aramean? The story of the east bank altar built by the two and a half tribes (Josh 22:10) forbids such constructions on the part of Israelites; the story of Naaman, while reinforcing the Jordan as the border between permitted and forbidden practices, makes another point about the nature of a border. The porousness of the Jordan prevents it from definitively differentiating nations.

---

14. Douglas, *Leviticus as Literature*, 14.
15. Rofé, *Prophetical Stories*, 131.

The national posturing of Naaman is parodied and then subverted when the little Jordan cures him and restores his youth. Instead of distinguishing Israel, the Jordan serves as a kind of magical gateway to a holy land.

Anyone can enter through the gateway of the Jordan and recognize the sacred nature of the land to the west. There is no national prerequisite, according to the text, for appreciation of Israel's land and God. However, such appreciation doesn't change much of anything: Naaman's revelation has no impact on his position as Naaman the Aramean (5:20). Although it has no extant text or precedent to contest, the story of Naaman can be labeled an anti-conversion tract. Revelation and even recognition of the God of Israel exert no influence on identity.[16] This theme is consistent throughout the Hebrew Bible.

> Just as there was no established mechanism by which to allow outsiders to become insiders . . . there was no established mechanism by which to recognize gentiles who had come to respect Israel's God. The narrators of the Hebrew Bible knew, or knew of, such gentiles . . . But in none of these texts, even in the eschatological visions, is there a sense that non-Israelites somehow become Israelites through acknowledging the God of the Israelites.[17]

Naaman is miraculously healed in the Jordan, where he becomes cognizant that God resides in the land of Israel alone, yet nothing changes. He returns home, occupies the same position, and the wars between Aram and Israel rage on. The story establishes the land of Israel's holiness as a kind of objective fact while maintaining that the exit of Israelites and the entrance of foreigners exert no effect. The land is holy irrespective of who dwells there. An Israelite is still an Israelite outside of the land and an Aramean still an Aramean inside of it, although only an Israelite can sacrifice to God and such sacrifices can be offered only west of the Jordan. Exilic logic and exilic longing are expressed in tandem.

---

16. Other scholars have failed to acknowledge this point. Reading anachronistically, most have taken Namaan's requests as signs of his conversion. "This is probably an early example of conversion to Yahwism." Webb, *John the Baptizer*, 104. "That this declaration (v 15) was a confession of monotheistic belief is clear from what follows: Naaman decides to serve the Lord alone, even in Aram (v 17), and requests His pardon for the times when his official duties will require him to participate in the worship of another god (v 18)" Rofé, *The Prophetical Stories*, 127–28. Robert Cohn calls the story's climax a conversion and sees it as balancing halves of the story: "Naaman has been both cleansed and 'converted' and his journey to Elisha is now balanced by his departure." Robert L. Cohn, "Form and Perspective in 2 Kings 5," *VT* 33 (1983): 171.

17. Shaye Cohen, *The Beginnings of Jewishness: Boundaries, Varieties, Uncertainties* (Berkeley: University of California Press, 1999), 131.

As if to balance the piety of an outsider with the corruption of an insider, Elisha's disciple Gehazi chases after Naaman to extort the tribute offered Elisha. Gehazi's plot sprouts like Shakespeare's Richard III with an interior monologue critiquing his master's behavior. After the self-justifying critique, Gehazi pursues Namaan. As Naaman steps out of his chariot to hear out Gehazi, the difference between the disciple who faithfully watched his master ascend in a chariot of fire and the disciple seduced by the luxury of an Aramean general's chariot comes into stark relief as proof that prophecy can rarely be transmitted. Gehazi spins a tale of two humble members of Elisha's prophetic band in need of one talent of silver and two sets of clothing. Naaman, no doubt perceiving the request as a manner of acceptance, happily provides two talents of silver along with the two sets of clothing. Gehazi stashes his booty and returns to Elisha.

Upon his return, Elisha asks Gehazi where he has been and Gehazi dissembles by claiming that he has not gone anywhere. In a sad echo of the disciple's vision of his master lifted to heaven in a chariot, Elisha informs Gehazi that he has seen the whole transaction. "Did I not perceive that a man stepped out of his chariot to receive you? Is this the time to take silver, to take clothing, olives, vineyards, sheep, cattle, servants, and maids?" (2 Kings 5:26). One wonders about Elisha's time, in which extorting material goods is particularly egregious. Perhaps it is not the time because Israel is humbled by Aram or not the time because of the economic disparity in Israel, but all the same Gehazi is caught and condemned by the transfer of the very leprosy removed from Naaman. Gehazi is cursed with eternal leprosy that will cling to his descendants after him, never to be dispelled by the waters of the Jordan. As a healed Naaman departed from Elisha's presence, so does an afflicted Gehazi. The point brought home by this final scene reinforces the notion that insider and outsider do not correlate with good and bad; the fearsome general can be God fearing and the prophetic disciple scheming. Since leprosy seems to be a punishment for reaching beyond borders and norms, it is all the more curious that Namaan is granted a cure denied Gehazi.[18] Betrayal of a familiar

---

18. Yair Zakovitch reads Naaman's story in the context of other named lepers, Miriam (Num 12:1–17), Uzziah (2 Chr 26:17–21), and Gehazi (2 Kings 5:27), and suggests that all share a "distinctive punishment visited upon those who fail to recognize their subordination to those with higher status." Zakovitch, *Every High Official Has a Higher One Set Over Him*, 23. Zakovitch analyzes the story through the leitwort לפני (before) that appears seven times in the story. As the plot twists, the restoration to stability occurs only when each character learns his place in the elaborate hierarchy in which "every high official has a higher one set over him" and over which the Hebrew God reigns supreme. Zakovitch concludes that, in the biblical estimation, a healthy individual, and indeed a healthy society, is one that upholds rigid hierarchies. I argue that the story, in fact, subverts hierarchies and instead invests in the idea of an indelibly sacred land.

code appears graver than crimes of war. The failed prophet is worse than the successful enemy general. Gehazi's eternal leprosy becomes, like the land's innate holiness, a state that cannot be changed.

As received in early Christianity, the story of Naaman's miraculous immersion multiply influences ideas about how baptism affects the body. The connection between the Naaman story and the baptism ritual becomes apparent through the translation of the Hebrew verb "to dip" טבל used for Naaman's immersion into the Greek βαπτιζο (baptize) in the Septuagint.[19]

> The verb טבל 'to dip', is usually translated in the LXX by the verb βαπτω 'to dip'. However, 2 Kgs 5.14 is the only place where טבל is translated by the intensive form βαπτιζο. The employment of this verb in the LXX form of this story may have suggested the use of the Jordan River as a site for immersions in later Judaism. It also might have influenced early Christian thought to use this form of the verb to refer to its own rite.[20]

The translation of "dip" into "baptize" bespeaks an understanding of Naaman's immersion in the Jordan as a transformative experience and paves the way for the reading of Naaman as a convert. Since, as discussed above, Naaman's social position and identity are not transformed by his Jordan bath, its salubriousness becomes the evidence of baptism's transformative power.

As 2 Kings 5 interacts with and reinterprets Leviticus 13 and 14, so the Gospel accounts of John the Baptist interact with and reinterpret both the Priestly and the Deuteronomistic texts.[21] The Gospels establish a transition for the baptized from a state of impurity to one of purity while transforming the meanings of "impure" and "pure." "Pure" primarily indicates a state of moral righteousness that can be achieved through a combination of repentance and immersion. This radical recontextualization of Priestly ablutions that dissolve states of physical impurity marks an important stage in the transformation of the concept of purity. The association of impurity with sin

---

19. The Septuagint translates טבל "he dipped" in verse 14 "he went and dipped in the Jordan seven times" as εβαπτισατο from the stem βαπτιζο.

20. Webb, *John the Baptizer*, 103–4.

21. My sense of baptism's origins comes very close to that of Adela Yarbro Collins. Collins proposes an origin in the combination of the Priestly concept of ritual purity and the prophetic notion of cleansing from sin (Isa 1:16–17, Ezek 36:25–28). "The Origin of Christian Baptism," *Studia Liturgica* 19 (1989), 35. I agree with Collins that the ritual of baptism represents Priestly ideas refracted through a prophetic lens and understood according to concurrent notions of purification from sin, but also see Leviticus 13–14 and 2 Kings 5 as particularly influential. The apocalyptic dimension of John's description of his baptism can be attributed to both the prophetic tradition and to eschatological traditions surrounding the Jordan River.

and purity with atonement is already a trope of prophetic writing (Isa 1:16; Ezek 36:25; Zech 13:1), which has corollaries in Psalms (18:21–25, 24:3–4, 26:6, 51:4, 9) as well as the Holiness Code. Although the terms for ritual and moral purity are synonymous in the Hebrew Bible, they remain categorically distinct. Ritual impurity is an expected part of the Israelite condition that can be checked through prescribed symbolic acts where moral impurity is a state of imbalance that can impact the collective and must be attended to through repentance and the reinstitution of just systems. Jonathan Klawans has documented the conflation of moral and ritual impurity among Second Temple sectarian groups like that at Qumran.

> At Qumran, sin was considered to be ritually defiling, and ritual defilement was assumed to come about because of sin. Sinners had not only to atone, but also to cleanse themselves of the ritual impurity their sins produced... In short, what were, in the Hebrew Bible, the independent concepts of ritual and moral impurity have become, at Qumran, fully intertwined.[22]

Repentance involved claiming and annulling transgression while immersion dispelled the physical residue of sin. The slippage between the two kinds of impurity becomes characteristic in New Testament texts such as those describing baptism, where immersion offers ritual purification from the moral impurity of sin. As repentance is coupled with baptism, immersion is literally understood to dissolve past sins and the physical state associated with them.

In the contexts of the Gospels and early Christian interpretation, Naaman's affliction and sickness in general are understood as outward signs of sin. "The physical consequences of mortality—sickness, bodily decay, and disintegration—were the direct results of that sin and therefore always the indication of sin's presence."[23] Baptism provides the occasion for moral sanctification as well as physical healing. The Baptist, for example, enjoins his followers to undertake "a baptism of repentance for the forgiveness of sin" (Mk 1:4). Those who emerge from the Jordan have satisfied the Baptist's command to repent and achieve a new state of purity that allows them access to the Kingdom of Heaven. In this sense, baptism recalls the river ordeals of the ancient Near

---

22. Jonathan Klawans, *Impurity and Sin in Ancient Judaism* (New York: Oxford University Press, 2000), 88.

23. Susan Ashbrook Harvey, "Locating the Sensing Body: Perception and Religious Identity in Late Antiquity," in *Religion and the Self in Antiquity*, ed. David Brakke, Michael L. Satlow, and Steven Weitzman (Bloomington: Indiana University Press, 2005), 149.

East in which the guilty sink and the innocent emerge.[24] The alternative to salvation in water is punishment in fire. John the Baptist damns the Pharisees and Sadducees, whom he dubs "you brood of vipers," to the alternative fate of fiery obliteration (Mt 3:7–10).[25] Rendered impure through their actions, denied baptism and relegated to the wrong side of the eschatological divide, destruction is the destiny of these descendants of Abraham. Baptism serves as the dividing line between the bodies destined for eternity and those slated for death. Jesus makes the correlation between sin and sickness explicit: "Those who are well have no need of a physician, but those who are sick; I have come to call not the righteous but sinners" (Mk 2:17). In this formulation, the sick are sinners and the sinners are sick and both species of malady can be cured by the call of Jesus.[26] As Jesus travels through Israel, he confronts a society characterized by illness at every turn. Since the removal of sin by Jesus functions as a healing, the dissolution of sin in baptism may also serve to heal the infirm body of Israel or prepare it for another stage of healing.

The restoration of Naaman's body to "the flesh of a young boy" (2 Kings 5:14) also becomes incorporated as a baptismal motif. As Jesus informs his followers: "Truly I tell you, unless you change and become like children, you will never enter the Kingdom of Heaven" (Mk 10:15; Mt 19:14; Lk 18:17). In the Gospel of John and in Christian tradition, rebirth occurs over the course

---

24. In the Laws of Hammurabi, there are two instances in which the river judges the accused. In both cases, the ability to swim is a sign of innocence and drowning is a sign of guilt. An accused sorcerer "shall go to the divine River Ordeal, he shall indeed submit to the divine River Ordeal." If he drowns, then "his accuser shall take full legal possession of his estate." If he survives, then his accuser is put to death and the accused "shall take full legal possession of his accuser's estate." Laws of Hammurabi ¶ 2, Martha T. Roth, *Law Collections from Mesopotamia and Asia Minor*, ed. Piotr Michalowski (Atlanta: Scholars Press, 1997), 81. An accused adulteress must also "submit to the divine River Ordeal for her husband." LH ¶ 132, Roth, *Law Collections*, 106. Frank Moore Cross observes that "the cosmic river springing up from the underworld is also 'Judge River,' as in Mesopotamia the place of the river ordeal, the place of questioning or judgment, as one enters the underworld." Cross, *From Epic to Canon: History and Literature in Ancient Israel* (Baltimore: Johns Hopkins University Press, 1998), 88. In the Gospels of Matthew and Luke, where John the Baptist promises a future baptism of "Holy Spirit and fire" for those baptized in water and a burning "with unquenchable fire" for the nonbaptized, there is some sense that the Jordan anticipates a cosmic/apocalyptic river. In later Christian texts the association of the Jordan and such a fiery river become more explicit.

25. As scholars have noted, it is highly unlikely that the Pharisees and Sadducees would venture out to the Jordan together. The animosity between these groups often ran as high as that between the Jesus people and the Pharisees. The lumping together of these two groups is the author's way of rejecting the ideologies and practices of both groups. Neither is admitted into the kingdom of heaven. "Matthew's 'Pharisees and Sadducees' may be due to a Matthean redaction, because in a number of passages he shows a heightened interest in the Pharisees. Also, Matthew adds Sadducees in 16.1–12 where his source (Mk 8.11–21) only mentions Pharisees." Webb, *John the Baptizer*, 174.

26. When Jesus heals the paralytic of Capernaum, he says, "Son, your sins are forgiven" (Mk 2:5).

of baptism.[27] "Very truly, I tell you, no one can enter the Kingdom of God without being born of water and Spirit" (John 3:5). The Pauline assurance that "in Christ Jesus you are all children of God through faith" (Gal 3:26) and the rabbinic assessment that "a convert is like a new-born child" (Yebamot 22a, 48b, 62a, 97b) are perhaps also influenced by the language of rebirth used to mark Naaman's coincident healing and recognition of the God of Israel.

Along with refiguring the nature of purity, the Gospel accounts of John's baptism also innovate by establishing the Jordan as a kind of social boundary between "pure" and "impure" groups within Israel. The Jordan, the biblical boundary between Israel and Others, is refigured in the accounts of baptism as the dividing line between the damned and the saved within Israel as evident in a statement like "the one who believes and is baptized will be saved; but the one who does not believe will be condemned" (Mk 16:16).

> John's baptizing ministry, therefore, created a fundamental distinction between two groups of people: those who received the repentance-baptism and those who were unrepentant; those who were forgiven and those who were unforgiven; those who were purified and those who were unclean; those prepared to receive the expected figure's ministry of restoration and those who would be judged and face that figure's wrath. While John addressed his message to all Israel, the effect of that message was to divide them into these two sets of people.[28]

The difference between those who have and those who have not immersed in the Jordan becomes as fundamental a distinction as that between Israel and the nations. As the Jordan becomes the boundary between the baptized and the unbaptized, the river loses significance as the periphery of the land of Israel.

Naaman is named explicitly in the Gospel of Luke as an example of the outsider more deserving of redemption than the native Israelite. As proof that "no prophet is accepted in the prophet's hometown," Jesus expounds, "There were many lepers in Israel in the time of the prophet Elisha, and none of them was cleansed except Naaman the Syrian" (Lk 4:24, 27). For Jesus, the

---

27. "In taking over John's baptism the early Church was able to adopt its specific characteristic, 'a baptism of repentance for the forgiveness of sins,' in order to acknowledge the new beginning established through baptism. This new beginning was understood as a new birth, a rebirth or the beginning of life." Hermann Lichtenberger, "The Dead Sea Scrolls and John the Baptist: Reflections on Josephus' Account of John the Baptist," in *The Dead Sea Scrolls: Forty Years of Research*, ed. Devorah Dimant and Uriel Rappaport (Leiden: Brill, 1992), 341.

28. Webb, *John the Baptizer*, 197.

statement expresses his rejection by the community of his childhood at the very moment he is embraced by other Galilean communities. If the insiders will not accept him, then Jesus will take his healing elsewhere. For the author of Luke, the statement authorizes the mission to the Gentiles. As a prooftext, it is one among a number of devices that correlate Jesus and Elisha. Infusion with the Holy Spirit (Lk 4:1, 14) enables Jesus to heal outsiders (4:18 by way of Isa 61:1) in the manner in which Elisha healed Naaman (4:27). Jesus subsequently heals a leper, not ostensibly an outsider, through the laying of hands—a technique of which Naaman would have likely approved—and the words, "Be made clean" (Lk 5:12–14).[29] Jesus transmits word of his healings and cleansing of leprosy to John the Baptist as evidence that he is "the one who is to come" (Lk 7:18–22). The removal of sickness is a sign that Jesus has come to dispel the sin of the human condition. The Jordan crossed at baptism incorporates a singular body into a collective while at the same time designating it as a site of individual redemption. Baptism secures a symbolic border around the body that cordons it off from social ills and protects it from the contamination of sin. This trend in which the singular body "stand(s) for (the) bounded system" of a community defined by practice continues in earnest in Christian and Jewish exegesis.[30]

Interpreting 2 Kings 5

Naaman's body, the bodies of contemporary Israelites, and the body of Elisha become devices for differentiating among groups as well as practices. Early Christians and rabbinic Jews have much to say about one another's bodies. Both groups tend to represent the other's collective body as ailing in order to define themselves, in contrast, as healthy. These interpretations also portray Jordan immersion as a transition from sickness to health and read the physical status of the body as an indicator of identity. The focus on the construction of healthy vs. ailing bodies indicates something of a return to the early Boyarin thesis that "what divides Christians from rabbinic Jews is the discourse of the body."[31] Where Boyarin examined the distinguishing techniques of representing the body as a sexual agent or as a bulwark against sexual agency, here the representation of the body as either vigorous or afflicted is at question. As Boyarin has also argued, exegetical polemics about bodies are ciphers for

---

29. Bruce Chilton sees this as an explicit parallel between Elisha and Jesus. As Elisha could heal a leper because he possessed the spirit of Elijah, so Jesus could heal a leper because he was filled with the Spirit. Chilton, *Jesus' Baptism and Jesus' Healing*, 76.

30. Mary Douglas, *Purity and Danger: An Analysis of the Concepts of Pollution and Taboo* (New York: Routledge, 1991), 116.

31. Daniel Boyarin, *Carnal Israel: Reading Sex in Talmudic Culture* (Berkeley: University of California Press, 1993), 2.

interpretive practices. The ailing flesh of the Jew for the Christian embodies the drive to understand the letter of the Law and the physicality of midrashic engagement with language. The staging of the Christian body in midrash similarly works as a rabbinic mode of contending with allegoric practices of reading, perceived as dispensing with apparent meaning in the name of spiritual significance. The use of represented bodies as markers of borders assumes a kind of urgency in an atmosphere in which early Christians and Jews share so many physical sites. These sites include the holy land, cities, and the Jordan River as well as the texts of Hebrew Scripture.[32]

The rabbinic body, blessed and alive on earth, adheres to the purity laws and abstains from a litany of potential contaminants. The Christian body, poised for a more glorious avatar, is defined by the immersion of baptism. These practices that identify bodies simultaneously construct them as bearers of ideology. At the same time that water immersion is a shared practice of Christians and Jews, the immersions are distinguished as being either consistent with a christological pattern or consistent with Jewish law. Among the strategies of distinguishing between the two is the act of interpreting Naaman's Jordan immersion in 2 Kings 5. At stake in such interpretations is how much figuration can be discerned in the biblical texts or alternately how much the story supports the system of ritual purity. Naaman's freshly "baptized" enthusiasm and Elisha's distancing reluctance are reflected and reproduced in the early Christian and rabbinic responses to and interpretations of the story. While baptized interpreters celebrate Naaman's healing as their precedent, rabbinic readers echo Elisha's instruction to the newly baptized to "go in peace."

True Water, Saving Water

To Origen, the third-century Christian exegete, the Jordan River Valley is a location of redemption because the river runs through it, while the river in which the believer is inaugurated into the Kingdom of God is, allegorically, Christ. To Origen, Jesus is the Jordan that dissolves past sins and restores the soul:

> For, what other places ought the Baptist have made the rounds of except the neighborhood of the Jordan? If anyone wanted to do penance,

---

32. On the contest to claim the Holy Land and its biblical memories, see Andrew Jacobs, *Remains of the Jews: The Holy Land and Christian Empire in Late Antiquity* (Stanford: Stanford University Press, 2004); Hagith Sivan, *Palestine in Late Antiquity* (New York: Oxford University Press, 2008); Joan E. Taylor, *Christians and the Holy Places: The Myth of Jewish-Christian Origins* (New York: Oxford University Press, 1993).

a bath of water was available. Then, "Jordan" means "descending." But the "descending" river of God, one running with a vigorous force, is the Lord our Savior. Into him we are baptized with true water, saving water. (*Homilies on Luke* 21:3–4)[33]

The Baptist chose "the whole region lying along the Jordan" as the location of his ministry because there the intention to repent could coincide with baptism, the enactment of repentance. The Jordan's etymological significance as the river that "descends" (ירד) points to the descent of God incarnate to earth and its "vigorous force" that impacts landscape signals the force of divine descent on history. Where in 2 Kings 5, the Jordan, a porous boundary between the national bodies of Israel and Aram, displayed the ability to heal the human body, in Origen's estimation, the Jordan is Christ's body. "Holy space seems to have been 'transubstantiated' into a community of persons, the Body of Christ."[34] The decision to join one's body to Christ functions like the Jordan as a border demarcating redemption.

According to Origen, baptism is the transition from an inner landscape of "stain" to one of "salvation." Although the process is internal, the movement is as spatially significant as Abraham's migration from his home to the land shown to him by God.

> If anyone wants to be baptized, let him go out. One who remains in his original state and does not leave behind his habits and his customs does not come to baptism properly. To understand what it is to go out to baptism, accept the testimony and listen to the words by which God speaks to Abraham: "Go out of your land," and so forth. (*Homilies on Luke* 22:5)

"Going out" to baptism requires renunciation of a former state with its attendant "habits" and "customs." Such renunciation then is a precondition for baptismal transformation. The baptized can retain previous affiliation no more than Abraham could remain in his land, birthplace, or paternal abode. Like Abraham, the baptized must sever himself from a previous state of being in order to be accordingly blessed. There is an implicit correlation between the geography of the Promised Land and the state of salvation: demarcated by

---

33. Just as the Holy Spirit descends upon Jesus at the Jordan in the Gospels, Origen views Jesus himself as a "descending" river. All translations of Origen are taken from Origen, *Homilies on Luke, Fragments on Luke*, trans. Joseph T. Lienhard, S.J. (Washington, D.C.: Catholic University of America Press, 1996).

34. W. D. Davies, *The Gospel and the Land*, 185.

the Jordan, both require a departure from a prior location. Origen's Jordan, of course, is not the one set in the Syro-African Rift, but all the same it delineates a past and future heavy with the connotations of stain and salvation. The operative analogy is that as Abraham reached a physical site of redemption in the form of sacred territory, so the baptized achieve salvation in the body through the physical act of baptism.

Once the fusion with Christ's body becomes the definitive boundary crossed, the biblical motif of the Jordan as border is transposed to the next world, where Jesus stands in a river of fire near the flaming sword administering the baptism in fire promised by John the Baptist.

> At the Jordan River, John awaited those who came for baptism. Some he rejected, saying, "generation of vipers," and so on. But those who confessed their faults and sins he received. In the same way, the Lord Jesus Christ will stand in the river of fire near the "flaming sword." If anyone desires to pass over to paradise after departing this life, and needs cleansing, Christ will baptize him in this river and send him across to the place he longs for. But whoever does not have the sign of earlier baptisms, him Christ will not baptize in the fiery bath. (*Homilies on Luke* 24:1)

Jordan baptism is Origen's shibboleth, a password imprinted on the body to be deciphered at paradise's gates by Jesus himself. The unbaptized body, in Origen's view, will remain on the other side of the river, relegated to the wrong side of the eschatological divide. Redemption, in Origen's view, is a realm entered by way of three gates: water baptism in this world; the corollary baptism "with the Holy Spirit" (Luke 3:15); and baptism in a fiery river by Christ on the path to paradise. In all three instances, the rite is not a universal right, but is achieved through confession, repentance, and faith.

When Origen retells the story of Naaman in the Jordan, he emphasizes that there are two types of ailing bodies: those of Israel and those of "the gentiles." Specifically, the story illustrates the Lukan maxim that "no prophet is accepted in the prophet's hometown"; the supporting examples are Elijah, who performed miracles for a widow in Zarephath near Sidon, and Elisha, who cured the leprous Aramean, not the "many lepers in Israel" (Lk 4:27). Origen reads the verse: "Recall, too, the many lepers in Israel in the time of Elisha the prophet; yet not one was cured except Naaman the Syrian" (Lk 4:27) as proof that Elisha can only heal the foreign body, because it alone is sufficiently pliant to be transformed by spirit. In Origen's assessment, Naaman's healing in the Jordan provides proof that Israelite lepers went uncured and thus unredeemed by the Israelite prophet or by other means.

In his own time, Origen perceives a form of spiritual leprosy lingering on the obdurate body of Israel that refuses to plunge into baptismal waters.

> Consider that right up to the present day there are many lepers in "Israel according to the flesh." Realize, in contrast, that men covered with the filth of leprosy are cleansed in the mystery of Baptism by the spiritual Elijah, our Lord and Savior. To you he says, "Get up and go into the Jordan and wash, and your flesh will be restored to you." Naaman got up and went. When he washed, he fulfilled the mystery of baptism, "and his flesh became like the flesh of a child" (2 Kings 5:14). Which child? The one that is born "in the washing of rebirth" (Titus 3:5) in Christ Jesus, to whom is glory and power for ages of ages. Amen. (*Homilies on Luke* 33:5)[35]

By upholding the past and resisting the merger with Christ, the Jews remain embodied in a flawed state. In Origen's words, "many lepers" populate "Israel according to the flesh." The inherent malady of embodied Israel is further reflected in an understanding of the Bible that is "too 'fleshly'—that is, too focused on the literal, nonspiritual interpretation of the text."[36] In contrast, bodies made Christian slough off their "leprosy" through baptism. The baptismal removal of sin and illness enables the clarity of an allegorical understanding in which the physical Elijah of Scripture is assimilated by "the spiritual Elijah, our Lord and Savior."

By identifying the agent of this healing baptism as Christ, Origen clarifies the association of Elijah and Jesus, thereby weighing in on the question of whether John the Baptist is Elijah redivivus. Capacious enough to assume most roles in the story, however, Jesus also plays the voice of Elisha and the role of the Jordan. Naaman here signifies a collective gentile body cleansed of the filth of sin and rendered Christian. His healing in the Jordan is fulfilled when gentiles collectively go out for "the mystery of baptism." At the same time, Origen argues that when the biblical character, Naaman, immerses in the Jordan, he too is baptized. When in the text of 2 Kings Naaman's body

---

35. On Origen's closing doxology that quotes I Peter 4:11, see Marc Hirshman, *A Rivalry of Genius: Jewish and Christian Biblical Interpretation in Late Antiquity*, trans. Batya Stein (Albany: State University of New York Press, 1996), 68.

36. Origen, *De principiis* 4.3.2, 6–7 (SC 268:346–52, 364–68), cited by Jacobs, *Remains of the Jews*, 62. Jacobs points out that reliance on Jewish "philological and geographical expertise in his own interpretative efforts to produce a thoroughly spiritualized interpretation of the Old and New Testaments" accompanied Origen's critique. The infirm Jews, in Origen's perspective, could still bestow authenticity on Christian interpretation. One can imagine that Origen's proximity to "actual" Jews may be what necessitated the construction of a ritual barrier between them.

becomes like that of a child, it is actually "renewed by the Holy Spirit" as described in Titus. As Naaman's immersion makes him like a child, so the baptized are reborn in water/Christ. "The mystery of baptism" then works in a transtemporal fashion: those baptized in "the present day" are transformed, thereby ensuring that they will cross the eschatological Jordan; such baptisms also cause Naaman's Jordan immersion to be fulfilled and to itself become an instance of baptism.

Origen focuses on Naaman's physical transformation while reading the body as an indicator of identity as well as of spiritual status. Through the link between the reconstitution of Naaman's body and salvation through the Holy Spirit, rebirth is promised to all those who, like the Aramean, go to bathe in the Jordan. Origen dispenses with Naaman's testimony most likely because his Jordan bath stands as proof enough that he will cross the river into salvation. As the gentile body is healed and saved in Origen's paradigm of Baptism, what happens to the Israelite body? Furthermore, what becomes of Elisha's voice and its imperative to "bathe . . . in the Jordan" now that it has been assimilated into both Jesus's and Origen's instructions to Christian believers? As do the Kohanim in Leviticus 14, Origen banishes the leprous body from his camp. Yet where the exile of Leviticus is time-bound, the exile from Origen's camp lasts for as long as the Jewish body remains Jewishly embodied.

These remarks, when combined with broader questions about the fate of the body, the fate of the past, and the fate of Jews within Christian systems of baptism and salvation, require engagement with John Dawson's critique of various contemporary "Jewish" readings of Origen.[37] Defending Origen from Daniel Boyarin's critique of the ways in which the practice of allegorical reading dispenses with literal meaning and with embodied states in the name of a transcendent spiritual truth, Dawson insists that Origen "understood quite well that the body is the inescapable site of identity—it is exactly where all the important things take place."[38] This assessment seems consistent with Origen's interpretation of Naaman's immersion in his *Homilies on Luke*. It is Naaman's flesh that becomes like a young child's, and it seems to be the bodies of believers that Jesus checks for "the sign of earlier baptisms" at the fiery

---

37. The work deals with Daniel Boyarin, Erich Auerbach, and Hans Frei. "Jewish" is placed in quotation marks here because Hans Frei was "a Protestant theologian of Jewish descent." In his review of Dawson's book, Mark Vessey comments that the section dealing with Frei is the one "in which one feels least tension between the author and his fellow modern interpreter. When in due course Frei is brought into conversation with Origen, there is no suggestion of his being taught a lesson, and the two theologians emerge from their encounter with honors roughly even." Vessey, *Bryn Mawr Classical Review* 11.16.2002, http://bmcr.brynmawr.edu/2002/2002-11-16.html.

38. John David Dawson, *Christian Figural Reading and the Fashioning of Identity* (Berkeley: University of California Press, 2002), 10.

river surrounding paradise. The body does then convey identity, but what are "all the important things" that can transpire there?

In Dawson's mind and indeed in Origen's, the important movement is a kind of ascent toward baptismal transformation, merger with Christ, and a physical and perspectival state wherein all signs point to an ongoing christological drama. The body has validity "as a complex and rich psychosomatic medium of a person's divine transformation." This physical striving toward a certain kind of change entails its own hermeneutic, that of allegory that "leads the reader toward fuller, richer embodiment by illuminating the body's irreducible spiritual dimension." While maintaining that allegorical reading leaves literal meaning in place and that the "spiritual dimension" of embodied existence in no way reduces material experience, Dawson maps an analogy between "the allegorical reader's necessary departure from Scripture's literal sense" and "her resistance to the fall of her soul away from contemplation of the *logos* into body, history, and culture." Instead, the literal text must point to a "spiritual metanarrative" and the body must always be gesturing toward the "transformative promise of Christianity." The dangers of reading text as text and not as the process of Christological unfolding are either that "one will find the narrative satisfying as it stands and learn nothing truly divine, or one will find the narrative repellent and learn nothing worthy of God. In either case, the transformative promise of Christianity will be lost."[39]

As if anticipating the response that the biblical text (and more specifically the Hebrew Bible) is not the script for "the transformative promise of Christianity" alone, Dawson identifies the readers prone to such dangers: "Simpleminded Christians (and Jewish readers) are the typical victims of the first sort of literalism, Gnostics of the second." Dawson, elsewhere a careful exegete of the parenthesis, here slips a consciously alternate Jewish hermeneutic into a parenthesis associated with the Christians too obtuse to know how to read. It is through such statements that Dawson seems to support rather than refute Boyarin's reading of Origenist allegory. Indeed, the binary opposition between the literal and the allegorical, the body and the spirit may not be as definitive as some of Boyarin's early works make them out to be—binaries ultimately never line up rigidly as opposition—but Dawson and his subject, Origen, do not have a place in their schema for the Jewish reader or for the body whose transformations follow other trajectories.[40]

---

39. Dawson, *Christian Figural Reading*, 47, 54, 61, 58.

40. Dawson, 58, 127–28. By not submitting themselves to such change, Jewish bodies never quite "become what they are." Dawson explains "what is central to Origen's vision" as such: "personal and historical human possibilities are being realized through transformations in which things increasing become what they are." Ibid., 215. What things "are" here are embodiments of spirit that need to reclaim the spiritual state through the medium of Christ.

Dawson shows that Origen does not efface the body; rather, he deems it valid insofar as it acts as a site for Christian transformation. As we have seen, the "body" of the biblical text is similarly valuable insofar as it points to the plot of incarnation, baptism, crucifixion, resurrection, ascension, and parousia. This valuing of the biblical text presents a problem for those who read it differently, particularly when such readings are deemed the "fall of the soul... into body, history, and culture." What if the body, history, and culture are in and of themselves valued as sites of revelation and planes of transformation? Is this, as Dawson would have it, necessarily a limited, partial understanding? From a textual standpoint, such a reading is indeed a problem for "(Jewish readers)," as it involves their disqualification as readers and claimants of a biblical legacy. In other words, by not reading allegorically, the Jews are distanced (at least in Origen's estimation) from the true text. The corollary exclusion of the Jewish body from salvation can be seen in the *Homily*, where "one who remains in his original state" cannot receive the proper baptism. Origen reads the rejection of those who do so in the Gospel of Luke when John the Baptist addresses the crowds as "you brood of vipers" and enjoins them not to suppose that Abrahamic descent will save them from a fiery end (Lk 3:7–9). As the Baptist repels those unwilling to relinquish prior "habits and customs," so Christ staves off those without "the sign of earlier baptisms" from paradise. Where does this leave the Jewish body "in the present day"? "Covered with the filth of leprosy" and on the wrong side of two Jordans.[41] So Dawson is correct that Origen does not spurn the body as such, but only those bodies that fail to "recover" their "most authentic dimension" through Christian spiritual transformation.[42] In the passage at hand, Origen dispenses with such bodies by barring them from eternity.

Testing the Waters

The final section of Tractate Sanhedrin in the Babylonian Talmud concerns itself with the eventual entry of "all Israel" into the World to Come as well as the exceptions forever barred from this Jewish version of eternity. Gehazi, the wayward disciple of Elisha rendered forever leprous, is among the figures forbidden from redemption in the Next World. In justifying his exclusion, the Rabbis visit the biblical text concerning Naaman's healing and Gehazi's affliction. There is ultimately no way to know whether the Babylonian Rabbis

---

41. The two Jordans may also correspond to two leprous afflictions. Dawson quotes Origen's gloss on Matthew 8:3: "In the same way as in these instances Jesus touched the leper spiritually rather than sensibly [Matt. 8.3], to heal him, as I think, in two ways, delivering him not only, as the multitude take it, from sensible leprosy by sensible touch, but also from another leprosy by his truly divine touch." Origen, *Cels.* 1.48, quoted in Dawson, *Christian Figural Reading*, 57.

42. Dawson, *Christian Figural Reading*, 63.

were aware of Origen's interpretation of the Naaman story or other similar Christian interpretations; if they did, we don't know whether the talmudic retelling constitutes a formal response.[43] Questions of time, space, and contact preclude a confident assessment of polemic.[44] However, Christian appropriations of Elisha the Prophet and allegations that the Israelite body is incomplete or somehow flawed do seem to be countered in the rabbinic narrative. When the Rabbis interpret the story of Elisha and Naaman, they display an acute awareness of the association between Naaman and baptism as well as a desire to counter it. Furthermore the leprous body excluded from the World to Come is one antithetical rather than similar to Origen's portrait of the Jewish body.

When the Rabbis retell the story of Naaman, the drama of Naaman's conversion in the Jordan is perceptibly absent. Thus, if the interpretation aims to subvert readings like Origen's, then it would be an occasion of silent polemic in which nonacknowledgment operates in place of refutation.[45] Along with potentially spinning an antibaptismal polemic through silence, the Rabbis also audibly hint that they are taking a stand against the practice of baptism. Such a hint takes the form of context: the story of Naaman is told within the framework of a discussion about the degree to which a teacher is to blame for his apostate students. One example of such a student is Jesus and another is Gehazi. So Naaman, Elisha, and Gehazi materialize along with Jesus and his alleged teacher, Rabbi Joshua ben Perahyah.[46] Elisha and Jesus are cast

---

43. The point is taken from Andrew Jacobs that construing Christian and Jewish exegetical divergences as dialogue ignores power asymmetries, particularly after the rise of Christian empire. However, I do maintain that biblical exegesis constituted the central arena of Christian-Jewish theological contest.

44. For example, centuries probably elapse between Origen's account and the written version in the Talmud. Sanhedrin 107b, the page containing the Naaman story, contains both Tannaitic and Amoraic materials. This variant seems to be later than its parallel in BT Sotah 47a, and the story of Naaman only appears in the later version. For more on issues of dating this text, see David Goldenberg, "Once More: Jesus in the Talmud," *Jewish Quarterly Review* 73 (1982): 78–86; Stephen Gero, "The Stern Master and His Wayward Disciple: A 'Jesus' Story in the Talmud and Christian Hagiography," *Journal for the Study of Judaism* 25 (1994): 287–311; Jacob Neusner, *The Rabbinic Traditions about the Pharisees Before 70, Part I*, 82–87 (Atlanta: Scholars Press, 1999).

45. The history of censorship of BT Sanhedrin 107b points to the necessity of a muffled polemic against Christianity. In some cases, the Church censored the passage. In others, "supposedly incriminating passages were left out by the Jewish printers themselves in order not to jeopardize the publication of the Talmud." Schäfer, *Jesus in the Talmud*, 132. For the varying manuscript evidence for Rabbi Joshua ben Perahya pushing away "Jesus" or in turn "one of his [R. Joshua's] disciples," see ibid., 142.

46. For an analysis of the Joshua ben Perahya and Jesus story, see Daniel Boyarin, *Dying for God: Martyrdom and the Making of Christianity and Judaism* (Stanford: Stanford University Press, 1999), 24–25.

within a triangle of stories in which the Rabbis unravel the association between Elisha and Jesus forged in the Gospels by opposing Elisha to Jesus in doublet stories with a chiastic structure. Once the Elisha-Jesus connection is severed, Elisha is repatriated into the rabbinic camp, where he functions as a metonymy for both the Rabbis and the Jewish body. Jesus then serves as a foil to Elisha and as a parallel to Gehazi, the charlatan disciple. In addition, Elisha's recommendation of a Jordan bath does not transform Naaman, rather Elisha the healer heals himself.

Like Origen, the Rabbis state a principle, "Let the left hand repulse but the right hand always invite back," then provide two narrative examples to prove it true. Where Origen employed Elijah and Elisha as figures that affirmed the principle "No prophet is accepted in the prophet's hometown," the Rabbis draw on Elisha and Rabbi Joshua ben Perahyah as counterexamples. Both "stern masters," Elisha and Rabbi Joshua double-handedly spurn the "wayward disciples," Gehazi and Jesus.[47] Because the doublet stories describing the split between masters and disciples reads so tragically, the retold story of Naaman is easily lost in the pathos. A careful reading, however, reveals a three-way mirror. In the doublet, the Rabbis portray themselves metonymically as stern masters and, in the third story, they embody the role by acting as stern interpreters who divert the practice of baptism from the Jewish system of purity with no hands at all.

Let us look first at the healing of Naaman, which appears within the Gehazi narrative.

> Rabbi Isaac said: At the same time, Elisha was sitting and expounding upon the eight unclean crawling creatures. Naaman, captain of the King of Aram's army, was a leper. A certain young girl who had been taken captive from the land of Israel said to him: "If you go to Elisha, he will heal you." When he [Naaman] arrived, he [Elisha] said to him: "Go! Immerse yourself in the Jordan." He [Naaman] said to him: "Are you making fun of me?" His servants who were with him said: "What do you have to lose? Go and try it." Naaman went and immersed himself in the Jordan and was cured. He came [to Elisha] bringing all the things that he had with him, but Elisha had no desire to take the things from him. Gehazi slipped away from Elisha, went, took what he took, and deposited it.
>
> When Gehazi arrived, Elisha saw that leprosy had broken out on his face. He said to him: "O wicked one, the time has come to receive your reward for studying the laws of the eight unclean crawling creatures.

---

47. These terms are coined in Gero, "The Stern Master and His Wayward Disciple."

May the leprosy of Naaman cling to you and to your seed forever." He went from before him leprous as snow. (2 Kings 5:27). (Sanh. 107b)

The most striking omissions in this telling are the rebirth and conversion of Naaman. Naaman is healed, yet nothing in the story attests to a miraculous or transformative dimension. After unwittingly heeding Elisha's instruction, Naaman seems to feel no differently about the prophet and his God. In fact, upon his return from the river, Naaman says nothing.[48] Naaman's skin is healed, yet not purified and made youthful.[49] As a gentile, Naaman is not subject to the laws of purity and therefore cannot be purified in the Jewish ritual sense. The fact that Naaman's flesh does not become "like that of a young child" suggests that he does not experience any sort of rebirth.[50] It seems that he undergoes no "conversion" in the Jordan because immersion in these waters is not a valid mode of transformation.[51] By delegitimizing immersion in the Jordan, the Rabbis weigh in on the lack of efficaciousness of baptism. By excising the very verses that fertilize an interpretation like Origen's, they militate against the reading together of Naaman's story with that of Jesus in the Jordan or with accounts of Christian baptism in which believers become like children.

Where Elisha remains aloof from Naaman in both his leprous and his healed state, Gehazi, Elisha's disciple, chases after him seeking benefit. Gehazi takes what he can from Naaman, then returns to the Beit Midrash (House of Study), where Elisha is teaching a lesson on the laws of the eight unclean

---

48. This contrasts with the biblical story, where Naaman, after his Jordan immersion is characterized by "wordiness" Mordechai Cogan and Hayim Tadmor, *II Kings: A New Translation with Introduction and Commentary* (Garden City: Doubleday and Co., Inc. 1988), 65.

49. Word choice is indicative here. Naaman is cured as described by the Aramaic verb איתסי, rather than made clean, which would involve a variant of the verb טהר that we find in the biblical story.

50. Perhaps the rabbinic conversion ceremony with its absence of dramatic transformation offers a parallel: "In neither [rabbinic conversion] ceremony is anyone cleansed, unburdened, reborn, recreated, reimagined, or refreshed." Cohen, *The Beginnings of Jewishness*, 247.

51. The earliest attempt to distinguish the washings and immersions involved with maintaining purity from baptism may be the disqualification of Jordan water as a cleansing agent in Mishnah Parah 8:10: מי ירדן ומי ירמוך פסולים מפני שהם מי תערבות. According to the Mishnah, the problem with the Jordan as an agent of purification is the same as that with the River Yarmouk—they are two distinct bodies of water that are intermixed. Such a convergence, the Mishnah argues, sullies the resulting runoff only if one river is pure and the other impure. It is hard to distinguish which is the offending river in this case, but since the Jordan north of the Yarmouk is not deemed acceptable, it seems that the Jordan is rejected as a ritually purifying agent. The denial of the Jordan's efficacy on the grounds of its mixture/contamination may well imply that misconstrued principles and incorrect purifying agents lead to purported cleansings with no power in the realms of atonement or redemption. The passage may well stand as an early rabbinic refutation of baptism. In Bereshith Rabbah, a text significantly later than the Mishnah, the Jordan is said to pass through the lake of Tiberias without mixing with it (Bereshith Rabbah 4:5).

crawling creatures (Lev 11:29–37).[52] That Elisha's lecture concerns purity establishes that he is conducting normative rabbinic business at the very moment that Gehazi enacts his alternate designs. As Gehazi slinks in late to the lecture, leprosy erupts on his face.[53] His disease exemplifies the result of chasing after one's appetites and disregarding the study of law and the social hierarchy. Similar to Origen's portrait of the leprous Jewish body barred from paradise, the eternally leprous Gehazi is banished from the World to Come: "Four commoners have no portion in the World to Come . . . Balaam, Doeg, Ahitophel, and Gehazi" (Sanhedrin 90a).[54]

Gehazi, the disciple who breaks away from his master's teachings, disturbs the transmission of a continuous genealogy of knowledge that prevents the collective Jewish body from atrophying in exile. In rabbinic literature, Gehazi reads as greedy, secretive, and sexually forward.[55] These traits are also figured as stages in a plot to subvert Elisha's teachings. Gehazi represents the enemy within who outwardly conforms to convention while nurturing a secret agenda. He is furthermore a trickster who shuttles between camps. Rather than demonstrating that the camps or the categories are assailable, Gehazi as a trickster serves to reinforce the opposition between Elisha and Naaman, the Rabbis and the baptized. The trickster here is not the hero to be emulated, but the mediator who points to the danger inherent within interstitial zones. The implication of the trickster's ambiguity for the audience is that instability results from mediation while individual and communal stability is ensured by maintenance of binary opposition.

In the biblical passage (2 Kings 5:20) material appetites motivate Gehazi, but the Talmud suggests that his appetite for wealth bespeaks a heretical hunger. As he chases after Naaman, he also chases Rabbis away from the House of Study.

> Some say he [Gehazi] repulsed Rabbis from Elisha, as it is written, the sons of the prophets said to Elisha, "this place where we sit before you

---

52. This refers to the law of nonkosher crawlers: "The following shall be unclean for you from among the things that swarm on the earth: the mole, the mouse, the great lizards of every variety; the gecko, the land crocodile, the lizard, the sand lizard, and the chameleon" (Lev 11:29).

53. Note the shift from the biblical story, where Elisha's curse causes the outbreak of leprosy (2 Kings 5:27).

54. The mishnaic provenance of this declaration places Gehazi's disbarment from the World to Come closer to Origen's time period. For the substitution of Jesus for Balaam in a similar list of treacherous figures, see BT Berakhoth 17a–b, and Schäfer, *Jesus in the Talmud*, 30–31.

55. When the Shunnamite woman comes to Elisha in desperation because her son has died, Gehazi accosts her: *"And Gehazi came near to thrust her away.* R. Jose ben Hanina said: He seized her by the breast" (BT Berakhoth 10b).

is too cramped for us" (2 Kings 6:1), proving that until then it was not too cramped. (Sanhedrin 107b)

According to this passage, Elisha's quarters were vacant during Gehazi's tenure because he diverted inquiring Rabbis. As Boyarin argues, the study house is both site and symbol of rabbinic authority. Since Gehazi has repelled Rabbis from this space, it is possible to imagine a background story in which he invites them to another authoritative space, the entrance to which requires baptism. After Gehazi is banished, Elisha's quarters teem with Rabbis. Interestingly enough, in the Bible, the sons of the prophets (here Rabbis) move from Elisha's cramped quarters and resettle on the banks of the Jordan (2 Kings 6:1–7). Reading the biblical and the talmudic texts together implies a mode of Jewish reclamation of the Jordan River from its status as baptismal signifier.

Despite previous transgressions, it is only when Gehazi's betrayals become manifest that Elisha curses him: "You wicked man, the time has come for you to receive your reward for [studying the laws of] the eight reptiles. May the leprosy of Naaman cling to you and your seed forever" (Sanhedrin 107b). Gehazi is discovered to be the Rabbis' enemy only because a disease breaks out on his body. In other words, leprosy becomes the sign of a body operating at odds with rabbinic values and practices. As in Leviticus and in Origen's commentary, the leprous body must be removed from the camp. But what are the Rabbis really extricating when they banish Gehazi? Gehazi chased after whatever it was that Naaman offered and Elisha refused to accept. While the specific nature of this offering is left open for interpretation, the severity of Gehazi's punishment and his exclusion from the World to Come imply the potential danger of chasing after the "baptized."[56] In fact, chasing after the baptized renders the Jewish body impure. Gehazi, punished for the chase, is offered no dip in the Jordan River, no rebirth, no repentance. Gehazi's leprosy renders him impure in the categorical sense for an eternal duration.

Elisha tells Gehazi, "O wicked one, the time has come to receive your reward for the eight unclean crawling creatures." However, Elisha sat and expounded upon the laws of unclean crawling creatures at the very moment when Gehazi extorted goods from Naaman. Gehazi, it seems, missed the lesson and therefore is reprimanded both for his absence and his actions during this time. Irony is at work here, since Gehazi never studied the laws of unclean crawling creatures and his "reward" is physical affliction in this world and banishment

---

56. Allegations of Rabbis taking the Jordan plunge were certainly made, for Epiphanius' (himself a convert from Judaism) tale of the Jewish patriarch at Tiberias undergoing baptism prior to his death, see Adversus Haereses (Haeres. 100.30). For the connection to the biblical story of Naaman, see Hagith Sivan, *Palestine in Late Antiquity* (New York: Oxford University Press, 2008), 103–4.

from the next. The fact that Elisha espouses purity laws after Naaman's healing is particularly significant. By lecturing on the creeping animals that can render the Jewish body "unclean," the rabbinic Elisha distinguishes between the foreign body, healed by a dip in the Jordan, and the Jewish one, which must maintain its cleanness through adherence to the laws of purity. The fact that Elisha lectures on halakhah following his inadvertent yet wondrous cure of Naaman serves to repatriate Elisha and clear his name from the association with Christian baptism. Elisha, read by Christian exegetes as a figure fulfilled by Jesus, is a rabbinized character in the Talmud. Neither wonder worker nor agent of miracle, Elisha is a Rabbi who channels power by teaching Torah and gathering disciples.

Apostate Disciples

Naaman's story is set between parallel accounts of disciples spurned by their teachers' two hands. Where in the Gospels Jesus is figured as the fulfillment of Elisha, in the talmudic doublet Jesus and Elisha are construed as foils. Elisha is comparable to Rabbi Joshua ben Perahyah and Jesus to Gehazi.

> *And Elisha went to Damascus* (2 Kings 8:7): why did he go? — Rabbi Yohanan said: He went to bring Gehazi back into the fold [urge him to repent], but he would not repent. "Repent," he urged.[57] He replied, "this have I learned from you: one who sins and causes the multitude to sin is not given the opportunity to repent." What had he done? — Some say: He hung a lodestone above Jeroboam's sin [the Golden Calf], and suspended it between heaven and earth. Others maintain: He engraved the Divine Name [on the calf's mouth], and it would proclaim: *I am the Lord your God and Thou shall have no other gods before me.* (Exod 20:2–3)[58]

• • •

> When Yannai the king was killing the Rabbis, Joshua ben Perahyah and Jesus went to Alexandria of Egypt. When there was peace, Shimon ben Shetah sent to him: "From Jerusalem, the Holy City to Alexandria in Egypt: My sister, my husband is dwelling in you and I am sitting bereft." [Joshua] got up and left [taking Jesus with him], and came to a certain inn, where they honored him greatly. [Joshua] said: How beautiful this

---

57. Elisha's command contains half of John the Baptist's exhortation in the Gospel of Matthew.
58. The accounts of Gehazi diverting Rabbis from Elisha's house of study, the saying "let the left hand repulse but the right hand always invite back," and the story of Naaman follow this.

inn is . . .[59] [Jesus] said: Rabbi, her eyes are bleary. [Joshua] said: Wicked one. That's what you are busy with?! He brought out four hundred shofars and excommunicated him. Jesus came before him several times and said: Accept me! He didn't pay attention to him. One day, [Joshua] was in the middle of saying the *Shema Yisrael*, and [Jesus] came before him. [Joshua] wished to receive him, and made a sign with his hands [because he could not interrupt his prayer]. [Jesus] thought that he was rejecting him. He went and erected a tile and bowed to it. [Joshua] said to him: Repent! [Jesus] said to him: This is what I have learnt from you. Anyone who sins and causes others to sin is not enabled to repent. And our master has taught: Jesus performed magic, and misled and corrupted Israel.[60]

Although they are parallels, Jesus reads as a more sympathetic character than Gehazi. Gehazi's transgressions of diverting Elisha's students, betraying his teacher, and missing his lesson are direct attempts to subvert the order of the Jewish academy. Jesus's act of checking out his innkeeper, in contrast, seems rather benign. Part of the problem seems to be that Rabbi Joshua ben Perahyah comments positively about the inn because his rabbinic accomplishments are recognized therein. Like Gehazi then, when Jesus should be acknowledging his teacher, he expresses other designs. Jesus's sexual awareness then puts him at odds with the rabbinic establishment. One cannot help but detect some irony here since the Gospels (and all the more so Christian tradition) uphold Jesus as the paradigm of chastity while the Rabbis, for the most part, are recorded as being husbands and enjoying healthy sex lives. On a certain level, Rabbi Joshua's condemnation of Jesus seems to be based on Christian rather than Jewish standards. Where Gehazi's subversive designs become physically manifest, Jesus is banished though an elaborate ceremony involving "four hundred shofars." Following this excommunication, Jesus tries several times to repair the relationship with his teacher but again misreads the signs. Ultimately, the story does not fault Jesus for misunderstanding his teacher's gestures during prayer since Rabbi Joshua ben Perahyah is charged with unequivocally spurning him.

---

59. "A tragic misunderstanding is about to occur, because the word for 'inn,' אכסניא can also mean 'hostess.'" Boyarin, *Dying for God*. 24.

60. Translated by Boyarin, *Dying for God*, 24. Peter Schäfer shows that the identification of the wayward student with Jesus is a late tradition evident only in the Babylonian Talmud. "It is lacking in the Yerushalmi [Jerusalem] version and attested only in some manuscripts of the Bavli [Babylonian] version . . . the manuscript evidence clearly shows a tendency during the editorial process of the Bavli to identify the unknown student of Yehoshua b. Perhaya with Jesus." Schäfer, *Jesus in the Talmud*, 37.

Gehazi and Jesus become true parallels only after their expulsions. At this point, the two disciples engage in deviant magic and idolatry with the goal of engaging others in heresy.[61] Aware that their pedagogy has been perverted, both teachers pursue their students with the exhortation to repent. Even after their apostasy, both wayward disciples remember their Torah. As they tell their teachers: "This is what I have learnt from you. Anyone who sins and causes others to sin is not enabled to repent." As Torah is enlisted as a means of safeguarding heretical actions, expelled former insiders appear as the most dangerous of opponents, for in their hands Scripture becomes a dangerous weapon. Where Gehazi initially constituted the more rebellious student, following the banishment Jesus, who "misled and corrupted Israel" with his "magic," becomes more insidious. While Gehazi should certainly not be taken as an allegorical figure, some have cast him as a Jewish-Christian figure who chased after the "baptized" with the hope of gain and was accordingly punished.[62] What comes across more clearly is that temporary gains made in opposition to rabbinic practice ultimately entail the greatest loss.

Boyarin has discussed the episode of Rabbi Joshua and Jesus as a typestory in which a founder of heresy emerges from the very orthodoxy that the heresy challenges. "Jesus was at first a perfectly orthodox rabbinic Jew, and only because of the intransigence of an overly strict teacher and then a tragic misunderstanding did he found the great heresy of Christianity."[63] Rabbi Joshua ben Perahyah's rejection of Jesus also simulates the rabbinic rejection

---

61. Peter Schäfer observes that the labeling of Jesus as "a magician in a derogatory sense, is, therefore, an inversion of the New Testament, which connects him (positively) with magicians, with Egypt, and with healing powers." Schäfer, *Jesus in the Talmud*, 20. This observation supports Schäfer's argument that the Rabbis of Babylonia were familiar with the texts and not only the traditions of the New Testament. For his theories of textual transmission in this case, see ibid., 122–29.

62. Since the Jesus in this story, obviously, represents Jesus, the desire arises to read Gehazi as some other distinct Christian figure. R. Travers Herford struggles with this conundrum: "The connexion of a story about Jesus with a story about Gehazi suggests that there may be, under the figure of Gehazi, a covert reference to some person associated with Jesus. It is natural, therefore, to look amongst the followers of Jesus for the man of whom Gehazi is the type. I suggest that the man referred to is Paul." R. Travers Herford, *Christianity in Talmud and Midrash* (Clifton: Reference Book Publisher, 1966), 99. The association with Paul, while creative, is not substantive. Herford himself admits, "whether there is here any reference to Paul I am not prepared to say." Ibid., 102. Another approach would be to read the two stories chiastically with the two masters and two errant disciples being equivalent. A chiastic reading would associate Elisha with Rabbi Joshua ben Perahyah, and Gehazi with Jesus. Even so, I'm hesitant to reproduce a reading like Herford's that draws the analogy so tightly as to limit other readings. Whether or not Gehazi specifically represents Paul, John the Baptist, Jesus, or just a bad student, the Jesus/Gehazi parallel functions in tandem with the Naaman story as a rejection of Christian practices.

63. Boyarin, *Dying for God*, 26.

of Jesus, a rejection repeated each time a student of the Rabbis recites or reads the story. The two stories appear in two places in the Babylonian Talmud, Sotah 47a and Sanhedrin 107b. Only the Sanhedrin version includes the story of Naaman. I suggest that the later Sanhedrin version reflects the arrival at some kind of rabbinic conclusion to reject the practice of baptism.[64] With the story of Naaman functioning as a prooftext for Christian baptism, the Rabbis retell the story in a manner that supports their rejection of Jordan immersion. By linking a portrait of Jesus as a spurned Rabbi and a revised Naaman story that includes no conversion, the Rabbis catalogue both the teachings of Jesus and the practice of baptism in the category of incorrect interpretation. The stories, even while extending compassion for wayward disciples, assert that Jesus and his followers practice idolatry and that baptism or the pursuit of the baptized has the power to render the Jewish body forever impure. Context works together with a series of associations to counter interpretations of Elisha, Naaman, and Jordan immersion similar to Origen's.

Physician, Heal Yourself[65]

In the Talmud, Naaman comes and goes from Israel without experiencing miracle or conversion, and the Jordan is visited without much fanfare. Naaman's leprosy is transferred to Gehazi, whose disease-ridden body is extricated from the Beit Midrash and the World to Come. Once the foreigners have crossed back to the other side of the Jordan and the heretical students have been expelled, the focus shifts to the Jewish bodies that remain in the camp. Baptism's effectiveness has been disproved and in its stead halakhic observance is the proposed program. The Rabbis introduce the idea that upholding the purity laws keeps the Jewish body, in its individual and collective iterations, healthy.[66] I am not claiming that the Babylonian Rabbis anticipate the rationalistic interpretations of Jewish law of someone like Maimonides, but rather saying that the Rabbis ascribe to the purity laws the power to maintain the discrete nature of the Jewish community. In this way, upholding

---

64. Perhaps the Sanhedrin version is coterminous with or postdates the order of Heraclius (in 634) of "an empire-wide baptism—all Jews under Roman rule were to be baptized as proof of loyalty to emperor and empire." Hagith Sivan, *Palestine in Late Antiquity* (New York: Oxford University Press, 2008), 48.

65. Luke 4:23; Genesis Rabbah 23:4.

66. Again, I see a kind of silent polemic here insofar as the miraculous healing practices of Jesus are rejected through the contrast with Jewish law. The insufficient and indeed dangerous healings offered by disciples of Jesus are at issue in the stories of Rabbi Eleazar ben Dama's snakebite (Tos Hullin 2:22, BT Avodah Zara 27b) and the grandson of Rabbi Yehoshua ben Levi (Qoheleth Rabbah 10:5). See the opposing positions in Schäfer, *Jesus in the Talmud*, 52–62, and Boyarin, *Dying for God*, 34–41.

purity keeps Israel healthy. However, even those who adhere to the purity laws sometimes fall ill.

In contrast to Origen, who insisted that Israel's prophets could not heal the Jewish body, the Rabbis portray Elisha as the first prophet to be healed from a near-fatal illness. The rabbinic body, metonymically signaled by Elisha, does become sick; however, this body also possesses the inherent power to cure itself.

> Our Rabbis taught: Elisha was afflicted with three illnesses: Once when he incited the bears against the children, once when he repulsed Gehazi with both hands, and the third of which he died from; as it is written, *Now Elisha was fallen sick of his sickness whereof he died* (2 Kings 13:14). Until Abraham there was no old age. Whoever saw Abraham said, "this is Isaac," and whoever saw Isaac said, "this is Abraham." Therefore, Abraham prayed that there should be old age, as it is written, *And Abraham was old, advanced in age*. Until Jacob there was no illness, so he prayed and illness came into existence, as it is written, *And one told Joseph, behold, thy father is sick*. Until Elisha no sick man ever recovered, but Elisha came and prayed, and he recovered, as it is written, *Elisha had been stricken with the illness of which he was to die* (2 Kings 13:14). (Sanhedrin 107b)

Elisha brings illness upon himself insofar as sickness appears as a physical reflex of extreme actions. Such actions in turn seem to stem from an overexcited state of mind. Unlike Jesus in the Gospels, who heals the sick from moral decay or exorcises inhabiting demons from among Israel's population, Elisha's healing powers are focused within as a mode of self-betterment. This is an important rabbinic turn in the characterization of Elisha, since the acts of the biblical prophet and his master serve as a model for Jesus's acts of healing and resurrection of the dead. In this talmudic passage, Elisha does not miraculously heal others. He instead heals himself through the rabbinically legitimate means of prayer. In this evolutionary chart of sickness, Abraham introduces death, Jacob illness, and Elisha healing. In sickness and in health, the Jewish body can be transformed by prayer.

Elisha, the rabbinic metonymy, suffers from illness and conceives of its remedy. His body aches from the loss of his disciple as well as from holding an intransigent position of authority. Likewise, the rabbinic body aches from the loss of Jesus and his Jewish followers as well as from the scourges of exile and colonization. However, these states, like those of death and illness, were brought into the world for a purpose and mark stages in a cycle. Healing and redemption are promised in the cycle of fortune. Although the rabbinic/Jewish body may be pained or dismembered, it retains regenerative potential and can withstand many maladies before it expires. Since it innately possesses this potential, there is no cause for Jordan plunges or baptisms. Where Origen and indeed Paul insist that only baptism can correctly align the body and render it worthy of eternity, the Rabbis present Elisha as the physician who heals himself. The people of Israel and their Rabbis are shown to have long possessed the means of self-regeneration.

Conclusion

In *Illness as Metaphor* Susan Sontag addresses how "the subjects of deepest dread" become identified with disease and how civil disorder is analogized to illness. The dread at the base of Origen's and the rabbinic commentaries is that the collective Christian and Jewish bodies will lose their integrity by blending or too much resembling the collective body of the other. The sense of peril is heightened through mobilization of the illness metaphor—to get too close is to risk contamination and decay. A ritual regimen—baptism in one case and purity in the other—is prescribed as both treatment and prophylaxis. In this way, self-containment comes to define both individual and collective health. The formulation is different from those in the Hebrew Bible insofar as neither the early Christian nor the Jewish collective depends upon territorial limits as a means of definition. Therefore, the Jordan is attributed with the obstruction of ailing bodies rather than foreign ones.

The stakes in such a formulation become clear through Sontag's second point. "Illness comes from imbalance. Treatment is aimed at restoring the right balance—in political terms, the right hierarchy."[67] The claim to protect constituent bodies from illness functions to credit ritual with upholding the correct societal order while justifying the elevated status of the exegetes who command the collective body in an imaginary territory. By dictating the limits of a religious body politic, the exegetes put themselves at the head of the respective political bodies. The transfer of the Jordan, a territorial boundary in the Hebrew Bible, to the body enables the production of a Jewish-Christian binary as well as the establishment of a new elite for whom the body is sovereign territory.

The biblical border system becomes a conceptual template that can be realized through ritual and marked on the body. At the same time, it maintains the power to evoke "real" territory and "actual" miraculous events that transpired in the biblical past. After Origen, the process of creating a Christian holy land gets under way. As scholars have shown, this process required both the absorption of Jewish knowledge and the wresting of Hebrew Scripture

---

67. Susan Sontag, *Illness as Metaphor* (New York: Farrar, Straus and Giroux, 1978), 76.

and symbolic topoi (in some cases in the form of removing actual Jewish bodies) from the Jews. After Constantine, "the holy land itself was a prime locus for the elaboration of new modes of imperial Christianity in which a new Christian self could be fashioned and manipulated."[68] The power of the border system, as Jonathan Z. Smith has shown, rests in the multiple potential sites of its realization.[69] The geography of promise can be visited through the physical enactment of ritual, the observance of calendar, or constructive pilgrimage to the holy land. Again, the empire seems to set the map of the Jewish homeland insofar as imperial edicts determined where Jews could live and worship and Jewish exilic adaptations ran with the idea of the land's concurrent abstraction and materiality. To such adaptations we now turn.

---

68. Jacobs, *Remains of the Jews*, 2.

69. Although the territorial system was replicated and transposed in both Christianity and Judaism, J. Z. Smith draws attention to the different ways in which this occurred. The "synchronic structure" of the Temple based on difference rather than equivalence "could be replicated in a system of differences transferred to another realm or locale." In contrast, the Church of the Holy Sepulchre could stand only at the point of Jesus's burial fixed in Jerusalem. The sacred, in this case, requires equivalence. It is the reproduction of equivalence (the correlation of story to place, then to time) that comes to characterize the design of Christian liturgies and sanctuaries. Smith, *To Take Place*, 85–88.

Chapter Eight

TWO MORE MAPS OF ISRAEL'S LAND

The waters of the Jordan surround all the earth, half flow above the earth and the other half below the earth.
—Pirke de Rabbi Eliezer, 11

Rabbinic literature maintains the Jordan as a frontier. By the later rabbinic period and certainly in the Babylonian context, the land is not the 'real' space of Israel's requisite inhabitance, but rather an imagined (even when experienced) holy place that bears traces of the collective Israel. Here we will look at two mappings that cite the Jordan as a border. The two are parallel navigations of the land of Israel's waterways that arise in different contexts and arrive at different endpoints. The first mapping appears in the Babylonian Talmud Tractate Bekhoroth 55a, where a discussion elaborates upon the principles of animal tithing recorded in the Mishnah. The issue at hand concerns the permissible combinations of flocks when determining the correct tithing amount. Although much is said over the course of the discussion about the nature of rivers and borders, the context can be characterized as a halakhic (Jewish legal) conversation. The second mapping transpires amidst the tall tales of the Babylonian Talmud Tractate Baba Bathra 74b, in which the world's waters appear as a liminal space where features of creation and the eschaton become visible.[1] The tall tales, in a certain sense, represent

---

1. For the formal characteristics of these tall tales, see Dan Ben-Amos, "Talmudic Tall Tales," in *Folklore Today*, ed. L. Degh, H. Glassie, and F. Oinas (Bloomington: Indiana Semiotic Sciences, 1976), 29.

the outer edges of midrash, thus demonstrating the limits of narrative exegesis as they visit geographic peripheries.[2]

The relevant discussion in Bekhoroth 55a begins with a reiteration of Rabbi Meir's assessment that "the Jordan forms a division in the instance of tithing animals." Again, the question at hand involves the distances and barriers that can be spanned when enumerating the animals of a flock for the purpose of taxation. Questions about the nature of the Jordan as a divide arise from Rabbi Meir's statement that animals located on opposite sides of the Jordan cannot be combined when assessing the tithe. The first challenge to such an understanding of the Jordan comes from Rabbi Ammi, who takes Rabbi Meir's opinion to pertain only to places where there is no bridge across the Jordan. Where a bridge exists, animals from opposite shores can come into contact with one another and can therefore be grouped together for tithing, but where no bridge exists, the animals can neither be herded nor classified together. Rabbi Ammi's theory of the bridge is quickly discounted; since animals from different cities within the land cannot be tithed together, it goes without saying that animals in the land of Israel and outside of the land of Israel cannot be combined.[3]

While the tithing framework holds, the focus of the dialogue shifts to the nature of the Jordan as a border. The possibility that a bridge can unify the riverbanks is refuted, since the west bank remains inside and the east bank outside of the land. Furthermore, tannaitic tradition holds that despite the proximity of the banks, flocks from either side are not to be combined. Implicitly then the Jordan delineates the land from the outside. The divide deepens as Rabbi Hiya bar Abba supports Rabbi Meir's appraisal of the division by enlisting scriptural proofs that the Jordan is the border of the entire land. As Rabbi Hiya, following Rabbi Yohanan, recalls particular biblical maps, he puts a system of binding and nonbinding boundaries in place. He begins by citing the tribe of Benjamin's patrimony as stipulated in Joshua 18 in which the final coordinate falls where "the Jordan bounds it on the eastern rim" (18:20). Rabbi Hiya extrapolates from the verse that Scripture designates the Jordan as a border in its own right. A counterexample is then called into play that questions the binding nature of tribal boundaries. "Based on this then, when Scripture says 'the border turned' (Josh 18:14) and 'the border went up' (Josh 18:12), are these to be understood as borders in their own right?" If Joshua 18:20 is taken

---

2. See Stein, "Believing Is Seeing: A Reading of Baba Bathra 73a–75b." *Jerusalem Studies in Hebrew Literature* 17 (1999): 9–32 (Hebrew).

3. The Jordan, in this Babylonian text, is assumed to be a boundary. In texts compiled in the land of Israel, the eastern shore of the Jordan and its Jewish cities are included in the definition of the land, see PT Sheviith 6:1. Transjordan is a valid zone of Jewish residence and observance in Palestinian sources.

to mean that the Jordan serves as a border of the land itself, then do all of the tribal boundaries need to be understood in this way? In other words, do tribal divisions constitute legal and/or national boundaries?

The Jordan proves exceptional as both a tribal and a national border based on the map in Numbers 34, where it delimits the eastern edge of the land. Since the narrative mapping concludes with "this will be your land as defined by its borders" (Num 34:12), it can be safely assumed that the whole land of Israel shares a single border—the Jordan River. To review the stakes here, the Jordan is designated as a dividing line when it comes to the tithing of animals and as the definitive eastern border of the land of Israel. The stakes are somewhat lowered when one realizes that in the Babylonian Talmud the land of Israel is a geographic fiction. Even when articulated by Rabbi Hiya bar Abba, a sage who resided in Babylonia as well as in Palestine, the land of Israel in the Babylonian Talmud functions as a shifting memorial to a territorialized Jewish existence, as a structuring principle, and as a plane on which to understand biblical precedent. Because the land operates in such ways, it remains crucial to distinguish it from other locales in order to uphold the exceptional case of the exiled people of Israel. The larger project of maintaining and extending a legal framework that secures the uniqueness of the land then symbolically enacts the distinction of Israel among the nations where its members reside.

The revisitation of the land in talmudic discussion remembers Israel's place while creating a space for Jewish ritual performance in the many homes of a widespread diaspora. In the talmudic context, the biblical borders are useful memories that foster the concept of a unified people amidst a reality of dispersion. It is worth noting that even for rabbinic communities dwelling in the land following the destruction of the Temple and the Bar Kokhba uprising, the Jordan did not constitute a practical border since "the Land of Israel was defined as the Land upon which Israel lived in the context of its being the Promised Land, free of idolatry, not worked on the Sabbath or the Sabbatical Year."[4] The designation "land of Israel" was reserved for the specific locales within the biblical boundaries (according to the Jordan map) where Jews constituted enough of a majority to govern their lives through Jewish law. Despite their inclusion in a particular border system, the places where Jewish law did not prevail were not considered to be part of the land.[5]

---

4. Shmuel Safrai, "The Land of Israel in Tannaitic Halacha," in *Das Land Israel in biblischer Zeit* (Göttingen: Vandenhoeck & Ruprecht, 1983), 209.

5. Conversely, "the divinely ordained economy of the Land is not absolute but relative to the presence and agricultural activity of its Jewish inhabitants." That is, gentiles living in the land are not obliged to uphold Jewish laws. Richard Sarason, "The Significance of the Land of Israel in the Mishnah," in *The Land of Israel: Jewish Perspectives* (Notre Dame: University of Notre Dame Press, 1986), 118.

The continuation of Bekhoroth 55a is of further interest. Following the establishment of the Jordan as the eastern border of the land, a debate ensues about whether or not the Jordan itself constitutes part of the land. Based on differing scriptural precedent, Rabbi Judah ben Bathyra rules that the Jordan should not be included as part of land, while Rabbi Shimon bar Yohai insists on its inclusion.[6] The challenge to Rabbi Hiyya bar Abba's assessment that the length of the Jordan delimits the land comes not from these quarters, but rather from Rabbah bar Bar Hana's claim that "it is only the Jordan from Jericho and below."[7] This remark tempers Rabbi Hiyya's claim by implying that the Jordan only divides tithed flocks on its lower shores. However, the text maps a long span of the Jordan:

> The Jordan emerges from the cave of Paneas [Banias] and flows through the Sea of Sibkay, the Sea of Tiberias, and the Sea of Sodom, then continues until it feeds into the Great Sea [the Mediterranean]. Yet it is only the Jordan from Jericho and below.

The mapping recognizes the same river flowing from the northern mountains through the three seas (Sibkay/Samachonitis/Huleh, Tiberias/Galilee/Kinneret, and Sodom/Dead Sea) and ending at the Mediterranean. The anomalous twist is how the Jordan reaches the Mediterranean after it empties into the Dead Sea. The ensuing discussion of the interconnectedness of all waterways seems to answer this with the theory that all rivers empty into larger bodies of water themselves encompassed by the world ocean.

As evinced by Rabbi Isaac's statement that "the source of all water is the Euphrates" and Rabbi Yehudah's idea that the rivers of Eden feed all other bodies of water because "all of the other rivers are beneath the three rivers [of Eden: Pishon, Gihon, and Hiddekel] and the three rivers are beneath the Euphrates," the mythic geography discussed in chapter one is at work here. Rabbi Yehudah implies that the world's waters are all fed by sources in paradise. The talmudic discussion that begins with the question of districts in regards to tithing takes up the issue of the signifying border of Israel and ends by connecting all of the world's water to a source in the Garden of Eden. As in biblical texts, the very notion of a land bounded by water is linked to

---

6. Interestingly, Rabbi Shimon bar Yohai bases his rule of including the Jordan on Numbers 34:15 that speaks of the Transjordanian lands secured by the two and a half tribes.

7. Rashi seems to equate this opinion with that in Mishnah Parah that mixed waters contaminated by flowing through various lakes run along the Jordan's course. The alternate opinion of Rabbi Jonah that "the Jordan flows through the lake of Tiberias yet does not mix with it" and that this is "miraculous" appears in Genesis Rabbah 4:5.

the perception of a water-encircled world. The Jordan is recognized as the definitive boundary at the same time that the duration of its dividing force is called into question.

A parallel mapping of the Jordan's course in the Babylonian Talmud Tractate Baba Bathra 74b not only locates the river's source in the shared wellspring of Eden, but also tracks its endpoint to the eschaton. While this provides an example of sacred geography accruing apocalyptic import, it also evinces a trend apparent in Christian texts, where the Jordan becomes a gateway to an apocalyptic era. In this mapping, the Jordan terminates in the thirsty jaws of Leviathan and thus plays a role in the preparations for the messianic World to Come.

> Rabbi Judah further stated in the name of Rav: The Jordan issues from the cavern of Paneas [Banias]. It has been taught likewise: The Jordan issues from the cavern of Paneas and flows through the Sea of Sibkay and the Sea of Tiberias and rolls down into the Great Sea from whence it rolls on until it rushes into the mouth of Leviathan; for it is said: *He is confident because the Jordan dashes forth into his mouth* (Job 40:23). R. Abba b. 'Ulla objected: This [verse] is written of Behemoth on a thousand hills!—But, said R. Abba b. 'Ulla: When is Behemoth on a thousand hills confident?—When the Jordan rushes into the mouth of Leviathan.[8]

The Baba Bathra map omits the Jordan's flow into "the Sea of Sodom" (the Dead Sea) and, in place of the verb הלך, which I translated as "flow," in Bekhoroth, the Baba Bathra text uses the more evocative מתגלגל, "rolls."[9] Where the Bekhoroth map is invested in the nature of the Jordan as a border, the Baba Bathra map presents the Jordan as a connector that links periods of time as well as bodies of water. Leviathan, the beast subdued by God during creation, drinks from the Jordan as well as from the sea in preparation for the day of his annihilation after which he will be served as the fish course in the messianic banquet attended by the righteous in the World to Come. As it flows from terrestrial lakes into the world ocean and into the mouth of Leviathan, the Jordan models the basic structure of rabbinic time—a historical present that leads to an indeterminate messianic future.

---

8. In Bekhoroth 55a as well as in this passage in Baba Bathra 74b, the mapping is cited as a baraita. In this case, the additional levels of transmission from Rav to Rav Judah are also provided.

9. The verb shift occurs at exactly the point where the Baba Bathra version veers from that of Bekhoroth. So as the Jordan moves from the Sea of Tiberias, it rolls toward the Mediterranean and the mouth of Leviathan. In other words, the different course is marked by different locations as well as a different verb.

Along with the reemergence of the chaos monsters not seen since the creation, the messianic era seems to involve a reversal of appetites. The concept of a messianic banquet makes the point implicitly insofar as the food in the Garden of Eden was provided by God and then lost as a result of the first woman's appetite for knowledge. The food eaten in the present world is cultivated by humanity and involves great travail (Gen 3:17–19), but the future promises a great feast to end all daily meals.[10] The corollary between appetite and time becomes more explicit in the tales interspersed in Baba Bathra 74b–75a that provide something of a biography of Leviathan. Following rabbinic reports of Leviathan sightings on the world's waterways, Leviathan's origins are located on the fifth day of creation when God creates "the great sea monsters" (Gen 1:21). Rabbi Yohanan accounts for the plurality of these monsters as Leviathan the slant serpent and Leviathan the twisting serpent, the very monsters that God will slay "on that day" of the eschaton according to Isaiah 27:1. Rabbi Judah in the name of Rav—the very source that initiated the Jordan map—accounts for the Leviathan through a longer midrash.

> Everything that the Holy One Blessed Be He created in the world he created male and female. Thus Leviathan the slant serpent and Leviathan the twisting serpent he created male and female. If they had mated with one another, then they would have destroyed the entire world. What did the Holy One Blessed Be He do? He castrated the male and killed the female, salting her for the righteous in the future to come. (Baba Bathra 74b)

God curtailed the primordial sexual appetite of Leviathan in the name of protecting creation, leaving the beast to satisfy itself through mass consumption of food and water. As Leviathan's primordial sexual appetite is transformed into hunger and thirst during historical time, so the creature that spends the present eating will be eaten in the World to Come.

Leviathan's copious consumption in the present shows the depths of his appetites, affirming that had his mate been left alive, they surely could have overtaken the world. Rabbi Safra tells of a sea voyage during which those on board saw a fish raise its head above the water, revealing horns engraved with

---

10. The absence of food and attendant hunger that Adam and Eve experience following their banishment from Eden is the driving theme of the fourth-century Christian text *Life of Adam and Eve*. After standing in the Jordan River for forty days, Adam is awarded seeds by God (*Life* 20:1b). The life that Adam gains from this penitential act is of a mundane nature involving planting, harvesting, and feeding his children. The fact that Adam and his descendants continually hunger for something more indicates a loss that can be partially recuperated through penitential rituals in the present era, but only fulfilled in the eschatological future.

its fate: "I am a small sea creature 300 parasangs long and I am on my way to Leviathan's mouth" (Baba Bathra 74a).[11] The very dimensions of Leviathan's prey and the certainty of its fate instill fear about the sea monster's unrelenting hunger. What comes out of Leviathan's mouth is as frightening as what goes into it; Rabbi Dimi in the name of Rabbi Yohanan enumerates the consequences of Leviathan's hunger:

> When Leviathan is hungry he emits a breath from his mouth that boils all the waters of the abyss, as it is written, "he makes the deep boil like a pot" (Job 41:23) . . . when he is thirsty he causes great ripples in the sea . . . Rabbi Aha bar Jacob said, the deep does not return to normal for seventy years, as it is written, "he makes the deep seem old" (Job 41:24). and nothing old is under seventy. (Baba Bathra 75a)

Not only does the satisfaction of Leviathan require vast amounts of the world's resources, but the eruption of his hunger also leaves an enduring impression on nature. He boils the deep and disturbs the sea, illustrating that the need to sate Leviathan is a destabilizing component of the present. The world's waterways are places of chaos and potential threat because of Leviathan's abiding hunger.

Indeed, Leviathan's appetite for food and water is less portentous than the sexual appetite curbed by God to protect humanity, but it remains among the ominous features of the present. As the shift from the primordial past to the present involves the transformation of appetite, so the shift from the present to the future era will transform Leviathan from eater to eaten. Following the discussion of Leviathan's hunger, Rabbah, in the name of Rabbi Yohanan, describes how in the future God will prepare a banquet for the righteous from the flesh of Leviathan. The righteous are further stipulated to be the Rabbis themselves. The enormous mass of Leviathan will fill them up with enough left over to be distributed as a delicacy in the markets of Jerusalem. The division of time into the era of creation, the historical present, and the future World to Come corresponds with the different natures of appetite. This is reflected in the appetites of the chaos monster, Leviathan, whose sexual desire morphed into a disproportionate hunger that will cease to threaten humanity only after it is slain and served in a messianic banquet. The correspondence between time and appetite likewise characterizes humanity, once perfectly

---

11. Dan Ben-Amos recognizes "a conventional formula" of Talmudic tall tales here: they open with a variation of "'once we travelled on board a ship and saw . . .' The verb at the end of the formula 'we saw,' signifies the narrative position and point of view of the Talmudic storyteller." Ben-Amos, "Talmudic Tall Tales," 30.

nourished in Eden, now forced to labor for its repast and later—after the potentially dark days of apocalypse when, for example, "parents shall eat their children in your midst, and children shall eat their parents" (Ezek 5:10)—entertained by God at a lavish feast.

At the end of Baba Bathra's Jordan map, R. Abba b. 'Ulla raises an objection to the interpretation of the prooftext, "*He is confident because the Jordan dashes forth into his mouth*" (Job 40:23). In the context of Job 40, the Jordan verse stands at the midpoint between the description of the land monster, Behemoth, and the sea monster, Leviathan. R. Abba b. 'Ulla insists that the verse applies to Behemoth rather than Leviathan seemingly because Leviathan has not yet been introduced in Job 40:23. However, the next verse "Can he be taken by his eyes? Can his nose be pierced by hooks" (Job 40:24) clearly applies to Leviathan before he has been formally named. The mention of Behemoth in the talmudic discussion does not come as a surprise since a story about the sterilization of the male and female Behemoth precedes the Jordan map.

R. Abba b. 'Ulla's objection leads to an even more suggestive proposal: "When is Behemoth on a thousand hills confident? When the Jordan rushes into the mouth of Leviathan." R. Abba b. 'Ulla's explanation that Behemoth feels most secure when Leviathan is sated points to a tradition fully articulated in Leviticus Rabbah that Leviathan and Behemoth are to engage in a wrestling match to the death at the opening of the future era.[12] The current dispute over the Jordan's waters signals a more momentous eschatological competition.[13] During a historical moment when the Jordan no longer functions as an operative geographic border, its importance is refigured as a preparatory component of the World to Come. At a significantly later date, rabbinic literature arrives at a figuration similar to that in Christian literature of the Jordan as a border between this world and the next. In both cases, the geographic force of the river becomes a means of illustrating the distinction as well as the connectedness of the present and the future. The reservation of the fish and flesh

---

12. "R. Judah b. R. Simeon said: Behemoth and the Leviathan are to engage in a wild-beast contest before the righteous in the Time to Come, and whoever has not been a spectator at the wild-beast contests of the heathen nations in this world will be accorded the boon of seeing one in the World to Come. How will they be slaughtered? Behemoth with his horns will pull Leviathan down and tear him apart, and Leviathan with its fins will pull Behemoth down and pierce him through" (Lev. Rab. 13.3).

13. Later midrash grants the Jordan, when deemed sufficient, to Behemoth. Pirke de Rabbi Eliezer, for example, claims that: "The waters of the Jordan give him [Behemoth] water to drink, for the waters of the Jordan surround all the earth, half thereof [flow] above the earth and the other half below the earth ... This [creature] is destined for the day of sacrifice, for the great banquet of the righteous" (PRE 11). Numbers Rabbah, in contrast, states that because Behemoth swallows twelve months worth of Jordan water in one gulp, his thirst can be quenched by the paradisical river, Yubal (Num Rab. 21:18).

of Leviathan and Behemoth for the righteous in the World to Come further implies that the river operates as a symbol of restriction determining who will and who will not attend the messianic banquet.[14] This recalls some of the distinctions put in place by John the Baptist and elaborated upon by Origen.

Another sort of aquatic map follows Rabbi Judah's mapping. After endtime realities are suggested by the Jordan's termination in the mouth of Leviathan, a tradition that evokes the biblical maps of the land of Israel as well as the connection between nation and creation is cited.

> When Rabbi Dimi came, he stated in the name of R. Johanan: The verse, *For He founded it upon the seas and established it upon the rivers* (Ps 24:2) speaks of the seven seas and four rivers that surround the land of Israel. The seven seas are: the sea of Tiberias, the Sea of Sodom, the Sea of Helath [Eilat], the Sea of Hiltha,[15] the Sea of Sibkay, the Sea of Aspamia, and the Great Sea. The following are the four rivers: the Jordan, the Jarmuk, the Kirmiyon, and the Pigah.

Rabbi Dimi relocalizes the Jordan and brings it back into the realm of geographic familiarity. Rabbi Dimi describes a land encompassed by water that mirrors the cosmological order of the Jordan and Euphrates maps. Without citing the various scholarly and theological attempts to locate and fix the seven seas and four rivers, their number can be appreciated. The seven seas reflect the division of seven heavens, thereby rendering the land of Israel a kind of celestial mirror. The four rivers align Israel's territorial patrimony with the Garden of Eden, suggesting that the land is a model as well as an intermediary step toward paradise. Placed in the context of talmudic tall tales that locate apocalyptic signs at the margins of the world, Rabbi Dimi's map need not conform to known coordinates. It operates more as image of the revival of nation amidst the eschaton. Redemption, as we saw in the prophet Ezekiel's Temple visions, does not involve a return to Eden but rather the restoration of a boundaried land of Israel to a state with Edenic features. Forecasts of a renewed land permeate prophetic visions of what will follow the time of God's totalizing destruction. The appearance of Rabbi Dimi's map amidst tales infused with apocalyptic imagery suggests that the longing for national

---

14. The idea of restricted salvation is further elaborated in Baba Bathra 75a in a discussion following the description of the banquet that distinguishes among the various shelters to be provided in the World to Come. The most restricted category is that of the righteous, for whom tabernacles of Leviathan skin are constructed; the most inclusive category is that of the nations, who will bask in the light that shines from Leviathan's flesh splayed upon Jerusalem's walls. No one is here relegated to a fiery punishment.

15. Likely the sea of Ulathat mentioned in Josephus, *Antiquities*, XV, 10, 13.

restoration also finds a place in rabbinic ideas about the World to Come. The end-time return is not to paradise, but to a land reestablished according to a paradisical order. As we will see in the next chapter, similar dreams of a nationally rejuvenating return to the land of Israel as defined by particular sets of boundaries migrate into modernity, where they inspire the adherents of the early Zionist movement.

Chapter Nine

MY HOME IS OVER JORDAN

River as Border in Israeli and Palestinian National Mythology

Deep River, my home is over Jordan,
Deep River, Lord,
I want to cross over into campground.
Oh don't you want to go
To that promised land
Where all is peace.
　—African-American spiritual

The issue of borders rests at the heart of the Israeli-Palestinian conflict. To begin with, both Israelis and Palestinians define the symbolic whole of their respective nations according to the same set of boundaries. Thus, the very conceptions of Israel and Palestine overlap and conflict. Even in the absence of politically definitive state territory, border stories define geography as object of desire and constitutive of the national subject. Here we will consider how the Jordan became the border of Israeli and Palestinian national aspirations as well as how it has been continuously enforced and contested within their national myths. The analysis parallels that of the biblical maps in the first chapter by considering the degree to which imperial geographies influenced the map of the Jewish and the Palestinian national homes and then illustrating how national lore domesticates the lines drawn across the world by empires.

The utopian borders of the imagined Israel and the imagined Palestine both span from the Jordan River to the Mediterranean Sea. These aquatic borders signify territory and identity alike and render the map a highly charged and emotional symbol whose limits suggest distinction and separation. The very map that signifies the different homelands of Palestinians and Israelis corresponds with an internally divided space where the two peoples engage in an armed contest with arenas of demographics, media representation, and international legitimacy. The fantasy of sovereign, contiguous territory somewhat incongruously results in barriers, concrete and invisible, erected by the Israeli government to contain the Palestinians and to carve out zones of a distinctly Jewish Israel. Palestinians, meanwhile, actively wish the Jews away.

This chapter focuses on the Jordan River as a utopian and a political border between Jews and Arabs. In many ways, the Jordan is a deep border drawn in the final armistice between Israel and the Hashemite Kingdom of Jordan. Israel had fought Jordanian and other Arab forces at the Jordan in 1948 and fulfilled biblical myths when it captured the Golan Heights, a mountainous sliver of the east bank, and the entire West Bank in 1967. Palestinians have crossed the Jordan in large numbers throughout this tumultuous history. Crossing the de facto and ultimately recognized border entails particular obstacles for the stateless Palestinians to whom the Jordan symbolizes both collective exile and national aspiration. Even when the Israeli-Arab war is dormant, Jordan stories bolster the connection between the nation and its borders by reiterating a cultural difference between Jews and Arabs that overrides their proximity. In the stories, the Jordan separates the homeland from exile and those who cross the Jordan figure as Joshua-like heroes credited with national revival. In these ways, both Israeli and Palestinian national myths represent modern-day variants of the Joshua myth. Israeli national mythology strikes a distinctly Deuteronomistic tone, often employing terms taken directly from the Bible and, more surprisingly, some of these same terms appear in Palestinian national mythology.

The conception of the Israeli and Palestinian nations coincides with the fact that "the land is not yet fully acquired as a stable given" and that the borders remain undeclared, unofficial, and unstable.[1] With such uncertainty in the background, nationalists deploy mythologies of the border to hallow specific sites as stable, sacred frontiers that irrevocably define the nation. While counterbalancing political fluidity with claims of stability and perpetuity, the border mythologies perform another mediating role. A circumscribed terrain facilitates the formulation of national identities that overlap with religious and ethnic designations. Because national identities are neither natural nor unitary, the sense of a border enables their fluctuation. In other words, so long as a collective shares a sense of limits, the identities of its members can vary. Border stories assign a place to a people and index who belongs and who is excluded from it. For example, there are Jews, Arabs, and Muslims deeply invested in Israel and Palestine, but unless they claim origins or reside in the

---

1. J. C. Attias and Esther Benbassa, *Israel, the Impossible Land*, trans. S. Emanuel (Stanford: Stanford University Press, 2003), 219.

land, they are affiliates, not members, of the nation. Palestinian and Israeli identities depend on fixed territory despite their wide-reaching diasporas and the nomadism of both groups;[2] while the national identities continually shift, the competition over space sustains their ethno-territorial dimension.[3]

The symbolic codes at work in Palestinian and Israeli Jordan stories exhibit structural similarities and parallel themes, pointing toward the homologous rhetoric employed by competing groups. I will highlight the shared images that evoke common landscapes throughout. At the same time, I do not intend to elide the asymmetrical experiences of Israeli colonization and Palestinian displacement in the comparative enterprise. Binary formulations of symbolic codes like "home"/"exile" exert a powerful unifying force within Palestinian and Israeli societies that crosses the lines of class, political party, ethnicity, and religion by promoting putative nationalist underpinnings.[4] The invocation of a specific homeland rouses disparate factions to a common cause while the symbolism at play, however thematically parallel and internally unifying, mobilizes contention and motivates the clash at borders. Experiences at the Jordan become enshrined in national myth as moments of inception or transition when the national character underwent a transformation due to engagement with the enemy. In the narratives considered below, a group construed as paradigmatic breaks with previous behaviors of accommodation and redefines its environment through heroic acts. The Jordan functions as a site of beginnings that gives birth to the projects of collective struggle and where pioneers, refugees, and freedom fighters articulate the very character of the nation. In nationalist myth, the Jordan River demarcates a stark contrast between the past and the present that facilitates the construction of essentialized self-conceptions and definitive opposition.

Two Nations, One Map

The British created the map that operates twice over as a national icon. Biblical precedent exerted influence on European surveys of the region as did the administrative districts of the Ottoman Empire, but circumstance had the

---

2. On Palestinian diaspora see Helena Schulz with Juliane Hammer, *The Palestinian Diaspora: Formation of Identities and Politics of Homeland* (London: Routledge, 2003), 14; on Jewish diaspora see Jonathan Boyarin and Daniel Boyarin, *Powers of Diaspora: Two Essays on the Relevance of Jewish Culture* (Minneapolis: University of Minnesota Press, 2002), 11.

3. George Bisharat, "Exile to Compatriot: Transformations in the Social Identity of Palestinian Refugees in the West Bank," *Culture, Power, Place: Explorations in Cultural Anthropology*, ed. A. Gupta and J. Ferguson (Durham, NC: Duke University Press, 1997), 205.

4. For the class struggles embedded in Palestinian nationalism, see Ted Swedenburg, *Memories of Revolt: The 1936–1939 Rebellion and the Palestinian National Past* (Minneapolis: University of Minnesota Press, 1995).

greater hand in shaping what became Mandate Palestine. By 1873 Ottoman authorities had taken all of the steps they would toward subdividing the area (see figure 7). The Jordan served as perhaps the most definitive border, separating the districts (sanjaks) of Acre, Nablus, and Jerusalem to the west from the districts of Damascus and Ajlun to the east. The definitive nature of the Jordan, however, may say more about the fluidity between other Ottoman districts and subdistricts than about any barriers posed by the river. In fact, travel, trade, and migration across the Jordan characterize this period.

As the Ottomans finalized their administrative districting, the London-based Palestine Exploration Fund (PEF), a subscription-driven organization with missionary roots, was formed for the purpose of excavating and mapping the Holy Land.[5] The PEF's cartographic projects from the outset were subsidized by the British War Office and staffed by members of the Royal Engineers. In this marriage of convenience, members of the PEF remained largely unaware of the group's increasing dependence on the War Office, while the War Office ignored the PEF's archeological endeavors and pressed for a scientific map. That the PEF produced the map used, with some modifications, by the British general Edmund Allenby in ousting the Turks from Palestine and that this same map defined the initial colonial claims to Palestine shows how the measurement and representation involved in cartography also function as means of exerting control and absorbing space. This sort of quantified imperial knowledge that can be reproduced, cited, and disseminated then has the force to override historical or indigenous claims.[6] In the case of Palestine, scientific technologies substantiated biblical truth, which in turn inflected British military and imperial endeavors.

At an early stage of the PEF's survey, the Jordan River became recognized as a border. However much it would fit my argument if such recognition had a scriptural basis, I must admit that the reasons were of a more practical, logistical nature. The sparsely populated, less urbanized terrain east of the Jordan;[7] the competition between British and French explorers over the Holy Land; and the limited funds available to the Palestine Exploration Fund all contributed to

---

5. John James Moscrop, *Measuring Jerusalem: The Palestine Exploration Fund and British Interests in the Holy Land* (London: Leicester University Press, 2000), 70.

6. See Timothy Mitchell on Britain's map of Egypt completed in 1907. Mitchell, *Rule of Experts: Egypt, Techno-Politics, Modernity* (Berkeley: University of California Press, 2002), 9.

7. The Ottoman districts west of the Jordan hosted cities and sedentary agricultural communities, while those east of the Jordan were mostly populated by migratory Bedouin tribes. In 1881, after the Survey of Western Palestine had been published, Claude Conder, one of the lead surveyors, undertook an expedition east of the Jordan. "The unsettled state of the country and opposition from the Turkish authorities" prevented his mapping more than "500 square miles." Moscrop, *Measuring Jerusalem*, 135.

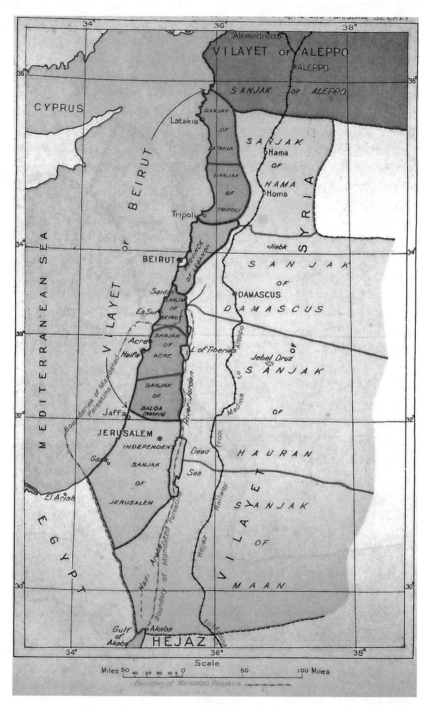

7. Ottoman administrative districts in Syria and Palestine. The Hejaz Railway runs through the eastern areas. Courtesy of the National Archives (UK).

the limitation of the survey to "Western Palestine." The PEF's Survey of Western Palestine invented the place that would become British Mandate Palestine and established the Jordan River as something of an indelible border in the colonial imagination. After the PEF committee on mapping drew the hypothetical line, it divided surveying duties between the PEF and the American Palestine Exploration Society (APES). Suspecting that the Americans were likely to fall short of English standards, the PEF gave them the less significant—from a biblical as well as military perspective—Eastern Survey in order to thwart French or Prussian inroads.[8] Indeed, the APES maps were disregarded by the PEF and their War Office sponsors and never included as part of the map of Palestine. This effectively led to a Palestine restricted to territory west of the Jordan.

The War Office published its version of the maps in 1879, with publication of the PEF maps delayed a year by agreement. The twenty-six maps published in 1880 presented a Palestine conforming to the biblical formulae "from Dan to Beersheba" for the north-south axis and "from the Jordan to the Sea" for the east-west axis; thus was born the Holy Land as a potential holding of a Protestant empire (see figure 8). The logic of mapping suggested that just as the territory could be contained and abstracted, so could it be ruled and administered as a colony. From a strategic perspective, a map of the entire Jordan River Valley remained essential: the French and Russians contemplated attacks on the Ottomans along this route, the Germans were building a railroad from Damascus to Haifa, and the British knew that the Jordan River Valley offered a route to the Suez Canal and their holdings in Egypt and India. In 1884 the PEF published an eastern map of "only 500 square miles of the land east of the Jordan," with additional knowledge acquired from the maps of Gottlieb Schumacher, a German railway engineer friendly to the PEF who surveyed the area from the eastern shore of the Galilee to just south of the Yarmouk River.[9] The contrast between the totalizing knowledge produced by the Survey of Western Palestine and the piecemeal familiarity with Eastern Palestine impacted the sweeping success of General Allenby's offensive west of the Jordan as well as his failed attempt to conquer the city of Salt to the east.

While fighting the Ottomans, the European powers began drawing new lines across the Middle East that ultimately resulted in the enduring shape of Middle Eastern nations. The hypothetical divisions in circulation included T. E. Lawrence's liberated Arab world marked by a "Greater Syria" ruled by the Hashemite heir Faisal, with whom Lawrence had led the Arab revolt, and zones of British influence and control administered by Faisal's brothers, Zeid and Abdullah (see figure 9), and the Sykes-Picot Agreement in which the

---

8. Moscrop, *Measuring Jerusalem*, 95.

9. Moscrop, *Measuring Jerusalem*, 146.

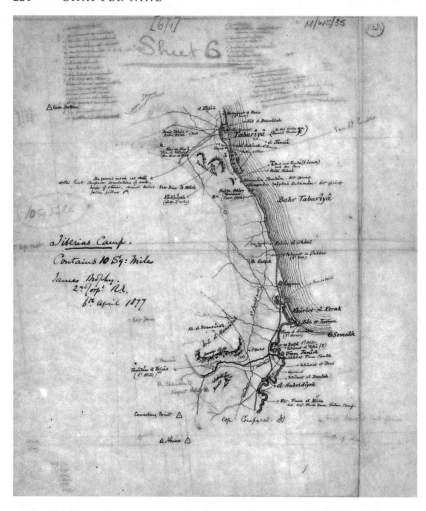

8. Tiberias Camp, sheet six of the Palestine Exploration Fund Map. Courtesy of the Palestine Exploration Fund archives, London.

British and French diplomats Sir Mark Sykes and François Georges-Picot shaded the map with gradated zones of British and French control and influence (see figure 10). A Holy Land, a jewel in the crown of Muslim lands, was designed by the Sykes-Picot Agreement with multiple Christian influences in which the northern Galilee region fell under direct French authority and French, Italian, Russian, and Greek control traversed a southern zone encompassing Jerusalem. A reflex of this model later appeared in the idea of an internationalized Jerusalem outlined in the 1947 partition plan, but neither scheme ever materialized. The British who mapped and conquered Palestine had little intention of allowing Catholicism and Orthodoxy to prevail in the Holy Land and even less intention of allowing another standing army

9. The hand-drawn map of T. E. Lawrence reflecting his vision of the Middle East following World War I. Courtesy of the National Archives (UK).

near the Suez Canal—"the 'lifeline' of the British Empire."[10] After the British army under General Allenby effectively took control of the region by 1918, the British rejected the Sykes-Picot configuration and established temporary demarcations of British administrative control.[11] While British actions made it clear that the Sykes-Picot map was no longer operative, the precise borders

---

10. Gideon Biger, *The Boundaries of Modern Palestine, 1840–1947* (London: RoutledgeCurzon, 2004), 42.

11. On the eve of the capitulation, specifically in October 1918, General Allenby reported to the War Office that he had divided Syria, Lebanon, and Palestine into three administrative areas called Occupied Enemy Territory (OET): North (Lebanon and the Syrian coast), South (Palestine), and East (Transjordan and the interior of Syria). Responsibility for the military government rested in the hands of Allenby until December 1919, when the French were given control of OET-North and OET-East was transferred to a provisional Arab government under Emir Faisal. The boundaries of these administrations, it must be pointed out, departed from the Sykes-Picot Agreement concluded in October 1916 between Britian, France, and Russia for the partitioning of the Ottoman Empire. Note that the Jordan divides zones in Allenby's scheme as well.

10. The personal map of Sir Mark Sykes used in negotiating the 1916 Sykes-Picot Agreement. The darker gray area along the Mediterranean coastline fell under direct French control. The Haifa Bay, the dark half-circle beginning at the coast, was part of the sphere of direct British control. The light gray area between the Jordan River and the Mediterranean Sea was designated an international sphere with French, British, Italian, and Russian influence. The A area fell under French influence and the B area under British influence. Courtesy of the National Archives (UK).

of the new Middle East remained undetermined during the postwar era of international conferences and local negotiations.

The ultimate British aspiration seems to have been direct control over oil-rich Mesopotamia and its Persian Gulf ports, political and economic influence over a semi-independent Arab state in the desert between the Euphrates and the Hijaz railway and a Jewish client state at the end of an oil pipeline near the Haifa port.[12] A beholden Jewish citizenry would, from the British point of view, be the group most likely to ensure that oil shipments reached European destinations and to develop infrastructure and industry around the pipeline. This vision of a self-sustaining colony that would front its own capital and labor contributed to the British Cabinet's acceptance of the 1917 Balfour Declaration that pledged "the establishment in Palestine of a national home for the Jewish people." Creating a Middle East conducive to the British pipeline, however, required direct negotiations with the French and other European

---

12. British National Archive, FO 608/278, FO 371/3385.

powers as well as a delicate balancing act in which Arab and Jewish national movements would be at once stimulated and reconciled. Since a binational state or a Jewish state with full Arab enfranchisement seemed the most tenable configurations (the Jewish state remained hard to fathom for some colonial officials either because of the greater Arab population at the time or because of an innate aversion to Jews), the British strategy of promising both groups national independence proved disastrous.

For the Hashemite sons of the sharif of Mecca, Arab nationalists, and Zionists, the period of indeterminate boundaries presented a golden opportunity to advance their claims along with their proposed boundaries of Palestine. Zionist representatives first submitted a confidential proposal outlining a basic boundary scheme to the British Foreign Office in November 1918:

> On historical, economic and geographical grounds, it is proposed that the boundaries of Palestine should be as follows: —
> In the East, a line close to and west of the Hedjaz Railway.
> In the West, the Mediterranean Sea,
> The details of the delimitation should be decided by a Boundary Commission, one of the members of which should be a representative of the Jewish Council for Palestine hereinafter mentioned.
> There should be a right of free access to and from the Red Sea, through Akaba, by arrangement with the Arab Government.[13]

From the Zionist viewpoint, it was essential to include the entire Jordan River Valley and a portion of the Yarmouk tributary in the nascent state (figure 11). The anticipated immigration of the Jewish masses required extensive agricultural and industrial development, and water was necessary for both. Plans to develop agricultural communities and hydroelectric power constituted the "economic grounds" for including the entire Jordan River Valley. Cognizant of the biblical romance motivating different levels of British support, the proposal also emphasized "historical grounds," meaning the two and a half Transjordanian tribes and the Jewish communities east of the Jordan during

---

13. British National Archive, FO 371/3385. British resistance to this boundary scheme is evident in the notes accompanying the document in which Arnold Toynbee notes, "These [boundaries[should not be published and it would be better if the following were instituted: An integrated Palestine, including the whole country from and inclusive of Dan to Beersheba and including the Jordan valley and the economic control of the water of the river and its tributaries." Another official hazards, "I think [it] is a somewhat controverted point. Sir Mark Sykes objected strongly." Major Ormsby-Gore comments on a later version of the proposal, "The Eastern boundary proposed here is in my opinion too far east and I do not believe that Akaba can be usefully developed as part of Palestine." FO 608/99/5.

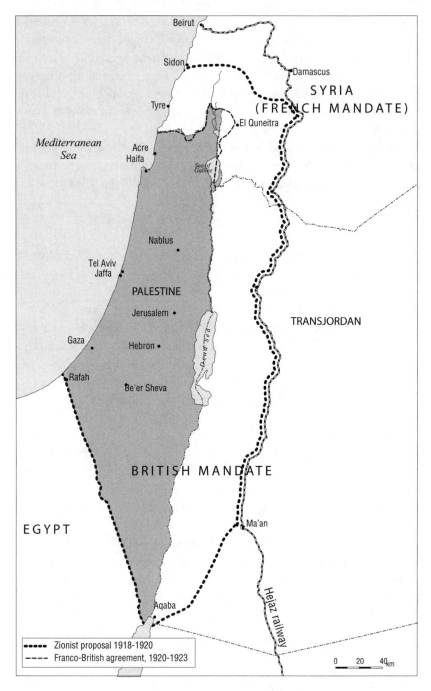

11. The Zionist territorial proposal of 1918–20 and the border that resulted from the Franco-British agreement of 1920–23. The British Colonial Office drew the borderline along the Jordan River in 1921 when it separated Transjordan from the Palestine Mandate. Soffer Mapping.

Hellenistic, Roman, and Byzantine periods. This was not simply a utilitarian act of political exegesis or something that the Zionists submitted to the British without believing themselves, but part and parcel of a project in which Zionists turned to biblical and postbiblical texts anew in order to reclaim a Jewish national history and establish a political charter. At the same time that certain texts were read with specific goals in mind, the biblical texts exerted an influence of their own. So as the Zionists cited the two and a half Transjordanian tribes as authorizing forebears, their more flexible claim to land east of the Jordan replicated the Bible's ambivalence about the Jordan as a border. In political terms, the Hijaz railway built on the footpath of Muslim pilgrims to Mecca set the limit of the Jewish national claim.

The impending peace conferences in Paris and San Remo lent a new urgency to the assertion of historical claims. The European powers planned to divide the spoils of the former Ottoman Empire; prior to the conference, frontiers were literally open. British officials were aware of but not convinced by Zionist boundary proposals.

> Palestine, for the purposes of this memorandum, may be best described as the Palestine of the Old Testament, extending from Dan to Beersheba. There is a question as to the exact boundaries, and these will have to be settled by commissioners. For the present it is sufficient to take the northern boundary as the Litani river on the coast and from there across to Banias, north-east of Lake Huleh, in the interior ... The eastern boundary is more difficult to determine. The Zionists are naturally looking eastwards to the Trans-Jordan territories, where there is good cultivation and great possibilities in the future. There is a general desire to get out of the steaming Jordan Valley and on to the uplands beyond; and we are undoubtedly face to face with a movement which is growing on the part of the Zionists that Palestine is now to include what it has not included for many centuries—if it ever did—and what would be regarded by the Arabs as part of their domain. It is assumed, however, for the purposes of this memorandum, that the Jordan and the Dead Sea will form the frontier on the east.[14]

As Zionists probed the Bible for the coordinates of the national home, the British limited Jewish territorial aspirations to regions recorded as belonging to Jews in antiquity. In the memorandum, Sir Erle Richards concedes to the difficulty of determining the eastern boundary. He grasps the economic

---

14. Palestine: Memorandum by Sir Erle Richards submitted to Lord Curzon and circulated for the consideration of the Eastern Committee, 4 Feb. 1919, British National Archive, FO 608/99.

and geographic logic of the Zionist proposal, but rejects it on political and exegetical grounds. If ancient Israel "ever did" include lands east of the Jordan, Richards reasons, then the many centuries without Jewish presence signaled a lesser degree of attachment. This constitutes an example of how the Priestly vision of Israel's land through its incarnation in the Palestine Exploration Fund map guided British policy.

The Zionist Organisation filed an interpretive rebuttal that maintained the vision of a British Palestine encompassing both banks of the Jordan.

> The fertile plains east of the Jordan, since the earliest Biblical times, have been linked economically and politically with the land west of the Jordan. The country which is now very sparsely populated, in Roman times supported a great population. It could now serve admirably for colonisation on a large scale. A just regard for the economic needs of Palestine and Arabia demands that free access to the Hedjas Railway throughout its length be accorded both Governments.[15]

In contrast to the colonial sense of the Jordan as a dividing line, the early Zionists stressed its connective qualities and hewed to a Deuteronomistic line in arguing for the multiple links between the banks "since the earliest Biblical times." History attested to the economic potential of an integrated valley, and should not England follow the example of Rome?[16] In another bold rhetorical twist, the rebuttal asks for more—access to the Hijaz (Hedjas) Railway—in response to the British memorandum restricting the area of the Jewish National Home.[17] Apparently, Chaim Weizmann, head of the Zionist Commission, tem-

---

15. Ibid.

16. As noted, references made to "the historic boundaries" by Zionist or British leaders always connoted the boundaries of the Bible. In the long sweep of history, the Jordan had both been and not been a border. "The Romans had formed two provinces on the two sides of the Jordan river and the Arabs ruled the eastern bank with a different and separate name, Urdun, as early as the 7th century CE. The First Crusader Kingdom (1099–1187) had controlled all the lands west and east of the Jordan river and the Arava valley, although its rule did not last long. During the centuries that preceded the determination of the boundaries of modern Palestine, the area east of the Jordan was usually controlled by the ruler of Damascus, and not by whomever was ruling western Palestine." Biger, *The Boundaries of Modern Palestine, 1840–1947*, 159.

17. A memorandum submitted to the Foreign Office weighs the ethnic and economic factors: "The Zionists have aspired to include part of Trans-Jordania, the ancient land of Reuben, Gilead, and Manasseh in the new Palestine, but this claim will undoubtedly be disputed by the Arabs, and by the local population. Either the Jordan river or a line drawn a few (not exceeding 10) miles east of the main stream of the Jordan should form the eastern boundary. From an economic point of view the latter is to be preferred as the Jordan could in skillful hands become a great source of industrial water power and intensive perennial irrigation and thus should be under one control. It should be laid down that the water of the Jordan belongs to Palestine." BNA 161 588 24 January 19 FO 608/99.

pered the claim by assuring British negotiators prior to boundary discussions that "the Jordan as an eastern frontier would suffice as a commencement."[18]

Arab nationalists also lobbied the European powers attending the Paris and San Remo peace conferences. Like the Zionists, Arab nationalists possessed a promise of independence from British officials and an authorizing scriptural-geographic precedent. The 1915 correspondence between Sir Henry McMahon and Sharif Hussein of Mecca committed the British to an Arab state abutting the Mediterranean in exchange for concerted Arab revolt against the Ottomans.[19] The Hussein-McMahon correspondence entailed its own set of boundary restrictions, yet Arab nationalists began thinking in terms of Greater Syria, a vast Arab kingdom mirroring the Levantine portions of the medieval Islamic empire stretching from the Euphrates to the Mediterranean. And like the Zionist representatives, the ruling Hashemites of Mecca were willing to make concessions while insisting upon their historical rights.[20]

After Sharif Hussein's son Faisal marched into Damascus as the leader of Greater Syria, he repeatedly reminded British officials of the pledge made to his father and gained the support of the Syrian and Palestinian National Congresses. For example, a 1919 decision submitted to the peace conference in Paris by all the delegates of the districts of Palestine and Southern Syria states: "We desire that our district Southern Syria or Palestine should not be separated from the Independent Arabic Syrian Government and to be free from all foreign influence and protection."[21] The Syrian National Congress crowned Faisal king of Syria just one month before he learned at San Remo that his kingdom fell under the direct jurisdiction of France, which felt bound by no promises and promptly expelled him. For Palestinian nationalists, "Faysal's

---

18. A telephone call with Weizmann, for example, is recorded by Major-General Thwaites, FO 608/99.

19. The 1915 pledge states that "Great Britain was prepared (with certain exceptions) to recognize and support the independence of the Arabs within the territories included in the limits and boundaries proposed by the Sherif of Mecca; and Palestine was within those territories. This pledge was restricted to those portions of the territories in which Great Britain was free to act without detriment to the interests of her Ally, France." BNA FO 608/99.

20. Despite the relatively parallel scrambles to assert the boundaries of independent states, the British handled the two groups in different ways. For example, where the British always made clear to the Zionists that the Jewish state would have to include and economically benefit Arabs, they never admitted to the Arabs what they promised the Jews. BNA FO 608/99.

21. BNA FO 608/99. The Syrian Congress in Damascus issued an official statement to Faisal and asked that he present it to the various representatives at the peace conference: "The Great Danger of Zionism that threatens the southern part of our country and the rest of Syria in general from an economical, political and social point of view, intending to make Palestine a Native residence for the Jews, has caused the Syrians to unite and refuse the separation of Palestine from the rest of the Syrian countries, (being a natural and inseparable portion of land connected with Syria)."

(temporary) government (in Syria) represented a crucial step for the realization of the dream of Arab independence. More importantly, a strong and independent Arab government in Damascus was from their perspective a great source of strength for their struggle against Zionism."[22] When Faisal, more invested in the urban centers of Syria and more concerned with his standing vis-à-vis colonial administrations, reached an agreement with Chaim Weizmann not to interfere with Jewish immigration as long as Zionists purchased and developed only those lands not being cultivated by Arab farmers, Faisal's Palestinian supporters split from his movement and articulated a nationalist position "identified, first and foremost, with a specific territory—Palestine—and against a specific enemy—Zionism."[23]

With the exception of an internationalized Holy Land, the San Remo conference produced a Middle East that resembled the Sykes-Picot Agreement. The French received mandates in Lebanon and Syria and the British mandates in Mesopotamia and Palestine. The exact location of these places remained unspecified, and so measuring, mapping, and negotiating continued following the conference. Transjordan had started to take shape at the Anglo-French Convention in December 1920, where the Yarmouk River was designated as the border between French Syria and British Transjordan. Transjordan fell under the Palestine mandate approved at San Remo, but the British modified the mandate in 1921 through Article 25, which stated that the provision for a Jewish National Home did not apply east of the Jordan. Splitting the mandate, reasoned British officials, would allow them to fulfill promises made to Arab and Jewish nationalists alike. Approving the modification in 1922, the League of Nations granted international recognition to the Jordan boundary and, one could add, to the Palestine Exploration Fund map.[24] A single mandate governed Palestine and Transjordan, yet the Jordan River border distinguished the two "countries" in British as well as international eyes.[25] With his brother Faisal crowned king of Iraq, Abdullah ibn Hussein became the emir of Transjordan. Abdullah kept desirous watch over affairs in Palestine,[26] and Zionists

---

22. Muhammad Y. Muslih, *The Origins of Palestinian Nationalism* (New York: Columbia University Press, 1988), 119.

23. Muslih, *The Origins of Palestinian Nationalism*, 202.

24. Mr. Amery: "On the 23rd September, 1922, the League of Nations approved a definition of Transjordan as territory lying to the east of a line drawn from a point two miles west of the town of Akaba up the centre of the Wady Araba, Dead Sea and River Jordan to its junction with the River Yarmouk; and thence up the centre of that river to the Syrian frontier. It is clear from this definition that Transjordan territory was recognized at that time as extending as far south as the town of Akaba." BNA FO 608/99.

25. "The eastern borderline was defined in writing, although it wasn't marked in reality, and it appeared on maps only two years later." Biger, *Boundaries*, 183.

26. "The obscurity regarding the borderline's location, and especially the fact that it remained unmarked, allowed Emir Abdullah, the ruler of Tran-Jordan, to contest the legality of the line and

from time to time petitioned the British for the right to develop lands east of the Jordan, but by 1923 the Jordan River had force as a boundary.[27] While it remained easy enough to cross, the boundary substantiated both the McMahon pledge and the Balfour Declaration and suggested that the British could satisfy both Arab and Jewish national aspirations. The immediate effect of the demarcation was the birth of Transjordan, but it also set the first boundary contested by the two national groups now found between the Jordan and the Sea—Palestinians and Israelis. A delimited territory now had to satisfy two movements growing up in opposition to one another.

One Land, Two Maps

Because no other comprehensive maps of their prospective nations existed, the Palestinian and the Jewish national movements absorbed the British imperial map.[28] The strange effect of two aspirants adopting the same map was that despite their co-presence, each movement perceived the map as theirs alone. Mandate Palestine—a colony where British sympathies vacillated between the local Arabs and the immigrant Jews—set the template for two separate homelands. So long as the British remained uncertain of where precisely Palestine began and ended, the invested nationalists perpetuated multiple ideas about the scope of the nation and its relation to other political entities. Once the British promulgated a definitive map of Mandate Palestine, Jewish and Palestinian nationalists looked to a defining territory with certainty.

Zionist writings attest to the accommodation of imperial standards in national mythology. Prior to the 1922 White Paper, documents concerning the historic land of Israel often mention the two and a half tribes and use biblical

---

its location. He raised a demand for passing the Semah triangle and the Negev area to Trans-Jordan, shortly after the borderline was officially announced, by claiming that these areas belong to the *vilayet* of Syria (A-Sham), during Ottoman rule . . . Abdullah's demand was rejected." Biger, *Boundaries*, 184.

27. From 1921–1932 Transjordanian sheikhs, leaders and Emir Abdullah voiced and even published their readiness to sell and jointly develop land with the Jews. In the October 23, 1921 edition of the newspaper, Al-Jazeera, an article expressed the impatient wish of many Arab landowners on the East Bank to sell their lands to Jews. The article also alerted certain readers about the "new danger" that they faced, see Zvi Ilan, *Attempts at Jewish Settlement in Trans-Jordan 1871–1947* (Jerusalem: Yad Ben Zvi, 1984), 363. Due to the dire economic state of Transjordan during this period, the sheikhs and landowners saw such partnerships as their only available means of providing for their tribes and people. The British posed the greatest obstacle to the fruition of such desires citing concerns about transportation and security, although the real motivation was likely the desire to prevent an Arab uprising and limit the Jewish sphere of influence.

28. "It is clear that British map-making had a greater direct impact on the Zionist population of Palestine than on the Arab one. It certainly influenced the Arab population as well, but in more subtle and indirect ways." Efrat Ben-Ze'ev, "The Cartographic Imagination: British Mandate Palestine," in *The Partition Motif in Contemporary Conflicts*, ed. Smita Tewari Jassal and Eyal Ben-Ari (New Delhi and Thousand Oaks: Sage Publications, 2007), 104.

names for Transjordanian regions.[29] During 1917–20, Zionist leaders "put forward their own maximalist interpretation of the Balfour Declaration," in which the Jewish homeland spanned both banks of the Jordan.[30] Working toward this vision, they opened negotiations with Emir Abdullah and Bedouin sheikhs regarding the purchase and development of east bank lands. Despite the fact that Transjordanian leaders were more amenable to Jewish settlement than the Palestinians, no such arrangement came to pass due to insufficient Jewish funds, growing Arab resistance, and British opposition.[31] Some biblical ambivalence seems to have been replicated in the tepid approach and easy surrender of Zionist attachment to the east bank.

When the 1922 White Paper bestowed Transjordan on Emir Abdullah, the Zionist leadership acquiesced and concentrated its efforts west of the Jordan. As the river gained force as a natural delimitation of Zionist aspiration, Ze'ev Jabotinsky split from the mainstream movement and founded the World Union of Zionist Revisionists, that insisted, in the words of its popular song, "There are two banks to the Jordan, this one is ours and so is the other."[32] With the exception of the Revisionists, who would later come to power as the Likud party, Zionists then operated under the assumptions that the Jordan had been and would be the eastern border and that there was a qualitative difference between the two riverbanks. Thereafter, the Jordan map justified expansionist tendencies that halted at the Jordan. Discussions about a possible transfer of Arab Palestinians to Transjordan reveal the desire that the Jordan separate the Jewish state from the Arabs.[33]

The Palestinian homeland was similarly based on a combination of national conviction and imperial accommodation. A sense of territorial unity had survived Ottoman administrative divisions in both Christian and Muslim spheres through the religious acknowledgment of the interconnection of holy sites.[34] Although the Palestinian Arab Congresses held in the early

---

29. In a letter written by Chaim Weizmann to Winston Churchill just before the 1921 Cairo Conference concerning more precise definitions of the mandates, Weizmann presses for the inclusion of "Gilead and Moab" in Palestine. "This was the last public appeal made by a Zionist representative for the eastern boundary of Palestine." Biger, *Boundaries*, 167.

30. Avi Shlaim, *The Iron Wall: Israel and the Arab World* (New York: W. W. Norton, 2001), 8.

31. Yitzhak Gil-Har, "The Separation of the East Bank from the Land of Israel," *Katedra* 12 (1979): 47–69 (Hebrew).

32. The refrain rhymes in Hebrew. שתי גדות לירדן/זו שלנו זו גם כן. In 1923, Jabotinsky resigned from the Zionist Executive, charging that its policies, especially its acceptance of the 1922 White Paper, would result in the loss of Palestine.

33. See T. Reinhart, "The Second Half of 1948," in *The Other Israel*, ed. R. Carey and J. Shainin (New York: New Press, 2002), 16–19; Chaim Simons, *International Proposals to Transfer Arabs from Palestine, 1895–1947: A Historical Survey* (Hoboken, NJ: Ktav, 1988).

34. For Muslims this sense of Palestine as a country went back to the *Fada'il al-Quds* (or "merits of Jerusalem") literature, which described Jerusalem and holy sites and places of note throughout

1920s rejected the British Mandate as well as the 1922 White Paper, striking and rioting against both, the borders of the demanded Palestine were constituted by the Mandate. Mandate borders became fixed in Palestinian consciousness, with the 1968 Palestinian National Charter making the case most bluntly: "Palestine, with the boundaries it had during the British Mandate, is an indivisible territorial unit."[35] The period of opposing the British and rioting against the Jews became enshrined as a paradigmatic period of Palestinian nationalism.

With Palestine under their control, the British continued surveying in order to establish internal boundaries for districting and taxation.[36] Because it was a work in progress, the nationalists inferred that their efforts could impact the map. In historical terms, this means that between 1922 and 1937, Palestinian and Jewish nationalists experimented with ways to draw themselves into the map. Each group eventually landed on a technique that it continues to employ in the twenty-first century. For the Palestinians, uprising became the means of influencing policy; for the Zionists, settlement constituted the act of self-assertion par excellence. Palestinians rioted in 1920, 1921, and 1928, when the Zionist bent of the early mandate became clear. From 1936 to 1939, the Palestinian countryside and city alike participated in a general strike and revolt against Jewish immigration and land transfer. The revolt aimed to destabilize British control and thwart the Zionist enterprise. At the same time, the Jews in Palestine established collective settlements as borderlines and frontlines in the conflict. Through development, defense, and expansion, the Zionists sought to subdue Palestinian revolt and implement a Jewish state.

The 1937 Palestine Royal Commission Peel Report recommended partition as a solution to the ongoing conflict (see figure 12). It specified that the largest area including the central mountain region and Negev desert would belong to an Arab state, and a long coastal area connected to the Jezreel and Jordan Valleys would make up a Jewish state. A reduced mandated area around Jerusalem would interrupt the contiguity of both states, and a population transfer, resembling that between Greece and Turkey, would ensure that all Jews and Arabs resided in their designated territory. Although the Peel Commission recommendations were not implemented, partition came into vogue as viable solution to the conflict. In 1938 variations on the Peel map circulated among

---

Palestine. Rashid Khalidi, *Palestinian Identity: The Construction of Modern National Consciousness* (New York: Columbia University Press, 1997), 29.

35. In 2001, the PLO leader Faisal Husseini described the dream of a "Palestine from the [Jordan] river to the [Mediterranean] sea." Quoted in Efraim Karsh, *Palestine Betrayed* (New Haven: Yale University Press, 2010), 249.

36. On British cartographic endeavors during the Mandate, see Dov Gavish, *A Survey of Palestine under the British Mandate, 1920–1948* (London: RoutledgeCurzon, 2005).

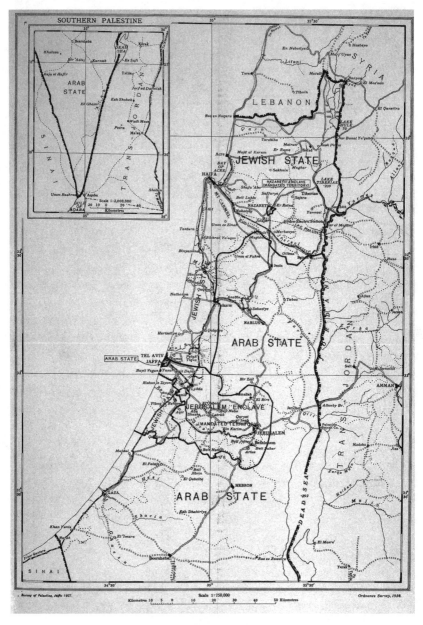

12. A plan of partition that accompanied the 1937 Peel Commission Report. Courtesy of the National Archives (UK).

British and Zionist officials,[37] and when the British ultimately departed in 1947, they left a Palestine partitioned by the United Nations (figure 13). The two-state solution theorized at diplomatic talks in the early 1990s represents to many a viable outcome. Although the separation barrier, discriminatory zoning of the West Bank and Gaza Strip, and Israeli checkpoints effect a de facto partition, the two-state solution imagines separate Israeli and Palestinian states adjusted to the boundaries that preceded the 1967 war.

Partition proposals have the good intention of giving Palestinians and Israelis lands of their own and drawing a line behind which armed forces will retreat, but such blueprints also have the effect of fetishizing boundaries. The diplomatic suggestion of borders as the solution to the conflict promotes a sense that the attainment of the "original" borders is, in fact, the only resolution. Among certain sectors of Israeli and Palestinian society, the popularization of partition plans stokes an apocalyptic attachment to the borders of the homeland. Needless to say, these borders are not thought of as colonial lines in the sand, but as the divinely ordained limits of the Promised Land or the land of the Prophets. Moreover, the threat of having to relinquish—in real or imagined terms—land belonging to the map of Greater Israel/Palestine strengthens the attachment to its borders and intensifies the border mythology. Exactly as world powers begin thinking in terms of subdivided zones, Israeli and Palestinian leaders weigh whether it is better to deliver a reduced homeland, or whether to wager that war or demographic outnumbering may one day achieve the national dream. And so—without an endgame—aggressive settlement, uprising, and partition plans cycle through the decades.

No matter how many lines are drawn in order to produce a pure Israel or a pure Palestine, the two peoples sit on the same land and depend on the same dwindling resources.[38] That partition actually increases attachment to the

---

37. The theory of partition was also abandoned in 1938 by the British government's Woodhead Commission, which admitted, "We have been unable to recommend boundaries which will afford a reasonable prospect of the eventual establishment of self-supporting Arab and Jewish states." Quoted in Shabtai Teveth, *Ben-Gurion and the Palestinian Arabs: From Peace to War* (Oxford: Oxford University Press, 1985), 183.

38. For example, the Lower Jordan River "is expected to run dry by the end of 2011. The Lower Jordan River today is a highly degraded system due to severe flow reduction and water quality decline. Over 98% of the historic flow of the LJR is diverted by Israel, Syria, and Jordan for domestic and agricultural uses. The remaining flow consists primarily of sewage, fish pond waters, agricultural run-off, and saline water diverted into the LJR from salt springs around the Sea of Galilee. The river has lost over 50% of its biodiversity primarily due to a total loss of fast flow habitats and floods and the high salinity of the water. Long stretches of the LJR are expected to be completely dry unless urgent action is taken by the parties to return fresh water to the river." *Towards a Living Jordan River: An Environmental Flows Report on the Rehabilitation of the Lower Jordan River*, Friends of the Earth Middle East, http://www.foeme.org/index_images/dinamicas/publications/publ117_1.pdf.

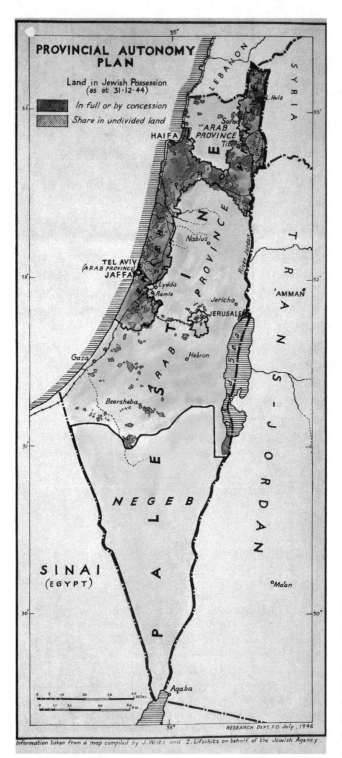

13. The proposed partition of Palestine (1943–46) from the British prime minister's confidential file. Courtesy of the National Archives (UK).

mandate borders becomes evident in the wars of 1948 and 1967. Viewing the United Nations 1947 partition plan as an anathema, the Palestinian leadership immediately began organizing attacks on Jewish communities and urging the states belonging to the Arab League to dispatch their armies.[39] The goal was to end the possibility of the Jewish state and to erase the very lines set by the United Nations. From a territorial standpoint, the war ended in Israel's favor. Other than the West Bank—absorbed by Jordan—and Gaza—which fell under Egyptian control—, Israel claimed most of the mandate map (see figure 14). Yet the 1949 Armistice Lines—particularly the Green Line delimiting the West Bank—now seemed to many Israeli leaders like an unnecessary constraint. The armistice partitions stoked the Israeli desire for a revival of "Biblical Israel," in which the entire span of the Jordan formed the border.[40] This version of biblical Israel/mandate Palestine became a territorial reality when Israel occupied the West Bank and Gaza Strip as a result of its victory in the 1967 war.

In the realm of national mythology, stories of pioneers, refugees, and freedom fighters insist upon the Jordan as the eastern border of the homeland (*ha'aretz* and *il-balad*). The pioneers in question are Arabs from east of the Jordan who establish cities and towns in the West Bank and Jews who found the kibbutz movement on the banks of the Jordan. The refugees are, of course, the Palestinians who fled or were expelled by Israeli forces during the wars of 1948 and 1967 as well as the Jews who fled European pogroms or survived the Holocaust. The freedom fighters are Israelis and Palestinians who breached the borders in the name of national assertion. In the stories, pioneers, refugees, and freedom fighters all function as national heroes whose actions redraw cease-fire lines and elevate the Jordan River to the level of national symbol.

The lore of the border indicates an initial stage of collective expression that generates national identity. Border stories set the limits of group identity as they order geographical imagination.[41] Admittedly, in reference to Semitic traditions, "folklore" is something of a suspicious term with an orientalist past that includes ethnographic studies of Palestinian village life by Western scholars in order to reconstruct the customs and mores of biblical times,[42]

---

39. On the Transjordanian and Iraqi plans to occupy the whole area of mandate Palestine, see Karsh, *Palestine Betrayed*, 89.

40. Moshe Dayan, for example, "held the view that Israel's borders had to be expanded so as to rectify the omissions of the 1948 war. In particular he felt that the border with Jordan, which he himself had helped negotiate, was impossible to live with and had to be replaced by a natural border running along the Jordan River." Shlaim, *The Iron Wall*, 100.

41. National lore as well as identity exist in a state of flux and undergo a constant process of change despite the claim of tradition as fixed and perpetual. See Bisharat, "Exile to Compatriot," 203–33.

42. See Mun'im Haddad, "The Relationship of Orientalism to Palestinian Folklore," in *Folk Heritage of Palestine*, ed. Sharif Kanaana (Tayibeh: Research Center for Arab Heritage, 1994).

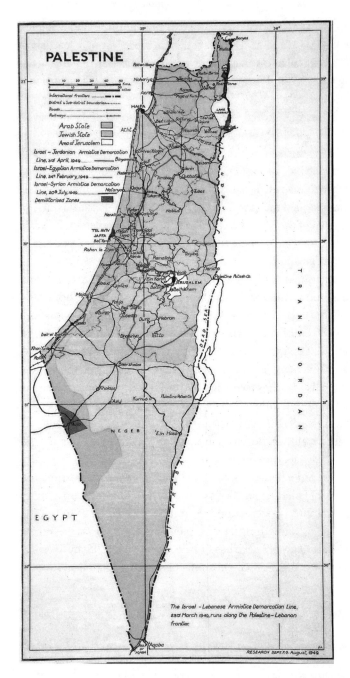

14. The shaded regions indicate the Arab state, the Jewish state, and the international area of Jerusalem as approved by the United Nations in 1947. The Armistice Lines of 1949 show the area conquered by Israel in the 1948 war. Following the 1948 war, the Gaza Strip fell under Egyptian control and the West Bank became incorporated into Jordan. These areas were occupied by Israel following the 1967 war. Courtesy of the National Archives (UK).

as well as the colonial claim that where the Palestinians had lore, the British, and later the Israelis, had history. At the same time, the Palestinian folklore movement has played a central role in forging collective identity and building nationalist sentiment, and occasionally in resistance.[43] During the first Intifada, for example, the performance of traditional music and dance at cultural festivals functioned as an open form of resistance against the occupation and the Israeli army presence. The creation of a distinctly Israeli culture relied on the synthesis of the traditional lore of Jews from vastly different host countries, the revitalization of tales of the land of Israel from Hebrew sacred texts, and the popularization of new traditions, many of which were borrowed from Palestinian or Bedouin culture. While creating a new Jewish culture, many immigrants to Israel also sought to distance this culture from the traditions and folk customs that defined life in the villages of the Diaspora. Thus both Israeli and Palestinian cultures have forged a relatively novel body of folklore that draws from the past while signaling the difference between the ancestral transmitters of the old traditions, who proved weak or passive in the face of enemies, and the "new Jew" and "the new Palestinian," who pledge to defend themselves, resist, and fight.

Pioneers

The city of Ramallah's founding begins with a Jordan River crossing by the Haddadins, an Arab Christian clan that resided east of the Jordan during Ottoman rule in the sixteenth century. Ramallah, interim capital of the Palestinian state, hosts the governing bodies of the Palestinian Authority and other important political, social, and cultural institutions. Six miles north of Jerusalem, cosmopolitan Ramallah has the status of a center in Palestine (or at least the center of the West Bank), although the movement of its residents and the sphere of its influence are restricted by the Kalandia checkpoint and the Israeli separation barrier. The account of Ramallah's founding is transmitted through various means. As it extols their ancestors and explains their interconnection, the story is a staple of Ramallah's Palestinian Christian culture.[44] In interviews conducted with multiple generations present, members of the

---

43. "The Palestinian national heritage is taking shape in accordance with the needs of the nationalist movement. As poet Sameh al-Qassem points out: 'We have folklore—so we exist.'" Meron Benvenisti, *Sacred Landscape: The Buried History of the Holy Land since 1948* (Berkeley: University of California Press, 2000), 262.

44. A man in his sixties from Ramallah residing in San Francisco explained that the story has been passed down as the central identity narrative of Ramallah Christians: "My father told me this story and so did my grandfather. When we were little, our great-grandmother told all the children, 'we are from so and so.'" Interview, San Francisco, April 17, 2001. All interviews were conducted in confidentiality, and the names of narrators are withheld by mutual agreement.

youngest generation told me, "we were first in Ramallah," "we are descended from the seven brothers," or "we are descended from the Haddadin," yet used the occasion of my interview as an opportunity to brush up on the story by querying older relatives who narrated with detail and animation. Genealogical records of Ramallah Christians are meticulously maintained and available in historical studies, posters, and websites. The Haddadin clan division according to seven families continues to organize Ramallah Christians, who host annual reunions in the West Bank and abroad. The importance of the narrative is not restricted to Christians; the Haddadin migration across the Jordan opens accounts of Ramallah's history as narrated in monographs and websites. Palestinians from Ramallah, as well as from other regions, cherish the origin of a purely Palestinian Arab city whose rise predates the Western colonial reinvention of the Middle East and, in certain instances, marshal it as a counterclaim to the Israeli pioneer myth.

The story begins east of the Jordan in the region of Kerak and Shobak in 1550. The Haddadins had settled in the area, farming and increasing their wealth by shepherding cattle and sheep. They live "with al-Qayasilmah Bedouins around them," and they gain "the trust and love of the Bedouins because of their kindness and charity."[45] The balance of power is tenuous as the Haddadins, along with the other tribes in the region, are ruled by a despotic Bedouin prince, Dhiab Ibn Qaysum. One day when Rashid Haddadin, chief of his clan, hosts Ibn Qaysum, news reaches him of the birth of a daughter. Ibn Qaysum proposes that after the girl matures, she be wed to his son. Rashid, not wanting to offend his guest and assuming the arrangement to be either false formality or a joke, agrees. From the Christian perspective, such intermarriage is anathema to the point of being a profound impossibility. As one narrator explained, "He thought that it was just a joke because no way do Christians and Muslims marry."[46]

For Ibn Qaysum the arrangement is no joke; when Rashid's daughter reaches marriageable age, he calls for the girl to be brought forward and the marriage to be consummated. When Rashid initially resists, the prince responds with violence. The nature of this violence differs in the versions. In one account Ibn Qaysum kidnaps two Haddadin children and kills them by rolling them down a mountain beneath a stone.[47] In another, hostilities erupt between the tribes, two Haddadins are killed, and, upon prevailing, Ibn Qay-

---

45. Khalil Abu Rayya, *Ramallah: Ancient and Modern* (Ramallah: American Federation of Ramallah, 1980), 12–13.
46. Interview, San Francisco, April 17, 2001.
47. Abu Rayya, *Ramallah*, 12.

sum threatens to kill all the men and take the girl.[48] Whatever the nature of the threat, it motivates Rashid to employ a different strategy. He requests that Ibn Qaysum grant the family some time to prepare his daughter for marriage and, having bought himself a few days, alerts all Haddadins to pack what they can and prepare for flight. Under the veil of darkness, they load their possessions on animals and head in the direction of the Dead Sea. Crossing a narrow bend in the Jordan, they reach safety and secure themselves through a clever ruse. Rashid and his brothers scatter sharp shards of metal at the crossing point so when Ibn Qaysum and his men pursue them on horseback, the horses' legs are cut, rendering them unable to reach the other side of the Jordan. Free of threat, the west bank opens to the Haddadins as a pristine sanctuary distant from the dangers that lurk in the more familiar east bank. The Haddadins tame the imposing landscape by felling trees, building homes on steep inclines, and establishing agriculture.

Here the Jordan operates as a spatial border whose traversal entails a break with a past in which the proximity of neighbors leads to inevitable tensions. When the tensions flare into hostilities, the Haddadins exhibit intrepid resolve and a pioneering spirit. To a certain degree, the recitation of the story has the effect of conferring these qualities on Ramallah itself in the sense that stories of founding fathers are used as a means of making space meaningful by endowing it with personal qualities. The different versions of the Haddadin story air the complexity of the relationship between Palestinian Christians and Muslims. The story begins on the uneasy note that the Haddadins are encircled by the Bedouins and lorded over by a despotic prince; instability results from the fact that the lines of distinction are not clearly drawn. Such lines become clear, however, when Muslim neighbors initiate a marriage with a young Christian woman and the Christians in turn assert the Jordan as a boundary between them. The story does not end on a note of religious distance or spatial separation, but rather by pointing toward the potential peace between Christians and Muslims in a city established by Christians who fled persecution. In Khalil Abu Rayya's printed version, Rashid Haddadin has a strong alliance with his Muslim neighbor, Husayn Banawiyah. As Husayn likewise feels oppressed by Ibn Qaysum's tyranny, the two organize their families and flee together. In Naseeb Shaheen's version, the wronged party is Rashid's brother, Sabra, who eventually returns eastward and reconciles with Ibn Qaysum.

Another context in which I heard the story of the Haddadins was in Christian households following their harassment or the destruction of their prop-

---

48. N. Shaheen, *A Pictorial History of Ramallah* (Beirut: Arab Institute for Research and Publishing, 1992), 10.

erty by Islamicists. Older family members would tell younger ones or, in my case, foreign guests the story of the Haddadins as a means of contrasting an initial Christian hospitality to Muslims in the Ramallah region with occasions of Muslim intolerance toward Christians. It should be emphasized, however, that Palestinian Christians feel bound to Palestinian Muslims due to the collective national struggle and that in one instance a Christian narrator told me the story of the Haddadins as well as the description of how her family car had been torched by Islamicists in the company of a Muslim friend and landlady who disparaged the "shabab" (youth active in Palestinian uprising) who burnt the car and took pride in the qualities of the Haddadin.[49] Recalling the idyllic beginnings of Ramallah as a refuge fosters nostalgia at the same time that it suggests that periods of threat and oppression can come to a definitive close. As Palestinians perceive no practical end to such a period in the near future, the story offers comfort and hope for the future in the form of praiseworthy forerunners.

A parallel story tells of the founding of Bir Zeit, a college town suburb of Ramallah known for its church and olive trees. The fame of Bir Zeit University, a crossroads of various Palestinian ideological movements, spread in the first Palestinian uprising as a place of clashes with Israeli soldiers and closures. A thirty-year-old narrator who grew up in Jordan with a mother from Bir Zeit recalled that he had heard a similar story prior to the first Intifada at a summer gathering where the people of Bir Zeit hosted a reunion for Bir Zeitis living in the Diaspora. After I mentioned the Haddadins to him, he responded, "Actually, many villages were founded this way. In those days, the West Bank was like a frontier." Solidifying the analogy, he compared the legend of Bir Zeit's founding to "the stories you have of Americans settling the West."

> This is the story of the Mudainat.[50] They came from Kerak a long time ago. It's the same story like the Haddadins. They moved in small groups of people using horses for all of their stuff and they crossed the river. I don't know the details of them crossing the river, but they chose Bir Zeit. It was empty then. They put up the tents and they started living,

---

49. Interview, Beit Hanina, February 10, 2000. The cause of Palestinian nationalism has to a certain degree united Christians and Muslims. For the tension that persists between the groups despite their common cause, see Anthony O'Mahony, *Palestinian Christians: Religions, Politics, and Society in the Holy Land* (London: Melisende, 1999).

50. The narrator explained that the Mudainat clan is made up of five families: "There are five main families of Bir Zeit, and from each family there are four smaller families. The main ones are Kasis, Naser, Abdallah, Samandar, and there is one more, I forgot the name. And from Kasis, there are four main families like Dar Jaser, Dar Harb, like that. I'm from Kasis, the main Kasis."

making life. It was desert then. That's how it goes with all the small cities in the West Bank, that's how it is. Everyone moved from point A to B and everyone chose his land, like here, in the States, with the pioneers. Remember how in the United States, there was the law that if you go to this point of land and put your flag, that was your land. This was the same thing.[51]

The Christian Haddadin and Mudainat families share a point of origin in Kerak. In the versions of the story of Bir Zeit's founding that provide a motivation for the emigration, it is always a forceful marriage proposal on the part of Muslims. Both the Haddadin and the Mudainat families flee the east bank in order to escape a perceived threat to the integrity of the family structure and find safety and freedom on the west bank. In most versions of the story, the pioneers find the west bank destinations empty. However, a female narrator of the Bir Zeit frontier story explained, "Well, actually, there were already Muslims there, but they made a treaty with them."

For Palestinian Christians, the Jordan is doubly connected to their origin as the site of the baptism of Jesus Christ as well as a barrier crossed in the migration of their ancestors.[52] Both Jordan "crossings" usher a wave of Christianity into the Holy Land. The Jordan plays an important religious role in the lives of Palestinian Christians as the site of baptism as well as an annual pilgrimage to honor the Feast of John the Baptist. Al-Shariah ("ish-shri'ah" meaning "the Watering Place" and also "the Law"), the Palestinian Christian name for the Jordan, is the widespread appellation of the Jordan in Palestinian culture. Palestinian Christian pioneer tales have assumed national importance: they inscribe Christian inclusion in the Palestinian collective in the lore of the land while also claiming forerunner status in the interim Palestinian capital of Ramallah. These pioneer tales, however, are also claimed as part of the narrative legacy of Muslim Palestinians. In the Palestinian as well as the Israeli context, territorial lore is no quaint activity for history buffs, but rather a means of establishing and reinforcing the claim to the land. While Occupation/Intifada stories have overwhelmed much of the narrative activity of contemporary Palestinians, both Christian and Muslim, the pioneers still function as part of the Palestinian story of national origin.

The pioneer enjoys a privileged place in Israeli national mythology as well.

---

51. Interview April 17, 2001, San Francisco.
52. For Palestinian Christian theological positions on the land, see Monika Slajerova, "Palestinian Church reads Old Testament: The Triangle of Ethnicity, Faith, Land—and Biblical Interpretation," *Communio Viatorum* 46, no. 1 (2004); 34–62, esp. 56.

The Jew as farmer, soldier, and founding citizen indicates a radical break from Jewish history and an appeal to biblical precedent.[53] In the myth, the Jewish pioneers represent the modern-day generation that crossed the Jordan and motivate Zionist action. "Halutzim," the word for Zionist pioneers, is derived from the book of Joshua where the *halutz* is a military vanguard that crosses the Jordan ahead of the other tribes (Josh 1:14, 4:12–13) and serves as the infantry in the battle of Jericho (Josh 6:7, 9). According to mythic logic, the establishment of the Jewish state represents an occasion of Jewish national rebirth and a renaissance for the Jews of the world. By analogy, those Jews who remain in the wilderness of exile are consigned to death. The eerie resonance between the death of an entire generation in exile followed by a return to homeland and the annihilation of European Jewry as precursor to the rise of the State of Israel makes the myth all the more persuasive. The pioneers, bent on settling the land against all odds, took furious notes, recording their daily lives in journals, autobiographies, meeting notes, and memory books. The archive became simultaneous with the action as the dissemination of the pioneer tales coincided with the influx of immigration. The tales from the frontier instructed the immigrants that pioneer action alone could bring Moses's—read Theodore Herzl's—dream into being by settling in kibbutzim, enrolling in clandestine military units, and withstanding hardship of heroic proportions.

No one did more than David Ben-Gurion to perpetuate this myth, which expressed the Labor ideology of Jewish metamorphosis through agriculture, territorial acquisition, and national assertion. Ben-Gurion seems to have identified with the biblical Joshua: he characterized Joshua as "the true father of the people," studied the book with the preeminent Bible scholars at the prime minister's residence, and cited its boundaries as pertinent to the modern state.[54] Following the 1948 war, Ben-Gurion publicly analogized Joshua's warriors

---

53. Attias and Benbassa, *Israel, the Impossible Land*, 155; Tamar Katriel and Aliza Shenhar, "Tower and Stockade: Dialogic Narration in Israeli Settlement Ethos," *Quarterly Journal of Speech* 76, no. 4 (1990): 366; Ze'ev Sternhell, *The Founding Myths of Israel: Nationalism, Socialism, and the Making of the Jewish State* (Princeton: Princeton University Press, 1998), 20; Yael Zerubavel, *Recovered Roots: Collective Memory and the Making of Israeli National Tradition* (Chicago: University of Chicago Press, 1995), 215.

54. See Anita Shapira, "Ben-Gurion and the Bible: The Forging of an Historical Narrative?" *Middle Eastern Studies* 33:4 (1997): 645–74; Michael Keren, *Ben-Gurion and the Intellectuals: Power, Knowledge, and Charisma* (DeKalb: Northern Illinois University Press, 1983), 100–117; Shabtai Teveth, *Ben-Gurion and the Palestinian Arabs: From Peace to War* (New York: Oxford University Press, 1985), 34–38; David Ben-Gurion, *Ben-Gurion Looks at the Bible*, trans. Jonathan Kolatch (Middle Village, NY: Jonathan David Publishers, 1972), 57–98. The conversations of the book of Joshua study group are recorded in *Studies in the Tanakh by the Study Group in the Home of David Ben-Gurion*, ed. Hayim Rabin, Yehuda Elitsur, Hayim Gevaryahu, and Ben-Tzion Luria (Jerusalem: Kiryat Sepher, 1971) (Hebrew).

and Israel's soldiers: "Not a single biblical commentator, Jewish or Gentile, medieval or contemporary, would have been able to interpret the Book of Joshua in the way the Israel Defense Forces did this past year."[55] After the Israeli victories in the 1967 war, Moshe Dayan mythologized himself as the biblical Joshua and bumped David Ben-Gurion to the place of Moses.[56] While the myth has lost ground among secularists and liberals due to increased national cynicism in the wake of the Palestinian uprising, the birth of new historicism, and the economic crisis of the kibbutzim,[57] it maintains a presence in Israeli national consciousness. The myth is ritualized through school trips to pioneer sites, museums, and interactive exhibits and by the establishment of retreat centers and guesthouses that offer urban visitors contact with the natural environment and accomplishments of the kibbutzim.[58]

The first pioneers established Kibbutz Degania, "the mother of the Kibbutzim," in 1910 on the eastern edge of the Jordan River.[59] The enterprise of "establishing an independent settlement of Hebrew workers on the national land" was staged at a highly symbolic location in order to link the modern national movement with biblical history, highlight the Torah as a mandate for Zionism, and transpose text on landscape so that immigrant Jews could forge a sense of familiarity.[60] The Degania pioneers resemble other nationalist elites who initiate nation-building projects at the peripheries while asserting themselves as conveyers of progressive modernization and at the same time as the heirs of

---

55. Cited in Shapira, "Ben-Gurion and the Bible," 651.

56. "David Ben-Gurion was the Moses of our time." Moshe Dayan, *Living with the Bible* (New York: William Morrow & Company, 1978), 77. In the same volume, Dayan understands exiled Palestinians in terms of "Arab clans who have some members living on the West and others on the East Bank, like the Israelite tribe of Manasseh in biblical times," 212, and himself as Joshua at the end of the battle assigning boundaries, 225–26.

57. In February 2007, the members of Kibbutz Degania voted to end the cooperative and privilege private interests. Daniel Ben Simon, "From Kibbutz Secretary to Board Chairman," *Ha'aretz*, February 23, 2007.

58. On the kibbutz museum, see Tamar Katriel, "Remaking Place: Cultural Production in Israeli Pioneer Settlement Museums," in *Grasping Land: Space and Place in Contemporary Israeli Discourse and Experience*, ed. E. Ben-Ari and Y. Bilu (Albany: State University of New York Press, 1997), 169.

59. Current members of Kibbutz Degania told me of the dilemma faced by the pioneers when choosing a burial ground. Although there was ample space to the east of the river, this territory has the status of diaspora in the Jewish legal context. Bodies buried inside the land of Israel do not need to be placed in a wood coffin, but can be wrapped in a shroud and buried in the earth. Outside of the land however, bodies must be placed in a wood coffin. Had the Deganians established their cemetery to the east, they would have had to use coffins and would have been discomforted by the fact that they buried their dead outside of the homeland. The cemetery was established to the west of the River.

60. "Degania, Pioneers' Courtyard Museum" (Kibbutz Degania Alef, Mother of the Kibbutzim: Association for the Commemoration of Degania Alef, 1990) (Hebrew).

ancient legitimizing traditions.[61] The Jewish National Fund (JNF) purchased the lands of Bab-al-Tum, where the Jordan reemerges from the Sea of Galilee with the intention that the first collective agricultural settlement would symbolize a twentieth-century Jordan River crossing to freedom.

Before expanding the analysis of this symbolic resonance, let me note that the JNF acquired lands on the western bank of the Jordan and in the Marj Ibn Amr/Jezreel Valley strategically near water sources so that the valleys would become sources of hydroelectric power and breadbaskets for the Yishuv to be incorporated as a future state. Settlement was the means of asserting borders that could later be upheld as nonnegotiable.[62] Kibbutz Degania and later the Naharayim hydroelectric plant were the only significant early Zionist developments just east of the Jordan.

Degania was situated in such a way as to proclaim the border and to showcase the transformation of the Jews of exile into the new Hebrews. Degania's setting implied that the first kibbutz was both a beginning and a revival: the metamorphosis of land and the Jewish body was broadcast as the very Jordan crossing that shuttled the new Hebrews into liberation.[63] Lacking familiarity or knowledge about the homeland to which they returned, the immigrants viewed the land through a biblical lens. The Degania memory book records a nostalgic pioneer address to the Jordan: "Did we not dream of you from the moment that we learned to read holy texts?"[64] Inaugurating a concerted nationalist project at the Jordan River sacralized a secular Jewish endeavor while defining a vital stretch of river as a rupture between the pioneers and both their diasporic past and their present Arab neighbors.

In daily life the Deganians perceived their community as straddling a line between safety and danger.[65] The complexity of relationships with Arab neighbors was simplified through "an Orientalist binary" in which violence

---

61. Tom Nairn, *The Break-Up of Britain: Crisis and Neo-Nationalism*, second edition (London: Verso, 1981), 33.

62. S. Reichman, "Partition and Transfer: Crystallization of the Settlement Map of Israel following the War of Independence," in *The Land That Became Israel: Studies in Historical Geography*, ed. R. Kark (New Haven: Yale University Press, 1990), 321–23.

63. In the words of a Degania founder: "I don't think we need to be so much afraid now. We are a different people from the people who lived in dispersion in ghettos; we even look quite different." Baratz, *A Village by the Jordan*, 70. The transformation of the Jewish body from ghetto to warrior physique has long been a standard of the Zionist movement. As Daniel Boyarin abstracts from Max Nordau's concept of the *Muskeljudentum*, "Zionism was, after all, explicitly designed to produce a Jewish version of the *Männerbund*, a culture of Muscle-Jews." Boyarin, *Unheroic Conduct: The Rise of Heterosexuality and the Invention of the Jewish Man* (Berkeley: University of California Press, 1997), 336.

64. Members of Kibbutz Degania Alef, *Degania through the Years*, 1962, 306.

65. Settlements on the border were considered so vital to national security that their residents did not have to provide any additional army service.

and opposition were attributed to marauding Bedouin bandits from the east while the Arab communities to the west with useful knowledge of farming, culture, and trade were construed as equally vulnerable to Bedouin raids.[66] In the records prior to 1948, the Arab threat is portrayed as a phenomenon of the east. Reality did not conform to the perceived dichotomy; coordinated Palestinian national efforts to stave off Jewish immigration and block development began in earnest as Degania was expanding.

The perception of a difference between west and east banks becomes apparent in the founders' accounts of their troubles with Bedouin bandits and thieves who find sanctuary in the mountains east of the Jordan River Valley following raids of their fields. The accounts present a picture in which peril lurks to the unknown east and the Jordan serves as a line between danger and safety. The Deganian accounts of traversing the river and venturing eastward are confrontations during which the pioneer identity is enacted and clarified. One of the turning points in Degania's early history is the murder of Moshe Barsky, the collective's "first sacrifice."[67] While crossing the river in order to obtain medical supplies from Menahamieh, a nearby Jewish town, Moshe is attacked. The written account describes how he descends from his horse and sends it back toward Degania in order that the collective not lose the precious animal and contends with his attackers in order to disprove the Arab perception of Jews as soft and easily killed "children of death." When Moshe fails to return from his mission, the Deganians go out as a search party and find him "late that night lying with a stick and a pair of shoes on his head: this was a sign of vengeance, it meant that in the fighting he had killed or wounded someone."[68] Focused on forging a new Jewish image characterized by physical strength, productivity, and truculence, the Deganians find solace in how Barsky met his end. The condolence letter from his parents in Russia encouraging the Deganians to work "with vigor and hope," pledging another son to replace Moshe, and assuring that "Moshe's death will bring all of us to the land" confirms their interpretation.[69] Degania's first loss and the resultant hostilities inspired a change in their sense of what it meant to be a pioneer.

The pioneer sensibility accrued an additional biblical image taken from the book of Nehemiah, where the rebuilders of Jerusalem hold work tools

---

66. Rebecca Stein, "'First Contact' and Other Israeli Fictions: Tourism, Globalization, and the Middle East Peace Process," in *Palestine, Israel, and the Politics of Popular Culture*, ed. Rebecca Stein and Ted Swedenburg (Durham, NC: Duke University Press, 2005), 275.

67. "Degania, Pioneers' Courtyard Museum."

68. Baratz, *A Village by the Jordan*, 80. As it turns out, Barsky wounded one of his attackers, who eventually died. *Degania through the Years*, 70.

69. Tanhum Tanfilov, *Stories of Tanhum*, ed. Yona Ben-Yaakov and Miriam Shion (Tel Aviv: Degania Alef, 1950), 15.

in one hand and weapons in the other (4:10–12). The Deganians adapted to the challenge of the frontier by becoming more vigilant in their guard and repossession of stolen property and by imitating Bedouin mores and cultural attitudes. The Deganians' sense of vulnerability, and that of the larger Yishuv, meant that each stage of expansion or act of defense was framed as a requisite move. The absorption of Jewish refugees from other regions of Palestine during World War I and the Arab revolt of 1936 as well as later refugees and survivors from Europe following World War II strengthened the Jordan River communities' dedication to Jewish survival and as a result saw their increased militarization.

Confronted with the indifference of the Ottoman authorities after the murder of Yosef Zaltzman and the theft of two mules, the members of the community decide to "take matters into their own hands." Degania member Tanhum Tanfilov finds an Arabic-speaking companion from a neighboring community and travels to the city of Salt in Transjordan. Received hospitably and fed generously by Bedouin hosts and local villagers alike, they are unable to learn anything about Zaltzman's murderers. Once they arrive in Salt, they covertly discover the stolen mules in dire condition and repossess them. Unable to determine a target for vengeance, they leave the increasingly menacing city and return home.[70]

Yosef Fine, Degania's Zionist cowboy, becomes famous in the region for his stealth missions to Syria in order to acquire weapons. On one journey, he hides guns beneath the carriage seats of an unsuspecting woman and group of children from Degania[71]; after another, Fine's mother resolved to transport weapons in a secret pouch stitched into a wide skirt.[72] Fine's purchase of weapons in Syria foreshadows the approach of a massive front of tanks, armored cars, and cannons from Syria on an unprepared Degania in the 1948 war. Where Transjordan figures as a point of origin as well as a place of impending peril in the Palestinian frontier stories, it appears in the Degania stories as both a locale of anti-civilization, associated with the danger of marauding thieves and hostile inhabitants, as well as one of advanced civilization hosting the well-developed cities of Damascus, Amman, and Irbid and the possibility of trade and commerce. While the Degania pioneers continue to farm and live east of the Jordan, they come to consider the lands just beyond their fields as the terrain of danger.

---

70. Tanfilov, *Stories of Tanhum*, 16.

71. "A Trip to Damascus (at the end of the 1930's)," *Ta'am Rishonim* 31 (1950) (Hebrew).

72. After one of his trips to acquire weapons in Damascus, Fine and his wife stop for the night at his mother's home in Mettulah. His mother suggests that since she is an old woman who is sick and he a young man who must stay alive, she would risk her life to smuggle the weapons. She gets medical permission for the trip to Damascus and sews herself the gun-smuggling skirt. Menuhah Fine, "Fine's Roots," *Ta'am Rishonim*, 31 (Hebrew).

Refugees

A critical number of the early pioneers were refugees who fled the restrictions and pogroms of Europe hoping to "cross the Jordan" to safety.[73] Degania's Joseph Baratz traces his commitment to Zionism to the pogrom that rocked his village in the Pale of Settlement: "In 1903 the quarter of the town in which we lived was sacked. We escaped unhurt but afterwards I saw the wounded and went to the funeral of those who were killed. Perhaps it was this that turned my thoughts seriously towards Zionism."[74] As they correctly perceived, the climate in Europe was becoming increasingly menacing, and a different scenario needed to be created for Ashkenazi Jews to stay alive. The Israeli pioneer rhetoric is replete with imagery of vitality and rebirth. The rebirth, understood to be collective, entailed the rescue and resuscitation of waves of Jewish refugees.

In the process of absorbing their refugees by the millions, the Jewish pioneers created the Palestinian refugees. The Jewish immigrants from Europe brought with them the notion that "war refugees seldom returned to their former places of residence if the victorious enemy had occupied their homes" and resettled their refugees in deserted Arab towns and on confiscated Arab lands.[75] To a certain degree, flight and the transplant of identity represent a level of normalcy to Jews and not the worst thing that could be done to a people. Refusing to recognize the claim of Palestinian refugees, Israel barred their return. The abandoned villages and lands surrounding Degania were quickly adapted as temporary absorption centers for new Jewish immigrants and then transformed into permanent Jewish communities.[76] Following the evacuation of neighboring Arab towns, the Degania pioneers had a sudden seniority that entailed participation in the project of absorbing and settling Jews. It should be noted that while the plight of refugees is a shared mode of justifying Jewish and Palestinian nationalism, both kinds of refugees have been mistreated in the national context. Many Holocaust survivors who immigrated to Israel were rebuked for not having stood up to the forces of fascism and for marching "like sheep to the slaughter."[77] Among West Bankers, Palestinian refugees were

---

73. "Just as the First Aliya (1882–1904) formed part of a wave of Jewish migration sparked off by a series of pogroms in Eastern Europe, so the Second Aliya was a proximate result of the persecution to which Russian Jews were subjected between 1903 and 1907." Near, *The Kibbutz Movement, A History*, vol. 1 (Oxford: Littman Library of Jewish Civilization, 1997), 11.

74. Baratz, *A Village by the Jordan*, 6.

75. Yoav Gelber, *Palestine 1948: War, Escape, and the Emergence of the Palestinian Refugee Problem* (Brighton: Sussex Academic Press, 2001), 8.

76. *Degania through the Years*, 311.

77. "Some welcomed only immigrants who came as ideologically driven 'pioneers,' while those who came as mere 'refugees' were despised." Tom Segev, *1949: The First Israelis* (New York: Henry Holt and Company, 1986). 133.

viewed with "a mixture of pity and contempt" stemming from "the sense that the refugees were 'defeated,' 'losers,' . . . complicit in their degradation."[78]

The members of Degania had a new sense of their role following the decimation of their relatives in the Holocaust. Realizing that there was no old world for them to go back to and perceiving that they were the only vital force in world Jewry, their position became intractable; the commitment to self-defense brought about a new readiness to take up arms and to settle as many refugees as possible. As a result, the relationship between the Deganians and their Arab neighbors worsened. This was the trend in the Jordan River Valley as well as almost every other region where Jews and Arabs had contact. The Arab disposition toward the Jews became acrimonious as the numbers of Jewish immigrants increased and larger tracts of land were purchased and settled. The Arab Revolt of 1936 had heightened the distrust on both sides.

The pioneer and the refugee, both shaped by the Jordan River as a border, are stock characters of Israeli and Palestinian national myth. The key years for these figures as well as for the mythic systems in general are the war years of 1948 and 1967. These are also the years when the Jordan actually became an Arab-Jewish political border as well as the dividing line between the Palestinian homeland and exile. It can be said that in 1948 the Jewish perception of this geography became the Palestinian reality. While all conflicts surrounding the withdrawal of the British and the partition plan can be seen as struggles over dividing lines, the Jordan served as a primary military front for Arab armies and assumed strategic importance as the eastern point of entry to Jewish communities and settlements. Thus as the Israelis fought in the name of their new homeland at the Jordan front, a dramatic percentage of Palestinians lost theirs through an eastward crossing.[79] 1948 is synonymous with the *Nakba*, the great disaster that befell the Palestinians; the collective expulsion from the Galilee and other regions is termed the *Nuzuh*, the "exodus" from Palestine.[80]

---

78. Bisharat, "Exile to Compatriot," 214.

79. Palestinians were also expelled in dramatic number to Lebanon. At least 750,000 people from across Palestine lost their homes and went into exile.

80. The nature of the exodus is central to Israeli and Palestinian national mythologies and therefore, not surprisingly, deeply disputed by the two. In the Israeli myth, a miracle occurs in which the Arab regular armies broadcast radio announcements urging their Palestinian brothers to move aside so that they can once and forever eradicate the Jewish roots in the land. This massive flight shocks the Jews but relieves them of the problem of Arabs in a Jewish state. They also beat their opponents and win the war. Jewish soldiers then intervene by urging and evacuating Arab residents. All tellers of the myth whom I have encountered will admit to the killing of the villagers of Deir Yassin, but understand the exodus to be proof that the Palestinians can find refuge in all Arab lands and originally saw themselves as citizens of a pan-Arab world, not of Mandate Palestine. In the Palestinian myth, the Jews are utterly ruthless and use any means possible to expel the Palestinians and seize their lands.

The cataclysmic loss involved with the exodus is a primordial event in Palestinian identity narratives; it brands the Jordan "as *the* border: the closest one spiritually, the one traveled across most painfully, the one that most fully characterizes the displacement and the proximity of its cause."[81] In the Palestinian imagination, the Jordan delineates the pre- and the post-1948 situation and stands for the distancing impediments that prevent a collective return.

The Palestinian exodus had several waves, beginning with the mass flight prior to May 15, 1948, and continuing with assumedly temporary emigrations combined with intimidation, expulsion, and deportation at the hands of the Israeli Haganah forces. At the beginning of the hostilities, buses went daily to Transjordan packed with those eager to escape military engagement; eventually the price of travel became so expensive that this mode of transit was available only to the wealthy. King Abdullah's Arab Legion Army opened fords over the river for those who could not find the money to travel by bus. As the war continued, Jewish military groups expelled Palestinians and forced the inhabitants of the Jordan Valley as well as other regions across the bridges and fords of the river. This was the case, for example, with the town of Beisan, close to the Sheikh Hussein Bridge, from which many inhabitants departed prior to the expiration of the British Mandate and those who remained were ultimately transported either to Nazareth or across the river by the Israeli troops.[82] In a first-person account, 'Issam Tahtamuni narrates the chaotic crossing:

The expulsion of Palestinians is the attendant dream of Zionism, which has no purpose except their harassment. The *Nakba* of 1948 changes nothing about their claim to ancestral land. The keys, deeds, and memories of original houses offer proof enough of ownership and prevent any future other than return to the way things were before the establishment of a Jewish state. "This 'other exodus' ... forms the core of the history and the legend which feeds today's Palestinian Arab nationalism." J. K. Cooley, *Green March, Black September: The Story of Palestinian Arabs* (London: Frank Cass, 1973), 40. The reversal of the biblical image has been noted: "While the Biblical Exodus could be viewed as a search for an identity or rootedness, the present exodus was a flight into exile and uprootedness . . . It was sudden and there was no prophet to lead it. Unlike the Biblical Exodus, there was no hope of a Promised Land ahead." Peter Dodd and Halim Barakat, *Rivers Without Bridges: A Study of the 1967 Palestinian Arab Refugees* (Beirut: Institute for Palestine Studies, 1969), 2.

81. Rosemary Sayigh, "Palestinian Women as Tellers of History," *Journal of Palestine Studies* 27/2 (1998): 42–58; Edward Said, *The Politics of Dispossession: The Struggle for Palestinian Self-Determination* (New York: Pantheon, 1994), 8.

82. For the dynamics that led to the removal of the entire Arab population of Tiberias in 1948, see Mustafa Abbasi, "The End of Arab Tiberias: The Arabs of Tiberias and the Battle for the City in 1948," *Journal of Palestine Studies* 37:3 (2008): 6–29. Beisan fell on May 12, 1948. After the surrender of Beisan, the Golani Brigade ordered its inhabitants to evacuate. See Benny Morris, *Righteous Victims: A History of the Zionist-Arab Conflict* (New York: Random House, 1999), 213; and Nafez Nazzal, *The Palestinian Exodus from Galilee* (Beirut: Institute for Palestine Studies, 1978), 17.

> We loaded two of our donkeys with a few of our personal belongings and left for the Jordan Valley. The road was full of panicked people, anxious to cross the river to Transjordan. My father joined us the next morning, while my uncle and aunt remained in the city for almost a month ... until the Jews ordered all the Arabs to leave, carrying them in trucks to the river and forcing them to cross to Transjordan.[83]

Panic spread across the Jordan Valley as it became apparent that only military victory would determine who would remain at home west of the river.

In a group memory book, the Degania members recall the unuttered thought that lurked behind their singing and exaltation following the proclamation of Israel's statehood: "What will become of us, who dwell at the border of Transjordan and Syria, tomorrow." On the eve of the war, dwelling beside the Jordan appeared to the kibbutz members as a position of extreme vulnerability.

> Degania is very close indeed to the borders of Transjordan and Syria; a little more than a mile from us, the Arab village of Samakh is actually on the frontier. We had always known this was the frontier, but it wasn't till the war began that we really knew it in our bones, knew exactly what it meant.
>
> As an objective for the Arab forces Degania was important. It is on the Jordan and we had built a bridge there ... If they could capture Degania they would have a hold over the Jordan and the way would be open to Tiberias and to Haifa—Degania was the gate to Galilee.[84]

In the Galilee, Palestinian resistance to the partition plan began in the winter of 1947, when the younger villagers of Samakh took to trying to destroy passing vehicles and firing guns openly.[85] Despite attempts by the elders of Samakh to prevent violent outbreaks, several Jewish villagers were killed while traveling on the roads. In late April 1948, soon after the British withdrawal, a battle erupted between kibbutz members from the Jordan/Sea of Galilee region and the villagers of Samakh over control of an abandoned British po-

---

83. Eyewitness account by 'Issa Tahtamuni, a tailor, interviewed at Dir'a, Syria, July 19, 1973, recorded in Nazzal, *The Palestinian Exodus*, 49.

84. Baratz, *A Village by the Jordan*, 156.

85. Members of Kibbutz Degania Alef, *Degania through the Years*, 1962. Abbasi, however, notes that these were isolated incidents uncharacteristic of the larger refrain from military activities of the Palestinian leadership of Tiberias and the surrounding villages. Abbasi, "The End of Arab Tiberias," 14.

lice fortress.[86] The engagement exhausted both parties, but when promised Arab reinforcements failed to arrive, the police fortress as well as the town of Samakh fell into Jewish hands on April 29.[87] The inhabitants of Samakh, numbering approximately 3,600, fled to the other side of the river, thereby losing their homes and entering the "wilderness" of exile.[88] Through an eastward crossing of the Jordan River, these Palestinians of the Jordan River Valley were transformed into refugees.[89]

Following the battle with the Samakh villagers, the Deganians' fear of danger from the east was realized as Iraqi and Transjordanian units moved across the Jordan and onto the Tiberias-Nazareth and Beisan-'Afula roads.[90] Interestingly enough, both sides were ill prepared for engagement; the Arab armies lacked a coordinated strategy to maximize their weapons and leadership, and the Deganians had sustained heavy losses at Samakh and did not possess sufficient weapons to defend themselves against the Arab armies. A group of representatives from Degania traveled to Tel Aviv to request the protection of the Israeli Defense Forces, but, as one Kibbutz member recalls, David Ben-Gurion told them, "I know you Deganians. If you meet face to face, you'll drive them out."[91]

As the indigenous Arabs were transported east, foreign troops moved west.[92] By the time the Deganians encountered the Syrians, they did so from positions taken in the abandoned village of Samakh. The Syrians began their

---

86. Arab Tiberias was besieged, defeated and evacuated by April 19. Nazzal, *The Palestinian Exodus*, 29–30.

87. This battle and the denial of weapons to the villagers of the Galilee by Arab armies are chronicled in Nazzal, *The Palestinian Exodus from Galilee 1948*, 16–27.

88. Efraim Karsh maintains that "the collapse of Tiberias's small community, less than 6,000-strong, triggered a flight from neighboring villages, including Kafr Sabt, Ma'dhar, Majdal, Nasr al-Din, Samakh, Samra, and Shajara." Karsh, *Palestine Betrayed*, 185.

89. Israeli intelligence briefs describe the crossing as a three-day affair during which long convoys were escorted to the river by Jewish military personal. See Gelber, *Palestine 1948*, 101.

90. With the exception of Egyptian forces and a few battalions that descended from the north, the Arab armies penetrated from the east side of the Jordan River. The ALA 1st Yarmouk regiment under Muhammad Tzafa entered Palestine, crossing the Jordan River at the Damia Bridge. A contingent from Damascus crossed the Jordan River on January 28, 1948, despite objections from the British, and scattered among the villages of Samaria. As the Iraqi forces prepared to infiltrate, the Israelis began to destroy the Jordan bridges in order to thwart their approach. See Yoav Gelber, *Palestine 1948: War, Escape, and the Emergence of the Palestinian Refugee Problem* (Brighton: Sussex Academic Press, 2001), 117–37.

91. Baratz, *A Village by the Jordan*, 157.

92. On April 18, the Arab community of Tiberias and the neighboring Nasir al-Din were "loaded onto buses and transported to the Transjordanian border at Samakh, while a small minority was relocated to Nazareth to the west." Abbasi, "The End of Arab Tiberias," 21.

campaign at Samakh with designs on seizing Tiberias and continuing to Safed, while the Arab Legion sacked the power plant at Naharayim and the Iraqi army besieged Kibbutz Gesher.[93] At Samakh the Syrian battalion clashed with the newly minted soldiers of Degania; both sustained heavy losses. The battle continued over a few days until the Syrian army took the former British police fort at Samakh, opened fire on the retreating Degania force, and pressed on to attack Degania itself.[94] From the accounts of kibbutz members, it was the victory of the weak over the strong as the Deganians repelled the Syrian tanks and armored cars with mismatched rifles smuggled over the years and Molotov cocktails.[95] Perhaps due to the visit paid by Yosef Baratz to Ben-Gurion, the Haganah lent the Deganians two cannons for twenty-four hours. After the destruction of their tanks and the sustaining of heavy losses, the Syrians pulled out of Degania and attacked two more kibbutzim in the region before moving to other fronts. After the Syrians fled, members of the kibbutzim in the Jordan Valley appointed their own "governor of Samakh," but were rebuffed by Ben-Gurion, who had not authorized the appointment.[96]

As Israel and the Arab states continued the war and negotiated a cease-fire, about 100,000 Palestinians found themselves as refugees in Transjordan.[97] Despite the fact that the refugees had lost their homes and property and now dwelled in the terrain of exile, they fared better than their counterparts in the Transjordanian-controlled West Bank, where the displaced huddled in orchards and lacked food and water. After arriving, the East Bank refugees

---

93. Although the Israelis destroyed the bridge between Naharayim (east side) and Kibbutz Gesher (west), the Iraqi army built a pontoon boat in order to cross the Jordan and attack Gesher.

94. "On May 15, the Syrian forces set out from El Hamma towards the Jordan Valley in three columns: the first was directed against Samakh, the second against the twin Jewish settlements of Deganiya and the third column was to cut the road connecting Beisan to Samakh . . . At dawn on May 20, the Syirans advanced toward Deganiya A, to capture the bridge across the Jordan River." Nazzal, *The Palestinian Exodus from Galilee*, 19.

95. Kibbutz Degania Alef, *Deraka Shel Degania* (Tel Aviv: Devar Publishing House, 1962), 209. Yosef Baratz tells the story of meeting Moshe Dayan, born in Degania, after the battle with the Syrians beside the Jordan bridge. Dayan tells Baratz, who had been meeting with Ben-Gurion and military officers in Tel Aviv, "Everything is alright, the old performed as if they were young." Ibid., 211. Historians have noted that the Syrian army itself was disorganized and lacked weapons. Morris, *Righteous Victims*, 232; Gelber, *Palestine 1948*, 141.

96. Gelber, *Palestine 1948*, 274.

97. The number of 1948 refugees in Transjordan is hard to determine. Some estimates put it at 160,000 and others at 60,000. The 1950 United Nations Economic Mission to the Middle East reported 100,905 refugees in the East Bank. The total number of 1947–49 refugees is also disputed. Benny Morris shows how difficult it is to arrive at a precise number. In light of estimates by the British Foreign Office in 1949, he sets the number between 600,000 and 760,000. Benny Morris, *The Birth of the Palestinian Refugee Problem, 1947–1949* (Cambridge: Cambridge University Press, 1987), 298.

lived in tents and received food rations and small quantities of money from the authorities and then moved to International Red Cross camps built outside of Amman, Zarqa, Salt, and Irbid. They would eventually be granted citizenship. However, when the news traveled of the superior conditions east of the Jordan, the Transjordanian government labored to distribute food to the refugees in Jericho and to prevent their movement across the river. Again the Jordan served as the border between uncertainty and stability, although in this case greater stability was to be found on the East Bank. When the West Bank was under Jordanian control, patterns of economic development and policy encouraged the migration, particularly of the wealthy and educated, to the East Bank.[98] However, from the point of view of the Palestinian refugees, their exodus would find no end until they returned en masse to their place of origin west of the river.

Freedom Fighters

What was the legacy of the pioneer trope in postwar Israeli and Palestinian national myth? What type of actions did these figures inspire after the borders were drawn, and what impact did these new pioneers have on the perception of the border's permeability? After the armistice lines were formulated in 1949, Israel's eastern border ran along the Jordan, dividing Israel and Syria north of the Sea of Galilee and Israel and Jordan south of the Sea for 40 kilometers (see figure 14). The "sinuous frontier" of the armistice—the Green Line—divided the central area of the nascent Israeli state from the now-independent Hashemite Kingdom of Jordan.[99] To the south the Arava, the Jordan Valley's dry counterpart, outlined Israel and Jordan as stipulated in the partition plan. During the 1950s, both Israel and Jordan "were independently engaged in formulating and implementing national water schemes to develop their economies for the absorption of immigrants on the one hand, and refugees on the other. Throughout this period, there were recurrent hostile and retaliatory incidents across the Armistice Demarcation lines between Israel and its four Arab neighbors."[100]

The other side of the Jordan, with its "alluring proximity," maintained a hold on the Israeli imagination and roused nostalgia for the era of open frontiers.[101] At the same time, the myth of the pioneer laboring at the periph-

---

98. Josef Nevo, "The Jordanian, Palestinian, and the Jordanian-Palestinian Identities." Paper presented at the Fourth Nordic Conference on Middle Eastern Studies, 1998, 4.

99. Helena Cobban, *The Palestinian Liberation Organization: People, Power, and Politics* (Cambridge: Cambridge University Press, 1984), 169.

100. Miriam Lowi, *Water and Power: The Politics of a Scarce Resource in the Jordan River Basin* (New York: Cambridge University Press, 1993), 79.

101. Stein, "'First Contact,'" 269.

eries along with the privileges bestowed on the pioneer generation caused some degree of resentment and frustration in Israeli society. Raised on the myth, the generation born in Israel yet too young to fight in the country's War of Independence experienced a sense of deflation and confinement.[102] In such a climate, penetrating enemy territory seemed a means of figuring the cease-fire lines as artificial barriers whose breach was a laudable means of both expanding and strengthening national identity.

The general fascination with what lay beyond the borders found a focus in Petra, the southern Jordanian red rock city hewn by the Nabateans. For young Israelis in the 1950s, Petra seemed a destination that represented the intrepid national character as well as one that could transport the Jewish traveler to the other period of time when Jews resided in a homeland. No matter that the Nabateans were enemies of ancient Israel and, at the time, the Jordanians enemies of modern Israel; the very antiquity of Petra offered a means of identification.[103] Contact with the ancient, perceived simultaneously as exotic and familiar, promised to contextualize Israelis in the Arab world and relieve them of quotidian modernity. Trekking eastward was a manner of infiltration that affirmed extant borders while testing the possibility of an enlarged space in which one could be an Israeli and an occasion in which a "point of contact with and separation from Arabs" became a "site in which Israeli identity and solidarity [was] acted out and reified."[104] In addition, such a venture beyond enemy lines suspended the pioneer myth and proved that the borders could still be redefined by Jewish initiative.

The challenge of reaching Petra was posed primarily in the ranks of the Palmach, the military wing of the Labor movement that aided the British during World War II, smuggled Jewish refugees to Palestine between 1942 and 1945, and constituted the military backbone during the 1948 war. After its units were folded into the Israeli Defense Forces, the Palmach persisted as an ideological and social organization. One of its central rituals was the *tiyul*, the trek into remote areas and sometimes beyond the ceasefire lines. The clandestine hiking expeditions tested the physical mettle and endurance of the participants and were intended as sacred occasions of communing with the

---

102. Anita Shapira, "The Kibbutz and the State," trans. Evelyn Abel, *Jewish Review of Books*, Summer 2010, 5–10; Y. Merkoviski, "The Red Rock: The Myth of the 1950s," *The First Decade: 1948–1958*, ed. T. Tzemrat and H. Yablonka (Jerusalem: Yad Ben Zvi, 1997), 337, 340 (Hebrew).

103. A woman of this generation, whom I interviewed at the Palmach archive, suggested that young people were encouraged at Palmach gatherings in the 1950s "to cross over to Petra and search for signs of Jewish antiquity." Interview, December 27, 2004.

104. Dan Rabinowitz, "Israel's Green Line, Arabness, and Unilateral Separation," in *Culture and Cooperation in Europe's Borderlands*, ed. J. Anderson, L. O'Dowd, and T. Wilson (Amsterdam: Rodopi, 2003), 219.

Jewish national past. Such pilgrimages offered psychological as well as physical means of establishing claims through presence while transforming the legacy of Jewish wandering into directed acts of reclamation in the present. Through interviews with Palmach members now into their seventies and eighties, I learned that in the 1950s reaching Petra was considered the trek of treks.

Two young, strapping members of Kibbutz Ein-Harod, Rahel Sevorai and Meir Har-Tzion, were the first to undertake the trek to Petra.[105] Sevorai described the trip to Petra as an Israeli rite of passage: "It wasn't that our generation lacked danger, but the trip was part of our settling the land: settling ourselves into the land and becoming settled with its dryness, its heat, the slope of its mountains."[106] For Sevorai, the border appeared an instrument of self-identification rather than a line of obstruction; transgressing it presented a way of exploring a potentially wider scope of Israeli sovereignty. As news of the Petra adventure spread, a second group began devising a follow up. During the autumn holidays in 1953, following a Palmach reunion, Arik Magar, Miriam Monderer, Eitan Mintz, Yaakov Kleifeld, and Gila Ben-Akiva set out in the direction of Petra. The three men and two women departed from a vacant wadi on the Israeli side and began walking along a navigated course. Five kilometers into Jordan, a barefoot Eitan Mintz was bitten by a snake and the group reversed its course, apparently in order to return to Israel.[107] Near the Bir Madkor police station, the five encountered a patrol of the Jordanian gendarmes and met their deaths in the desert; in some versions of the story, the five encounter a group of Bedouins who had been exiled in 1948. The fever pitch of reaction to the deaths quickly enshrined the event and conferred legendary status.[108] On one hand, there was criticism that five young people

---

105. A 2004 article in an Israeli magazine labels Har-Tzion and Sevorai as "the pioneers of Israel's hiking tradition" Yadin Roman, "The Lure of the Trail," *Eretz: The Magazine of Israel* 96 (November–December, 2004).

106. Quoted in Nesiya Shafran, "The Red Rock: A Retrospective," *Keshet. Documentation: Early Days in the Land of Israel*, ed. Aharon Amir (Ramat Gan: Masada, 1979), 8.

107. "His walking barefoot . . . becomes a symbol, and the land, the earth, is sanctified and rises to a level of supreme value." Haya Bar-Itzhak, "Walking Barefoot in the Land of Israel: Mythicization and Demythicization in Contemporary Kibbutz Narratives," *Contemporary Legend* 5 (1995): 87.

108. With the release of Arik Lavie's catchy pop ballad written by Haim Hefer, "The Red Rock of Petra," the myth became embedded in popular culture. The extensive radio play of a song in which the desert itself seems to swallow three valiant young heroes (the song dispenses with the two women) contributed to the construction of the 1950s border as "a fetshized entity" and "a space where the nation defines itself against the ultimate Arab Other." Rabinowitz, *Israel's Green Line*, 226. Ben-Gurion banned the song in 1958 because he feared that it would motivate other young people, lead to more deaths, and provoke the Jordanians. Between 1956 and 1957, eight young men were killed on the way to Petra. The five killed on the road to Petra were not officially commemorated by the Palmach until January 1, 2007, at the urging of Rahel Sevorai. See Yuval Azoulay, "Has the Debate over the Petra Trek Concluded?" *Ha'aretz*, January 2, 2007.

had sacrificed themselves not for the state, but for a reckless adventure; on the other, the Israeli public was captivated and saw the trek as expressing a national resolve to define the state rather than be defined by its enemies. The undertaking and its price disclosed a feeling of unnecessary constraint while reinforcing the ceasefire lines as partitions between relative safety and mortal danger. The fact that the five young people, remembered as being the best of a generation, fell on the other side of the border offered some psychological solace by locating the ever-present threat of Arab attack on external rather than internal terrain. In their memorialization, the Petra five were honored as having died for the collective and reaching Petra became the dream of new "pioneers."[109]

The mythification of the Petra five vindicated Israeli society by producing heroes who perished unarmed at the hands of Arab forces, thereby upholding a collective sense of victimization that in turn fed the reservoir of justification for Israeli provocation. It can also be seen as the cultural reflex of the Israeli military forays. The myth operated to obscure the Israeli incitements of diverting the Jordan River from the Israel-Syria demilitarized zone at Gesher Bnot Ya'aqov and brutally attacking the Jordanian town of Qibya in the name of reprisal.[110] Although the voyage to Petra was one incident among many that tested the borders in 1953, it was the one that best embodied how Israelis wanted to see themselves in relation to their Arab neighbors. By 1960, when two Israeli paratroopers, Shimon (Cushi) Rimon and Victor Friedman, reached Petra in purloined UN vehicles, the figure of the Israeli pioneer had been solidly refigured in a military context.[111]

The 1967 Israeli capture of the West Bank from Jordan was significant not only in strategic and territorial terms, but also because it realized the biblical map in which the Jordan forms the eastern border of the land of Israel (Num 34:12; Ezek 47:18). The figure of the soldier/scout/biblicist cut in the portraits of those fallen on the way to Petra became the new standard of national hero following the 1967 Six-Day War. The pioneer myth was easily coopted to condone the actions of settlers in the occupied West Bank and Gaza as the continuation of the work begun by the kibbutzim. While the secular kib-

---

109. Hadassah Avigdori-Avidov, "Arik," in *Until the Rock: The Five Who Went*, ed. D. Brash, A. Amiassaf, M. Berasilbaski, and M. Tal (Tel Aviv: Hakibbutz Hameuchad, 2001), 22.

110. Lowi, *Water and Power*, 87; Shlaim, *The Iron Wall*, 82, 86.

111. Meir Har-Tzion and Shimon (Cushi) Rimon, two of five who returned from Petra, fought with Ariel Sharon in the controversial Paratroop Unit 101, which carried out the attack on Qibya. Both were rewarded by Sharon with desirable properties close to the Jordanian border in the north and south, respectively. As those associated with the early kibbutzim comprised the Labor elite of Israel until 1977 (Sternhell, *The Founding Myths*, 79), so the new "pioneers" for whom borders were not limits formed the elite of the Likud party, which subsequently came to power.

butzim maintained affiliation with the left, the epithet "pioneer" became the rallying cry for right-wing religious nationalists staking claim and building fortress-like settlements in the "wild" West Bank.[112] "These new settlers regarded themselves as disciples of the early Zionist pioneers. And like their role models, many of them chose to farm the new land: agriculture was seen not merely as a way of life, but as a moral and patriotic calling."[113] The primary justification for continued settler presence is the biblical map of the Promised Land that reaches to the Jordan.

Although Jordan lost the West Bank in 1967, those who lost most were the West Bank Palestinians, some of whom lost their homes for the second time since 1948. At least 150,000 Palestinians were exiled to the East Bank of the Jordan and barred from crossing westward. For these Palestinians as well as for the larger global community of Palestinian refugees, the Jordan River came to symbolize the border between home and exile. To cross the Jordan from west to east was to descend into a shadowland of loss and negation and to imagine a westward crossing was to dream of the redemption of coming home.[114] This valuation of the East and West Bank parallels precisely that of the first Israelis who described ending their exile through immigration to Israel as "crossing the Jordan River."

During the Six-Day War of 1967, known to Palestinians as *al-Naksa*—the Setback—large numbers of Palestinians were transferred from the West Bank to Jordan. Some fled the war, but others were bussed to the Allenby Bridge by Israeli troops and others driven out by force.[115] A significant number of Palestinians found themselves in a state of exile with tenuous refuge on the Jordanian East Bank.[116] The Palestinians who crossed the Jordan in 1967 lost

---

112. It should be noted that Labor governments have not restricted the building and expansion of settlements. Leftist Israeli artists, however, have opened the founding myths as a mode of critiquing the continued Israeli occupation. The hip-hop group HaDag Nahash subjects the pioneer myth to ridicule in "Gabi and Debby"; the transsexual singer Dana International in "Traveling to Petra" "turns the Ashkenazi Israeli myth with its implicit sexual imagery of a heroic masculine penetration of a feminized Orient into an explicit sexual encounter with a desert that one returns to rather than from." Yael Ben-Zvi, "Zionist Lesbianism and Transsexual Transgression," *Middle East Report* 206 (winter 1998): 28.

113. Tom Segev, "A Bitter Prize," *Foreign Affairs* 85/3 (2006), http://www.foreignaffairs.com/articles/61726/tom-segev/a-bitter-prize. This stance, in many ways, is a continuation of the "agricultural fundamentalism" that characterized the Yishuv and early state. Attias and Benbassa, *Israel*, 157.

114. Although the terrain east of the Jordan epitomizes exile for the Palestinians, the Hashemite Kingdom of Jordan is, in fact, the only country where as citizens they are legally enfranchised.

115. Nur-eldeen Masalha, "The 1967 Palestinian Exodus," in *The Palestinian Exodus 1948–1998*, ed. Ghada Karmi and Eugene Cotran (New York: Ithaca Press, 1999), 63–109.

116. These numbers are disputed. The Jordanian government claimed that 361,000 people were displaced from the West Bank, UNRWA reported 162,000, and a 1993 study by the Israeli Institute for Applied Economic Policy Review counted 100,000 displaced persons in Jordan.

their homes, businesses, and belongings and became dependent upon UN aid and alms.[117] In Jordanian society, they were politically integrated at the same time that they were discriminated against and blamed for forfeiting their land to the Jews. Even the landscape of their resettlement proved a bleak shadow of West Bank terrain: "those who fled from the West Bank left behind them fertile areas for the semi-desert plains of the East Bank."[118] By the time the 1967 refugees reached the other side, all previous social structures were inverted or overturned. Although Palestinians traversed other borders during and after the 1967 war, the crossing of the Jordan came to symbolize Palestinian dispersion. Home, palpably close yet distanced by occupation, was revisited primarily through the reconstructions of ritual and memory.[119] For the global community of Palestinian refugees, the eastward exodus was recalled as a descent into a shadowland of negation and a future march westward stood for restitution and return.

The 1967 exodus is called "the Second Exodus" both because it was a replay of what had happened to Palestinians in 1948 and because many refugees from 1948, including those concentrated around Jericho, went into double exile.[120] The accounts of the refugees in the Second Exodus tend to focus on two aspects: panicked flight across the Jordan and confrontation with the bleak reality waiting on the other side. While the passage wounds the individual and the collective, the stasis on the East Bank proves haunting in its own right. In these narratives, the East Bank is a kind of shadow zone, an inverted mirror world where the refugees become specters of their former selves. The sense of exile overpowers them even in Jordan, a country of which they were already citizens.

---

117. By the Jordanian definition, the 1967 Palestinians are not classified as refugees "since they were living in Jordan up to 1967 and are currently 'displaced persons' rather than refugees who have crossed an international frontier." Marie-Louise Weighill, "Palestinians in Exile: Legal, Geographical, and Statistical Aspects," in *The Palestinian Exodus 1948–1998*, ed. Ghada Karmi and Eugene Cotran (New York: Ithaca Press, 1999), 16.

118. Dodd and Barakat, *Rivers Without Bridges*, 6.

119. For the delicate balance between Jordanian and Palestinian political power between 1949 and 1967, see Shaul Mishal, *West Bank/East Bank: The Palestinians in Jordan, 1949–1967* (New Haven: Yale University Press, 1978), 111–20. For refugee life on the East Bank, see Muhammad Siddiq, "On the Ropes of Memory: Narrating the Palestinian Refugees," in *Mistrusting Refugees*, ed. E. V. Daniel and J. C. Knudsen (Berkeley: University of California Press, 1995), 88; Schultz, *The Palestinian Diaspora*, 115; Slyomovics, *The Object of Memory*, 1998.

120. "Between 1949 and 1967, the Palestinian population in the West Jordan Valley was dominated by three huge refugee camps surrounding the town of Jericho: 'Ayn Sultan, Nu'aymah, and 'Aqbat Jabir. The residents of these camps had been driven out of present-day Israel in 1948–49. During the 1967 hostilities or shortly thereafter, virtually all residents of these camps, approximately 50,000 people, fled or were expelled to the East Bank." Nur-eldeen Masalha, "The 1967 Palestinian Exodus," 94.

One refugee who fled from Jerusalem to a camp in Zarqa, Jordan, recounted to me the collapse of the family structure and the humiliation of the refugee.

> We left because my father's oldest son scared him by saying, "Let's go, let's go, they're going to kill us." My father put all of us in the car and because he lost everything, he became blind. My sister was kicked from behind. We walked across the bridge. Other people walked under the bridge or swam through the water. The hardest part wasn't leaving, it was coming to Jordan. After we crossed the border to Jordan, they treated us like gypsies. Don't think that the Arabs like Palestinians. I've lived in Saudi Arabia, Kuwait, Iraq, Syria, and Lebanon, and no one likes the Palestinians. They say, "You sold your land."
>
> In Jordan, they threw us in a large camp, where we lived in a tent. The United Nations gave us some food and opened schools, stuff like that, but when the PLO started, the FBI started behind us. The Jordanians treated us like gypsies and put us in different schools. The PLO later opened a school for the children. When I crossed, I crossed with a lot of people, and the Israelis walked with us and fired their guns. We crossed the King Hussein Bridge. I know the first one who came to the Old City, Moshe Dayan. I remember that day when we knew that King Hussein was their partner. My father went blind when he lost everything. It was a terrible life. I was nine years old when they kicked us out, that's all.

The account begins and ends with the image of the blind father, a symbol of the emasculated patriarch and the refugee's loss of power and self-determination. In this interview, the narrator accounts for his father's blindness as a result of his losing everything, but in a later interview, he told me that it was a result of "the black water" in the refugee camp of Zarqa. The family's honor is eroded by the father's loss of vision as well as the sister's loss of honor when she is kicked "from behind." The degrading gesture not only shames her, but also disgraces her male relatives who find themselves in the position of being unable to defend her/their honor.[121] In this case, the Jordan delineates antitheti-

---

121. Note the corollary between the Haddadin flight from the east bank in order to protect the honor of family through the body of the young female relative and the disgrace experienced by the young female family member in this forced eastward crossing. "The fear of dishonor is also mentioned by a relatively large number of families. The honor of the womenfolk, *al-'ird*, is a central value to the Arab family. The defense of this honor falls upon the men of the family, and threats to honor are resented and punished. Honor is threatened not only by molestation of the women, but by insults and approaches to women made by strangers. The value of honor was seen as seriously threatened

cal experiences: residence with property and status in Jerusalem and homeless dependence and scorn on the East Bank. The act of crossing the Jordan is an experience of loss that ushers the Palestinian refugees of 1967 into a state of instability and international scrutiny. In the shadow of a definitive boundary, the refugees persist in the limbo of a Jordan River Valley refugee camp.

The family is part of a mass flight that the narrator recalls as a chaotic attempt to find sanctuary across the river. The sanctuary turns out to be of the most rudimentary kind, in that they are given a tent and small amounts of food. The sense of loss and disempowerment is emphasized in Jordan, where they are treated like gypsies and provided with only the most basic of needs. The only people who help the refugees are United Nations relief workers whose handouts are perceived as colonial merchandise.[122] After his family's flight from Jerusalem, the narrator confronts the impossibility of every situation: the Israelis bar a return, the Jordanians discriminate against him, and the Arabs of other countries blame him for his status as a refugee. Again the Jordan River is the border between stability and instability, and the East Bank constitutes a location of negation, the first stop in a long series of temporary "homes" in exile.

In the wake of the 1967 war, the Jordan also became the frontier where Palestinian identity was reconstituted as part of a resistance movement and the *fida'iy*, the guerrilla fighter, developed his tactical objectives and mythic status. Cross border attacks and subversion of Israeli development represented the restoration of national dignity and the possibility that the dispossessed Palestinians could succeed where the Arab armies had failed at humbling Israel. The *fida'iyin* movement contested the borders and with them the legitimacy of Israel through attack, sabotage, and the establishment of quasi-state apparatuses in exile while promoting a vigorous national identity that defied perceptions perpetuated by Israel and the Arab states. Resistance brought the new avatar of Palestinian identity into relief, and the Jordan River became a prime target as the resource that sustained Israeli communities and enabled immigration as well as the demarcation line beyond which Palestinians had been driven. Armed struggle at the Jordan announced that the Palestinians would define themselves through engagement with Israel rather than be forgotten through acts of removal and suppression. The resurgence and trans-

---

because the enemy was not Arab and would not observe Arab customs regarding the respect due to women." Dodd and Barakat, *Rivers Without Bridges*, 46.

122. "Resettlement plans were regarded as part of Western imperialism and a plan to eliminate the refugee problem—UNRWA was accordingly (in the early 1950s) seen as an agent of imperialism." Helena Lindholm Schulz with Juliane Hammer, *The Palestinian Diaspora: Formation of Identities and Politics of Homeland* (London: Routledge, 2003), 50.

formation of national identity at a frontier signaled a temporal break between an ignominious past of suffering and a new era of sacrifice in the name of collective redemption.

In addition, operating independently in border zones leveraged the Palestinian position within Arab states by creating an intermediary space for the formation of Palestinian institutions, asserting an identity distinct from pan-Arab or other Arab national formations and "political(ly) outbidding" other commitments to Palestine through persistent presence and attack.[123] Since the Arab regular armies had not only met with defeat in the 1967 war, but the Arab regimes, to varying degrees, had also immobilized the refugees through prohibitions, the rise of a distinctly Palestinian resistance was an assertion of self-determination amidst statelessness.

While several clandestine groups focused on Palestinian restoration via Israeli destruction formed around the post-1967 peripheries, it was ultimately Fatah operating in the Jordan River Valley that seized the reins of the Palestinian Liberation Organization (PLO) and assumed leadership of the national movement.[124] Through devotion to territorially focused guerrilla tactics and the creation of a victory narrative that countered historical circumstances, the Fatah leadership aligned the definitions of "warrior" and "Palestinian." Training camps were established just east of the Jordan, the longest continuous border with Israel, to offer rebellion-minded Palestinians an alternative to the stasis and despair of UN refugee camps. Similar to the way in which the Degania founders sought to reverse the image of the weak and humbled Jew at the Jordan, the early Fatah recruits endeavored to shift the image of the Palestinian from a dependent refugee to an impassioned revolutionary worthy of emulation by the whole Arab world. In both cases, image reversal required resistance, and resistance easily morphed into provocation.

Fatah's first operation in 1964, launched under the pseudonym "al-Assifa" ("the Storm"), was an attempt to blow up the station where Jordan water was pumped into the main pipe of the National Water Carrier of Israel. Although foiled by the Lebanese border patrol as part of a joint plan with Jordan to protect the states from Israeli reprisals, the sabotage aimed to counter the attempts of the 1964 Arab Summit to sequester resistance by establishing the PLO as Nasser's Palestinian puppet and its inaction regarding Israel's diver-

---

123. Yezid Sayigh, *Armed Struggle and the Search for State: The Palestinian National Movement 1949–1993* (Oxford: Clarendon Press, 1997), 174.

124. "Fateh, which is a palindromic [sic] acronym for *Harakat al-Tahrir al-Filastiniyya*, was established in the late 50s and early 60s, through the coalescing of various specifically Palestinian nationalist (as opposed to Arab nationalist) networks already active in the refugee camps, in diaspora groupings of Palestinian students, and in the embryo Palestinian communities of the emerging Arab Gulf states." Cobban, *The Palestinian Liberation Organization*, 6.

sion of Jordan waters. The second Assifa operation, planting explosives at the water carrier near Tiberias, was similarly unsuccessful.[125] Although the explosives never detonated and one operative was killed by Jordanian forces and the other captured by Israelis, these attacks stoked the motivation for cross-border raids and established the Jordan as the primary battle front. Assifa attacks were staged from the West Bank, but in the war's aftermath the guerrilla organizations regrouped on the East Bank, where they were policed by a Jordanian government forced into a level of permissiveness by the swelling public recognition and popular support.

Between January and March 1968, amidst constant negotiation concerning *fida'iyin* status in Jordan, Fatah launched seventy-eight raids into Israel. Public outrage about the explosion of an Israeli school bus by a mine laid by Palestinian guerrillas rallied support around a deterrence operation on Jordanian territory. Israel forewarned the United Nations of "a large search-and-destroy mission against guerrilla bases in and around Karama," and Jordanian intelligence, tipped off to the operation, encouraged the guerrillas to evacuate. The suggestion of flight, it seems, triggered multiple levels of resistance: resistance to being subdued by the Jordanians, to the pattern of Palestinian flight, and to the Israeli military. Karama, a refugee camp four miles east of the Jordan that served as Fatah's primary base, thus became the site of the inaugurating "success" of the Palestinian guerrilla movement. On March 21, 1968, Israeli battalions supported by air squads crossed the river anticipating that the guerrilla forces would abandon Karama and thereby facilitate a rapid operation. Instead, about 300 fighters held their ground reinforced by the Jordanian 1st Infantry Division stationed with tanks and artillery on the mountains above. Although the battle was primarily fought between Israeli and Jordanian troops and the Israelis gained control of Karma in less than a day, accomplishing the mission of its destruction, the battle became mythologized as a resounding Palestinian victory. Rashid Khalidi sees Karama, which became the "foundation myth of the modern Palestinian commando movement," as exemplifying Fatah's rhetorical strategy of portraying defeat at great odds as "heroic triumph."[126]

The myth of Karama overlooked the 120 guerrillas killed, the 100 wounded, between 40 and 66 prisoners seized, the steep Jordanian casualties, and the

---

125. Shlaim, *The Iron Wall*, 232; Cooley, *Green March, Black September*, 94; Sayigh, *Armed Struggle*, 107. For background of the two attacks, see Hart, *Arafat: A Political Biography* (Bloomington: Indiana University Press, 1984), 182–83.

126. Sayigh, *Armed Struggle*, 177; A. W. Terrill, "The Political Mythology of the Battle of Karameh," *Middle East Journal* 55, no. 1 (2001): 95; Rashid Khalidi, *Palestinian Identity: The Construction of Modern National Consciousness* (New York: Columbia University Press, 1997), 196.

destruction of the town and surrounding commando outposts. Such casualties were neglected amidst the emphasis on Israeli dead and wounded and by the choice of active defeat over passive retreat. The Palestinian guerrilla, hero of the myth, figured as the defender of borders and steadfast protector of the Palestinian identity. As *karama* means dignity in Arabic, the rallying cries of "Karama" were understood as a promise to Palestinian exiles everywhere that armed struggle would restore dignity and ensure the indelibility of Palestine's borders. The theme of the weak prevailing over the strong relied on the details of armaments: the Fatah fighters greeted the well-stocked Israeli brigades and battalions with, as Sayigh notes, "only a handful of anti-tank mines, seven anti-tank rocket launchers, and two 82 millimetre mortars." The Jordanian Infantry Division, which matched the Israeli forces and was responsible for most of the casualties, and its significant losses went unmentioned. The celebration of "a new Palestinian political identity" forged through heroism in a purely Israeli-Palestinian battle was upheld even in an environment where Jordanian commemorations narrated a more accurate version. Subsequent anniversaries of the battle, celebrated by Jordanians as "one of the symbols of Jordan's modern nationalism" and by Palestinians as their initiatory victory, highlight a rift in interpretation that widened into the standoff of Black September.[127]

While the Israeli soldiers lingered after the defeat of Karama to collect the bodies of the twenty-eight dead and shuttle the ninety wounded back across the river, they abandoned several tanks.[128] These vehicles became the relics of Karama, substitutes for the corporeal enemy, paraded through the streets of Amman and Salt as part of the ritual replay of the battle. The dissemination of the myth as well as the parades celebrating Karama boosted the number of recruits at guerrilla camps, inspired the proliferation of borderland outposts, strengthened the unity of East and West Bankers through common cause, encouraged a new audacity in attacks against Israel, and ultimately elevated Fatah to independent leadership of the PLO. At the 1968 meeting of the fourth Palestinian National Council, the myth of Karama underwrote the policy that direct attacks on Israel should be the dominant front in the struggle for liberation. Karama catapulted Yassir Arafat, who emerged from the battle yet again unscathed, to a position of fame as his visage began to appear on the cover of

---

127. Sayigh, *Armed Struggle*, 178; Terrill, "Political Mythology," 91, 105–6; Nevo, "The Jordanian, Palestinian, and the Jordanian-Palestinian Identities," 13.

128. The Palestinian version pits the guerrillas' bodies against Israeli tanks: "Our fighters, our children, they came up from their secret places and they threw themselves at the Israeli tanks. Some climbed onto the tanks and put grenades inside them. Others had sticks of dynamite strapped to their bodies," as narrated by Arafat in his biography, Hart, *Arafat*, 262.

magazines and one of prominence, as he soon became the official Palestinian spokesman.[129] The "romanticism, wishful thinking, and deliberate distortions" that characterized the Karama myth and answered Palestinian emotional and psychological needs, became Arafat's signature style.[130]

As a narrative, Karama was deployed to forge national cohesion in exile and bestow a new image on the Palestinians. The guerrillas were presented as "a different breed of men" whose stand against Israelis troops marked a new stage in Palestinian history.[131] Descriptions of martyr commandos igniting explosive belts and using their bodies as weapons against tanks, relished and embellished, defined the manner in which Palestinians would thereafter contend with Israel.[132] Iconic representations of Karama and its heroes proliferated in Palestinian visual and verbal art forms and, like the map of the land between the Jordan and the Mediterranean, became calls for remembrance and uprising. The Karama myth was put to several uses. As a charter, it granted Fatah the prime position in the hierarchy of the Palestinian resistance and placed the guerrillas above the law until the Jordanian crackdown. As a press release, it brought attention to Arafat as commander of the *fida'iyin* and contextualized the Palestinian struggle as a revolution worthy of international funding. As a "pioneer" story, it reverberated as a renaissance at the very border that symbolized Palestinian dispossession. As a narrative of beginning, it promised that future victory would be achieved only through armed engagement with Israel and self-sacrifice. Indeed, Fatah has been prone to reanimate the Karama myth as a means of resuming armed struggle when negotiations bore no results. Karama opened up new configurations of Palestinian identity while reinforcing their connection with national aspirations and the necessity of specific spatial coordinates. The battle's setting near the Jordan was mobilized as proof that resistance was the way back home.

Bridge over the River Jordan

On October 26, 1994, the Jordanian and Israeli governments signed a peace treaty that recognized a midpoint in the Jordan River as the border between the two countries and opened the border for crossings by Israelis, Jordanians,

---

129. "In May 1968 Fatah was granted a full 38 seats on the Palestinian National Council (out of 100 at the time) . . . In February 1969, 11 months after Karameh, 'Arafat was elected Chairman of the PLO Executive Committee, the most important Palestinian leadership position." Terrill, "Political Mythology," 103.

130. Terrill, "Political Mythology," 97–98.

131. Turki, *The Disinherited: Journal of a Palestinian Exile* (New York: Monthly Review Press, 1972), 60, quoted in Terrill, "Political Mythology," 98.

132. The Jordanian version does not attribute much assistance to the guerrillas: "for the most part, they simply got themselves killed." Terrill, "Political Mythology," 106.

and tourists.[133] In one another's eyes, the two countries now conformed to the lines drawn by the British Mandate (article 3 in Treaty of Peace between the Hashemite Kingdom of Jordan and the State of Israel). No boundary markers were placed in the Jordan from Beit Shean to Ein Gedi, the span of the West Bank, so that the border could remain symbolically open to defining a future Palestinian state. Palestinians have officially been able to cross into Jordan since 1967 under the open bridges policy of the Israeli government. However, neither the open bridges policy nor the peace treaty alleviated the combined burden of permits, stipulations, waiting, and questioning that characterizes the crossing of the Jordan for Palestinians. Palestinians cross into Jordan in order to travel to other Arab countries, receive medical care, reunite with relatives, or attend family functions. For Palestinians who can secure permission to travel between the banks, the passage proves humiliating and threatening to their sense of order, honor, and self. Stories of the crossing express the constrictions on Palestinian mobility and describe the Jordan River not only as a border, but also as an inverted liminal region where difference rather than unity is asserted and inscribed on those making the passage. Despite the conditions of occupation on the West Bank, it continues to function as the terrain of the homeland in the imaginations of exiled Palestinians and their children residing in Jordan. Most accounts of a post-1967 westward crossing focus on the trials of waiting and interrogation at the hands of Israeli officials. A female narrator's recollections of her westward crossings are characterized by the motifs of family destabilization, treacherous allies, and the confrontation with what it means to live in exile.

> It was very hard, very hard. We would sit in the car for many hours, then leave the car and go to the Jordanian crossing point. In those days, there wasn't any building even, so, forget it, we would bake in the hot weather. All of the people, even women and children, would stand in line for four or five hours and no one had the humanity to think about the heat and sun. We just went from checkpoint to checkpoint, first on the Jordanian side and then on the Israeli side. It was the same on both sides with the long, long wait and no building and no air conditioning. The officials have a desk that is shaded with a roof and the Israelis have a fan, but not us. Not the people, we just stand outside. It gets really hot by the Jordan, up to 40° C [100° F]. The worst part was the Israeli side where they take off everything, look at you, have you take off your clothes and check everything. Then you go to a room and then they order you like

---

133. Among the first Israelis to visit Jordan was a group of friends of the five who fell on the way to Petra. Yadin, "The Lure of the Trail"; *Until the Rock*.

> an animal to get out of the room. I shiver whenever I think about it. Then they check each of us individually, and do you know how much time is wasted and how bad the humiliation?

Part of the humiliation of the crossing is that elders are shown no particular respect, which comes as a blow to the Palestinian ethic of honoring older family members. Heads of families find themselves powerless to prevent the shaming of their children and grandchildren. Even symbolic attempts at maintaining some kind of order are thwarted.

> I remember how my mother spent like two days before the crossing ironing the clothes, putting everything together and wrapping the shoes individually in bags so as not to dirty the clothes. She would also wrap the medicine and put it to the side. When she came before the Israeli officer, he would take the suitcase, take it all out, throw everything on the floor, and put it upside down. My mother would scream "what are you doing?" at him. My mother is a neat freak, when she saw this her blood pressure would shoot up. The official didn't listen to her. She asked him, "Is there a bomb inside the medicine?" She was really mad and she was asking why are you doing this to us, you're not the only people.[134]

In the crossing from Jordan to Israel, Palestinians faced the fact that no matter who they were, they were suspects. After describing the impact of the crossing on her mother, the narrator recalled a crossing undertaken by her husband when he was incapacitated from serving as the family provider. The narrator provided the background that a Palestinian entering the West Bank from Jordan is only allowed to bring a specific amount of money determined by the trip's duration.

> There is a point where they check your money. One time my husband was crossing and had, I think, $2,500 with him. He was allowed to take $2,000 for the month and had no way of explaining the extra $500. He meant to give the $500 to his father as a present. His father was an old man and appreciated money more than clothes or material things. He met a lady when he was about five minutes from the checkpoint and he asked her, "Can you carry the five hundred with you just until we cross

---

134. Although the mother does not defeat the Israeli soldiers as do some of the women in the Intifada legends collected by Sharif Kanaana, his observation about the mother figures applies: "It is the adult women or mothers who appear to be the strongest figures—the pillars of the Palestinian Arab family." Kanaana, "The Role of Women in Intifadah Legends," *Contemporary Legend* 3 (1993): 58.

the bridge so that I won't have to pay a penalty or anything?" And she said, "okay." He gave her the $500, but after getting through the checkpoint, the lady disappeared and he lost it anyway. He was going to lose it anyway. The lady ran off with it and the Israelis were going to take it from him. So, do you see how hard?[135]

The crossing always entails some kind of loss, whether of money, order, or dignity, that reinforces the original loss of homeland and property. Each crossing of the Jordan, even a westward return to the landscape of home, causes a sense of homelessness and instability and results in exposure to the hot sun and the exposure of powerlessness. At the Jordan, the Palestinian realizes the degree to which she lives in exile. As scholars have noted about borders in general, the Jordan border serves as the place where the nature of Palestinian identity comes into undeniable relief.

Neither the Israeli-Jordanian peace treaty nor the Oslo Accords changed the bizarre nature of the border crossing for Palestinians; they rather multiplied the levels of scrutiny and obstruction.[136] An Israeli journalist for the newspaper *Haaretz* described the checkpoint at the Allenby/King Hussein Bridge as a study in mixed and misleading symbols.

> If you want to see a brilliant trompe l'oeil, a dazzling apparition, the ultimate phantasmagoria—the place to visit is Allenby Bridge on the Jordan River ... Dozens of Palestinians of all ages are standing in line and waiting to go through passport control, just as people do in any other country. But which country are they entering? Israel? Then what in the world are all these Palestinian policemen doing here? Palestine? Then what in the world are all these Israeli police doing here? A joint terminal? Then who in the world is the sovereign power here?[137]

Crossing a border triply guarded by Jordanians, Palestinians, and the vigilant Israelis fashions the contemporary Palestinian returnee a manner of pioneer staking a place in a homeland forever altered by the State of Israel. This type of Palestinian pioneer in the occupied West Bank figures as the protagonist in several contemporary literary works. Sahar Khalifeh's novel *Wild Thorns*

---

135. Interview in San Francisco, May 5, 2001.

136. "During the years of the peace process (1994–2000) it was possible (for some) to actually 're-turn.' However, it was not 1948 refugees, but persons displaced in 1967 or persons who had migrated for other reasons who were able to return, or rather resettle, in the West Bank or Gaza." Schulz, *The Palestinian Diaspora: Formation of Identities and Politics of Homeland*, 212.

137. Gideon Levy, "Twilight Zone: More than Meets the Eye," *Haaretz*, Friday, September 3, 1999, magazine, 7.

opens as Usama, the young hero, crosses the Jordan border in order to be reunited with his widowed mother and seek employment in Nablus. As he undergoes interrogation by Israeli officers and then gazes out at the barren landscape from a shared Mercedes taxi, he is haunted by a vision of "idyllic green meadows, the clear waterfall tumbling over the bottles of soft drinks in the green valley, the bags of almonds piled up in front of the waterfall, beneath the towering walnut trees" to which he expected to return. While the very unfamiliarity of his homeland makes him into a pioneer in a strange land, he confronts security zones, barriers and overcrowding rather than the vast tracts of land usually associated with pioneers. "Yes, heaven was here beneath his feet and before his eyes. But he was now a prisoner in the genie's bottle."[138] The restrictions he faces on his mobility and autonomy result in his development as a kind of anti-pioneer who is allowed no freedom to settle, work, and celebrate the passage of time and so turns to activities of rebellion and resistance.

The poet Mourid Barghouti's memoir of return to Ramallah also opens with a prolonged Jordan River crossing during which his mind is flooded with memories. He wonders how the narrow Allenby Bridge spanning the trickle of the Jordan has kept him from home for thirty years.

> Here I am, walking, with my small bag, across the bridge. A bridge no longer than a few meters of wood and thirty years of exile. How was this piece of dark wood able to distance a whole nation from its dreams? To prevent entire generations from taking their coffee in homes that were theirs? How did it deliver us to all this patience and all that death? How was it able to scatter us among exiles, and tents, and political parties, and frightened whispers?

On the other side, the difference between the banks becomes apparent with the sight of the first Israeli soldier as Barghouti struggles with the proper name to give to his destination. "And now I pass from my exile to their ... homeland? My homeland? The West Bank and Gaza? The Occupied Territories? The Areas? Judea and Samaria? The Autonomous Government? Israel? Palestine?" After passing Israeli and Palestinian passport control, he realizes that the conundrum of the land is more than a matter of names.

> I used to tell my Egyptian friends at university that Palestine was green and covered with trees and shrubs and wild flowers. What are these

---

138. Sahar Khalifeh, *Wild Thorns*, trans. Trevor LeGassick and Elizabeth Fernea (New York: Interlink Books, 2000), 19.

hills? Bare and chalky. Had I been lying to people, then? Or has Israel changed the route to the bridge and exchanged it for this dull road that I do not remember ever seeing in my childhood? Did I paint for strangers an ideal picture of Palestine because I had lost it?[139]

Return and reunion are acts of redemption, yet the redemption is called into question due to the element of humiliation and the ubiquitous signs of occupation. In fact, the very concepts of homeland, of return, and of Palestine itself seem to signify nothing to be found outside of memory. Rather than finding its fulfillment, the dreamt redemption of the West Bank dissolves into despair. Through inversion, the space of home becomes a reminder of exile. All the same, the Jordan always distinguishes between the two; to cross it is to enter a sacred past or a redemptive future. Like Zion for the Jews, Palestine for the Arabs is the place of ultimate belonging where one must pause to wonder if he really belongs.

The Jordan River Valley is a classic borderland zone characterized by displacement and reterritorialization; however, in this case, the experience of the border does not give rise to hybridized identities, but rather intensifies essentialist and national ones. Border crossings have the perhaps unanticipated effect of reifying the border. Neither their diasporic natures nor the fact that the Israeli identity came into being through participation in the mass immigration of Jews from across the globe and that the Palestinian identity emerged from the collective experience of displacement and exile results in a sense of the fluidity of boundaries. Instead, both peoples insist on indigenousness, primacy, and the inalienable right of territory. In the midst of mobility, the concept of the homeland becomes more entrenched.

The moral of this comparative story is that a symbol shared by two ethnic groups but diametrically interpreted leads to more strife than two distinct symbolic systems. A deep rift results when two groups delimit their identity along the same coordinates and the contest over space becomes a clash of self-definition. Because borders symbolize a collective whole, the overdetermination of a single border exerts a ripple effect of contentious engagement on other fronts. A borderland is a space where distinct conceptions of group identity come into contact producing either the celebrated hybrids of postmodern thought or essentialist identities reinforced through encounter with the Other.

Nationalist myth sculpts historical events into cultural traditions that determine collective identity while configuring systems of power. The territory

---

139. Mourid Barghouti, *I Saw Ramallah*, trans. Ahdaf Soueif (New York, Anchor Books, 2003), 9, 13, 28.

of homeland anchors the myths in space and delimits identity while naturalizing hierarchies. While outlining hierarchy and articulating the anxiety of contiguity, border stories establish new epochs of national development. The heroes of border stories are pioneers whose actions are enlisted to promote unyielding commitment to land and to require sacrifice in its name. The militarism of both Israeli and Palestinian cultures is traced to heroic origins when an improvising group strengthened by national conviction held its own against a state army. Both the account of the Deganians protecting their kibbutz from the onslaught of Arab armies and that of the Karama commandos facing the formidable Israeli Defense Forces are structured as a David versus Goliath battle of the weak against the strong. Another similarity arises from the fact that heroes distinguish themselves at frontiers, risking their lives for the nation's inviolability. Even as specific pioneers fade from national consciousness, the borders drawn by their tales persist as national signifiers. Two nations claim the Jordan as the ultimate symbolic border and so their mythologies exist in dialectic tension where the other is sometimes absent, sometimes negated, and always at some level operating as enemy. When aware of the other's national paradigms, Israelis and Palestinians interpret them through inversion so that Israeli pioneers are militarized colonizers of Palestinian land and Palestinian freedom fighters are terrorists bent on hunting down Israelis. Since each mythic context can neither support nor sustain the opposing perspective, shared symbols produce competition and perpetuate conflict. While feeding particular nationalist sentiment, the Jordan River stories maneuver within a single discourse of struggle.

Conclusion

THE BAPTISM BUSINESS AND THE PEACE PARK

The Jordan Valley is an integral part of the Palestinian land occupied in 1967, just like east Jerusalem.
  —Prime Minister Salam Fayyad, March 2011

Our security border is here, on the Jordan River, and our line of defense is here.
  —Prime Minister Benjamin Netanyahu, March 2011[1]

Contemporary theories of borderlands as well as transnationalism claim that despite the flags, guard towers, and, in some cases, walls, the hollow nature of the state and its subordination to borderless corporations and political movements become apparent at borders. As we have seen, such dynamics do not seem to be the case in the borderland of the Jordan River Valley. Here, at the line between the Hashemite Kingdom of Jordan and the State of Israel and the eastern border disputed by Palestinians and Israelis, no distinct borderland identity, no microeconomy based on tariffs or smuggling, no nonsovereign flow of labor or capital can be observed. Palestinians from the West Bank and Jordan frequently cross the border in the name of family visits, pilgrimage to ancestral lands, education, medical care, and access to the rest of the world via the Queen Alia International Airport near Amman, but as Edward Said has observed, the Jordan is *"the* border" where Palestinian identity is brought into relief and

---

1. Herb Keinon, *Jerusalem Post*, March 8, 2011, http://www.jpost.com/Defense/Article.aspx?id=211268.

policed.[2] The Jordanian state and, in particular, the State of Israel seem to assert themselves most strongly on the bodies of Palestinians crossing the river on the Allenby Bridge.

I want to explore why the state seems to persist in this particular borderland as I show how the apparently transnational business of Christian tourism constitutes an occasion of national assertion for Israel, Jordan, and, once upon a time in the days of Yasir Arafat, the Palestinians. Along with this line of inquiry, I want to ask a larger question about how authority over borders and resources is best determined at this moment of late capitalism. Or, in other words, what sorts of controls should be placed on the border—if we think that it should remain a border—and on the river and by whom? If we think that the resources properly belong to multinational corporate interests, then should the states move out of the way or at least realize their already co-opted role? If we oppose such multinational control, then should we see forms of ethno-national assertion—minority oppression and all—as a mode of resistance? Should we rather prioritize religious claims to sacred sites? Indigenous ones? And, if we listen to these claims, what roles do archaeology and its interpretation play in their authentication? Finally, in the clamor for oversight, where do we place transnational ecological appeals to focus on the disappearing resource of water in the Jordan River?

At today's Jordan River, the long history of Christian pilgrimage intersects with religious tourism; now, as ever, the act of pilgrimage sanctifies the very ground that it seeks. Despite the fact that its waters are no more potent than any others used for baptism, the Jordan is Christianity's holiest river. Its symbolic and ritual significance arises from the fact that here John the Baptist baptized Jesus (or, as some sources have it, Jesus baptized himself). The river is sacred because of the role it played in the early life of Jesus and because his going down into the river and rising up simulated his later death and resurrection. Christian pilgrimage to the Jordan is not undertaken in the name of expiation so much as in the name of retracing the footsteps of Jesus and thereby bringing one's life into greater alignment with Christ's. For many, it is an occasion for a secondary baptism and thus provides the opportunity for a volitional baptism or a different sort of rebirth experience.

There are three competing baptism sites on the Jordan River: Yardenit on Kibbutz Kinneret in Israel's eastern Galilee region, Al-Maghtas on the eastern shore of the Jordan north of the Dead Sea, and Qasr al-Yahud directly across from Al-Maghtas on the western shore. As the directors of the sites and the respective ministers of tourism work to definitely write the sites into the Gos-

---

2. Said, *The Politics of Dispossession: The Struggle for Palestinian Self-Determination* (New York: Pantheon, 1994), 8.

pel accounts of Jesus's baptism, they also compete for the devotion of a lucrative Christian tourism industry. At the same time that cornering this market hangs on biblical legitimacy, the attention of pilgrims and popes feeds the contest for national legitimacy among Israelis, Palestinians, and Jordanians. It is as if recognized sovereignty over a Christian holy place confers Christian favor on either the Muslim Hashemite Kingdom of Jordan or the Jewish State of Israel. During the Oslo Accords, the largely Muslim Palestinian Authority was a potential developer of a baptismal site and therefore another suitor of Christian favor, but currently Israel has developed and has full jurisdiction over the West Bank Qasr al-Yahud site.

Moreover, the competition for Christian recognition sustains a certain colonial dynamic in which Arab and Jewish national aspirations depend upon European consent. Just as the territorial and political form of the Hashemite Kingdom of Jordan resulted from British decisions, so the Israeli-Palestinian dispute is, in many ways, driven by questions of partition, population growth, and territorial rights that first arose during the British Mandate in Palestine. Since the British withdrawal and certainly since 1967, the Americans have served as arbiter and overseer of the conflict, and so a sense that Christian countries and Christian world leaders determine national sovereignty has persisted.

The opening of each baptism site has occasioned at least one form of national self-assertion and boundary drawing. The Yardenit baptismal site opened in 1981 on the grounds of Kibbutz Kinneret, the second kibbutz to be founded in Mandate Palestine in 1920. Its position on the western edge of the Jordan had symbolic resonance for its founders as well as for the larger Zionist movement as the paradigmatic site of Jewish national autonomy. Zionist mythology framed the kibbutz founders as intrepid pioneers who, like the ancient Israelite tribes under the leadership of Joshua, were crossing the Jordan to an era of independence and national settlement. Although Yardenit has been developed as a pilgrimage site for Christian tourists, its location attests to the sacred-secular hybridity that sanctifies Jewish national spaces. In other words, the avowedly Jewish secular endeavor of founding Kibbutz Kinneret garnered legitimacy in the eyes of world Jewry and in those of British Christians due to the biblical significance of its location. In developing the baptism site, the kibbutz members sought to further authorize their version of the Jewish national project by linking it with the biblical event of Jesus's baptism. While the baptism has little to no significance in Judaism and the kibbutz members are mostly secular, its antiquity establishes a line of continuity between first-century Jews like John the Baptist and Jesus and the current Jewish inhabitants of the State of Israel.

The Yardenit baptismal site opened in cooperation with the Israeli tourism ministry, which saw the site as a solution to the pilgrim practice of stopping "along the road and climbing down the river's steep, muddy banks to hold baptisms. The government built concrete tiers and handrails leading to the water; the kibbutz set up stands selling snacks, bottled water and T-shirts that say: 'I was baptized in the Jordan River.'"[3] The makeshift kiosks were upgraded in 2000 into a stone visitors' center shaped like a cross with a restaurant perched above the Jordan River. The renovation coincided with the Catholic Jubilee. Ubiquitous claims that Yardenit marks the site of Jesus's baptism are unsubstantiated by archaeologists or scholars of history and religion. Its legitimacy depends upon its position by the Jordan as well as on the biblical evidence for Jesus's early ministry on the shores of the Sea of Galilee. While never directly stated in tourist literature, the success of the Yardenit site results from its distance from more hotly contested areas and its proximity to other central Christian sites. In a single day, a Christian tourist can visit sites including Yardenit, Capernaum, the Mount of Beatitudes, the Church of Loaves and Fish, and Chorazin. It is likely that even with the increased exposure of the more historically plausible Al-Maghtas and Qasr al-Yahud sites, Yardenit will continue to enjoy robust business due to its location in the relatively quiet political terrain of eastern Galilee rather than at the literal edge of the West Bank.

Despite the fact that the Israeli Ministry of Tourism operates both Yardenit and Qasr al-Yahud, it is very likely that there are vested interests in maintaining the relatively large facility at Yardenit for economic as well as political reasons. The political calculus involves the sense that the Qasr al-Yahud site may eventually be turned over to the Palestinian Authority as part of a land transfer to the Palestinians. Should this be the case, Israel wants to be sure to have a recognized baptismal site as well as a place where Christian visitors tacitly support the Jordan River as the eastern border of the country, if only along its northern course. That said, Prime Minister Benjamin Netanyahu has folded the funds for the Qasr al-Yahud baptism center into the budget for West Bank development, meaning that money for Palestinian infrastructure has been diverted to the development of an Israeli tourist site. This sort of financial commitment bespeaks a proclamation rather than a planned retraction of the Israeli presence on the West Bank.

Kibbutz Kinneret, like the many kibbutzim established on the banks of the Jordan, began as an agricultural community dependent upon the fresh water of the Jordan. The agricultural focus later shifted to an industrial and, as has been made apparent, touristic one. At present, the kibbutz maintains

---

3. "Adding Symbolism to Baptism a Church Uses Water from the Jordan River, Where Christ Is Believed to Have Been Baptized." JoAnna Daemmrich, *Baltimore Sun*, May 25, 1997.

a large-scale date farm that is watered by the Jordan and an impressive emporium of date products. Another charming corner of the kibbutz hosts the grave of the Hebrew poet Rahel and a cabin where early Zionist heroes lived while developing the area.[4] Because of its location just south of the Sea of Galilee, Yardenit was the most practical place for a baptism site. The flow of clean water, unsullied by the wastewater dumped to the south, seemed guaranteed no matter the downstream state. However, the Sea of Galilee is currently so low that for the last two years the Israeli Water Authority has manually pumped water into the Jordan.

Our second site is Al-Maghtas, a state-sponsored Jordanian baptismal center on the eastern edge of the Jordan River just north of the Dead Sea. The development of the Al-Maghtas site and the surrounding baptism complex at Wadi Kharrar stands among the priorities of King Abdullah II. Abdullah's father, the late King Hussein, made the pursuit of Christian recognition a priority and set in place the conditions to enable such pursuit. Development along the Jordan River was possible only following the 1994 peace treaty between Israel and Jordan that effectively demilitarized the border. In 1995 Prince Ghazi bin Mohammed took an interest in Al-Maghtas and Wadi Kharrar as part of a project to promote Christian holy sites in Jordan. This interest led to a largely state-sponsored archeological dig led by Mohammed Waheed of the Jordanian Antiquities Authority.

For Jordanians, as for Israelis and Palestinians, the modern national project involves "'an obsession for archaeology' which makes use of historical remains to prove a sense of belonging."[5] In the case of Christian archaeological finds, the sense of belonging follows a convoluted path. On the one hand, antiquities attest to the long history of the recently nationalized territory. Yet since the antiquities do not point to the Muslim or the Jewish past, they are used as a draw for Christian support. The Christian support, in turn, is broadcast as political approval. As Prince Ghazi had suspected, Waheed's archaeological team made some wonderful finds of Byzantine churches and pilgrim stations as well as some water jars that may be pre-Byzantine. After initial excavations proved promising, King Hussein appointed a royal commission to oversee excavations and develop a baptismal park in time for the Catholic Jubilee.

The media life of an archaeological find tends to be more lively than the structures and implements dug up, and indeed the Wadi Kharrar area was

---

4. One of the celebrated early residents is Pinchas Rutenberg, who developed the Naharayim hydroelectric plant.

5. Haim Yacobi, "Architecture, Orientalism, and Identity: The Politics of the Israeli-Built Environment," *Israel Studies* 13 (2008): 94–118, 114. Quoting Edward Said, *Orientalism: Western Conceptions of the Orient* (New York: Random House, 1978), 12.

quickly identified as the place where John the Baptist set up camp, Bethany beyond the Jordan mentioned in the Gospel of John 1:28, the site of a church built by the Empress Helena, and the place where Elijah the Prophet ascended to heaven. This archaeologically substantiated mythology enabled the establishment of a Bethany beyond the Jordan baptismal park and the plan to donate Jordanian national land for the establishment of several church complexes.

By the year 2000, King Hussein had secured for Al-Maghtas the status of an official destination recognized by the Council of Churches of the Holy Land. The site opened on January 7, 2000, the day on which Middle Eastern Christians celebrate John the Baptist. The subsequent controversy helps to prove my point that national recognition and legitimacy are at stake in the competition for biblical sites and contemporary Christian pilgrims. In 2000, Pope John Paul II planned to end his Jubilee visit in Jordan at Al-Maghtas and for this to be his only visit to the Jordan River. The announcement of the pontiff's itinerary, suddenly and surprisingly, united Israelis and Palestinians around New Testament interpretation. Jesus most certainly did not enter the Jordan from the east, reasoned the Israeli and Palestinian political exegetes, but from the west. While archaeological traditions surrounding the St. John the Baptist monastery and the Qasr al-Yahud site lent force to the argument, Jordanian spokespeople openly disputed it.[6] The abandoned structures and scattered landmines no doubt made the West Bank site less desirable in Pope John Paul's eyes, yet Yasir Arafat appealed to the Vatican and the government of Ehud Barak for the pope to visit Qasr al-Yahud.

Arafat had earlier announced plans to develop a Palestinian baptismal site in time for the year 2000 Jubilee and had hoped that such a venture would draw more tourists to autonomous Jericho and thus confer international legitimacy on a liberated Palestine. Israeli impediments, insufficient funding, and corruption prevented the development of a Palestinian baptismal site by 2000, but Arafat and the Palestinian Authority still desired recognition of the place where such a baptismal site would or could be developed. After the pope agreed to fly over the river by helicopter and visit Qasr al-Yahud for

---

6. "To some Jordanians, the sudden inclusion of Qasr Al Yahoud in the Pope's itinerary caused some disappointment at the loss of an opportunity to receive a 'de facto' recognition of the absoluteness of Wadi Kharrar as the site of Jesus' baptism. Tourism Minister Akel Biltaji, who has defended the authenticity of the site and eagerly worked for its development since the early archaeological findings started emerging, in 1997, is of the same opinion. 'This is where John the Baptist lived,' Biltaji insisted, listing the many archaeological findings. 'We do not need a seal. This is where the first Christian community on earth emerged, because the baptism of Jesus was the dawn of Christianity,' he said." Francesca Ciriaci, "Controversy Aside, Wadi Kharrar Remains a Site Both Inspiring and Spiritual," *Jordan Times*, Tuesday, March 21, 2000.

fifteen minutes, the Barak government criticized Arafat's plans to greet the pope on territory under exclusive Israeli control. In the end, the pope visited Qasr al-Yahud under Israeli supervision and met with Arafat only when he traveled to Bethlehem.

Kimberly Katz has adeptly shown how the papal visits of 1964 (by Pope Paul VI, the first pope ever to undertake a pilgrimage to the Holy Land) and 2000 provided opportunities for Kings Hussein and Abdullah II of Jordan to come forward as the proper custodians of holy sites, thereby "promoting symbols of religion as symbols of national identification."[7] Deriving her evidence from tourist materials, newspaper articles, and postage stamps, Katz reveals how the two papal visits were symbolically linked as well as how both kings exploited the papal visits for their full political capital. Pope John Paul II's appearance at two baptismal sites and Pope Benedict XVI's recent delivery of his message to the Islamic world from Jordan point to the multiple currencies of such political capital. King Abdullah built on the political capital of the pope's visit by holding Jordan's National Arbor Day ('Id al-Shajara) at the baptism site.[8] At least two political maneuvers are elided in the symbolic setting of the national ritual. In the first, the pope's visit to a declared holy place is interpreted as a sign of political support; in the second, the now-legitimized holy place becomes sanctified as an inextricable national domain.

The competition between Al-Maghtas and Qasr al-Yahud is all the more direct since the Israeli Ministry of Tourism opened a $5 million baptismal center on December 18, 2008, with about six feet of river between it and Al-Maghtas. The argument that the Israeli and Jordanian baptismal sites function as ciphers for national self-assertion first occurred to me when I visited the Qasr al-Yahud construction site in August 2008 and saw Israeli and Jordanian soldiers facing off on opposite banks of the river while groups of European Christians self-baptized in the green and murky river. However, an official statement from the Israeli Ministry of Tourism claims that Israeli and Jordanian authorities will coordinate times of baptisms in order to "avoid overcrowding and to facilitate smooth and orderly baptismal ceremonies on both sides of the river."

The Israeli plans did not catch King Abdullah unaware, since the king and his tourism director had launched the plan of donating land for churches and baptism centers to the Russian, Coptic, Catholic, Armenian, and Anglican Churches. This mode of development increases the chances that members of those churches will visit the Jordan River from an eastern approach

---

7. Kimberly Katz, "Legitimizing Jordan as the Holy Land: Papal Pilgrimages—1964, 2000," *Comparative Studies of South Asia, Africa, and the Middle East* 23, n. 1/2 (2003): 184.

8. Katz, "Legitimizing Jordan as the Holy Land," 186.

and thereby support the authenticity of Al-Maghtas as well as the authority of the Jordanian kingdom as custodian of Christian holy sites. It seems that the leaders of these churches as well as other church leaders like Rick Warren are tacitly beholden to the king's position on the authentic site of Jesus's baptism since the official website of Wadi Kharrar and Al-Maghtas contains thirteen authentication testimonials.[9] Furthermore, on May 10, 2009, Pope Benedict XVI visited only the Al-Maghtas baptism site, where he blessed the cornerstones for the Catholic churches to be built on the donated land.[10] With the land donations, a certain loop is closed in which Jordanian national territory supports international churches that then endorse the authenticity of the site and, in turn, of the nation.

• • •

With church and state figuring out their relationship in some novel ways, it is worth asking anew whose authority determines the authentic. If authenticity for Christians depends upon the proven presence of Christ, then archaeologists and historical geographers are summoned to produce material signifiers of a god incarnate. For this reason in part, material culture has an out-sized influence on biblical studies. This comes with benefits including a look at settlement patterns and cultural diffusion that flies in the face of the Bible's more absolutist rhetoric, but at the same time gives rise to reactionary political claims. The baptism business, it seems to me, represents a mild form of religious materiality when compared with the expropriation of public land in Palestinian Nazareth for the privatized and costly Nazareth Village, an American Evangelical Christian production of Jesus's hometown with an alleged basis in archaeological truth, or with Jewish land grabs in east Jerusalem in the name of Davidic remnants. Who can say when, where, or whether Jesus of Nazareth went down in the River Jordan when the accounts are typologies of prophetic succession? And so, other than potential infection and transmission of viruses from the untreated wastewater in the Jordan, what harm is done if Christians immerse at any old twist in the Jordan?

If authenticity is a question of where the earliest Christians commemorated Jesus's baptism, then we can look forward to pilgrim accounts or backward to

---

9. http://www.baptismsite.com/content/category/4/12/22/lang,english/.

10. Nayef al-Fayez, director of the Jordan Tourism Board, cast the pope's exclusive visit as a "boost [to] Jordan's historic legacy as the cradle of civilizations and the land of the prophets." In this statement, the pope's visit attests to the antiquity of Jordan, which, in turn, supports its contemporary value and marks it as a place important to Catholics. If the 2009 statistics point to any kind of trend, it is that European Christians are more prone to visit the Jordanian site and American Christians more prone to visit the Israeli site.

suggestions in the Tanakh and Josephus of an ancient Jewish Jordan crossing festival with political ramifications. Egeria—one of the earliest travelers to the land of the Christian Bible—forded the river approximately at the Qasr al-Yahud/Al-Maghtas crossing, but made no mention of the baptism. Instead, she identifies it as the place where Joshua led the People of Israel into the Promised Land. Reversing their journey, she proceeds to Livias and visits their final campground in exile. In Egeria's view, John performed his baptisms at Aenon Salim, about eight kilometers south of today's Beit She'an. That a woman's authority on ritual geography has not been taken up as a development plan should come as no surprise. Egeria's hypothesis, however, has led scholars to conclude that no baptism site had been formally located in the fourth century. Mohammed Waheed's excavations have turned up a late ancient church with the unusual feature of an apse that opens to a causeway leading down to the river. There, a stone platform implies an early site of baptisms.[11]

If, as the saying goes, convenience is king in the world of consumption, then Yardenit rules. Church groups can step down into the water on a wooded bend visually apart from the conflict, then enjoy lunch in a kind of kibbutz dining hall shaped like a cross. It can be unbearably hot at Al-Maghtas/Qasr al-Yahud, and the ongoing nationalist rituals at the border mean that all visitors on either side are escorted to the edge, where armed Israeli and Jordanian soldiers stare one another down. In cases where church leaders command foremost authority, Al-Maghtas is the authenticated site; otherwise, people likely believe what it is in their best interest to believe about Jesus and his baptism. It is safe to say that although Christian tourists experience the Jordan River through Israeli or Jordanian facilities developed in the name of national benefit, these tourists interpret their experience within a religious and personal framework. That is, the tourists are unknowing players in the baptism business and its role in political competition. And while a diminished, torpid river (the lower portion of which carries wastewater, agricultural runoff, and saline waters) presents a disappointment, most tourists are unaware of Jordan's ecologic status and the imminent danger of its disappearance. One could observe that contemporary baptisms in the Jordan, no matter the point of entry, are initiations into a state of globalized pollution for which nations, not to mention industry, refuse to accept responsibility.

If we flip the question and pursue the water used in baptism instead of the place of entry, then questions of authenticity and authority have a different force. The question can be asked baldly: to whom does the water of the Jordan belong? On the basis of current use, the answer is that it belongs to industry and to states, with an unsurprising skew to the states. Since the 1950s, the

---

11. Archeologist Jennifer Tobin kindly shared her notes on Wadi Kharrar.

Israeli National Water Carrier has diverted Jordan River water to Tel Aviv and the Negev region. The Yarmouk tributary is largely diverted for use in Syria and Jordan before it ever reaches the river.[12] If we view resources like water as intended for the domestic and agricultural use of proximate citizens, then all three states would be within their rights to siphon off as much water as possible before it converges into an interstate boundary. If the resources are seen as belonging to future as well as to current citizens, then current practices can be called into question from the future-oriented perspective of the environmental movement. If we perceive resources as part of inalienable human rights irrespective of place and time, then we can perhaps be persuaded by bioregional theories of human organization that would disregard state boundaries and reimagine communities in terms of the waterways that support them. A community defined by the water that supports it, bioregionalism argues, would certainly behave differently than a state looking to advance its interests through resource extraction, not to mention differently than a multinational corporation pursuing a very different bottom line. Who should be the ultimate sponsor and overseer of environmental reclamation efforts, particularly those that cross state lines?

Such a question is not unrelated to the disappearing Jordan and the baptism business, inasmuch as the World Bank has commissioned global consulting firms to address the possibility of replenishing the diminishing water table in the Jordan-fed Dead Sea by pumping in water from the Red Sea, a case in which local needs are both addressed and determined by corporate and privatized economic interests;[13] the leading public proposal regarding the potential rehabilitation of the Jordan River comes from Friends of the Earth Middle East (FOEME), a consciously transnational environmental organization with Israeli, Jordanian, and Palestinian leadership. As a measure to restore the Jordan and establish an environmental basis for cooperation, FOEME is proposing the development of two transborder "peace parks" encompassing clusters of important archaeological and historical sites. The maps generated by FOEME remove the contested political boundaries to instead depict the Jordan watershed as a regional lifeline (see figure 15). Signaling postnational conceptions, the Jordan River peace park plan comes closest to the biblical countermyths.

---

12. According to the peace treaty between Israel and Jordan, Israel is supposed to receive 25 mcm of water from the Yarmouk per year. This does not always occur. I thank Gidon Bromberg for explaining current and proposed water sharing agreements.

13. The project has an estimated cost of $5 billion and requires extensive international financial support, where the World Bank, IFC, USAID, EU, and UNDP have all been approached to support this project. FOEME website: http://foeme.org/www/?module=projects&project_id=51

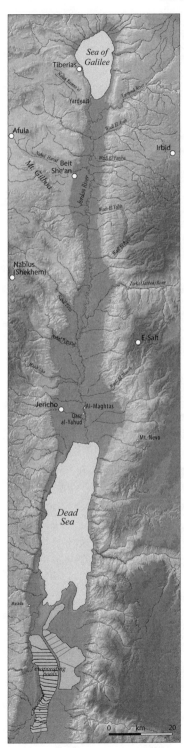

15. The lower Jordan River Valley with the Yardenit baptismal site just south of the Sea of Galilee and the baptismal sites of Al-Maghtas and Qasr al-Yahud to the north of the Dead Sea. Reproduced with the permission of EcoPeace/ Friends of the Earth Middle East, www.foeme.org. Redrawn by Soffer Mapping.

The peace parks would entail both the protection of the land and water and sustainable modes of treating and restoring the river. The northern peace park would encompass "two adjacent areas; Al Bakoora/Naharayim, where a small island was created at the junction of the Jordan and Yarmouk Rivers, and the Jisr Al Majama/Gesher site, known as the historical crossing point of the Jordan River Valley"; the southern park would include Qasr al-Yahud and Al-Maghtas. For some of the same reasons that make Yardenit a more viable baptism site, FOEME is currently focusing on the northern peace park. Their approach, laudably, is theoretically global and practically local, by which I mean that FOEME has adapted the concept of "Peace Parks" in "transboundary protected areas" from the World Conservation Union to the specific localities surrounding the Jordan.[14]

---

14. A "transboundary protected area," as defined by the World Conservation Union (IUCN), is a protected area that spans national boundaries, where the political borders that are enclosed within its area are abolished. This includes removal of all forms of physical boundaries, allowing free movement of people and animals within the area. A boundary around the area may, however, be maintained to prevent unauthorized border crossing. Such areas, also known as "peace parks," are formally dedicated to the protection and management of biological diversity, natural and associated cultural resources, and the promotion of peace

In one of the first practical steps toward the project, FOEME gathered the mayors of cities along the Jordan for a collective immersion and the writing of a memorandum of understanding regarding a peace park. The proposed development of the peace park includes the reflooding of the Rutenberg Lake to replenish the Jordan and create a bird sanctuary, the renovation of the Naharayim hydroelectric plant as a visitor's center, and the refurbishing of the workers' residences as eco-lodges. The overall concept is an eco-tourist haven with bird watching, hiking, biking, and kayaking attractions that can be reached from both Jordan and Israel. Stages of development have been planned in relatively realistic steps in which Jordan and Israel would respectively establish parks on Bakoora Island and at Old Gesher, agree to the reflooding of the lake, and, eventually, combine the areas into a cross-border park.

If there is precedent for any such development along Israel's borders, then I think it can be found at the very site of the proposed peace park. Bakoora or "Peace" Island was returned to Jordan in an item of the 1994 peace treaty with the provision that Kibbutz Ashdot Yaakov could continue to cultivate its land that fell on the Jordanian side.[15] On something of a regular schedule, the militarily administered gates to the island are opened to tourists. Understandably, the killing of seven Israeli girls visiting Peace Island on a school trip by a Jordanian border guard in 1997 goes unmentioned in the FOEME peace park proposal. In response to the killings, King Hussein returned from a trip abroad, visited the grieving families, and donated blood to benefit the injured girls. Such a royal response is said to have greatly moved the Israeli public.

The other precedent derives from the fact that the Naharayim hydroelectric plant was a rare site of Jewish-Arab economic cooperation in Mandate Palestine. Neither equal nor even, such cooperation was largely thwarted by the British colonial authorities. At present, the story of the Palestine Electric Company and the Naharayim plant circulates as a tale of either Zionist expropriation or victory, depending on the narrator, but there is a buried theme or at least an embedded possibility that industrial development in Mandate Palestine and Hashemite Transjordan could have yielded different political configurations.

Pinchas Rutenberg, engineer and head of the Palestine Electric Corporation, was a Jew from Ukraine who turned his efforts to Zionism following his participation in Russian revolutionary efforts. With a mind to build infrastructure for a future Jewish state and to secure jobs—a stipulation of Britain's

---

and cooperation. The parks encourage regulated tourism, sustainable development, and goodwill between neighboring countries.

15. Bakoora Island was formed as part of the rerouting of the Jordan and Yarmouk Rivers in order to create canals running into the Naharayim power plant.

policy on Jewish immigration to Palestine—for Jewish workers,[16] Rutenberg tirelessly lobbied British authorities and raised capital to build a hydroelectric plant at the confluence of the Jordan and Yarmouk Rivers. Because the Yarmouk flowed entirely under Transjordanian jurisdiction, Rutenberg and the Palestine Electric Company needed Emir Abdullah's grant of the right to harness it. Abdullah agreed, with the promise that Rutenberg would also supply electricity to Amman, but, as Renate Dietrich has argued, "strong nationalist and anti-Zionist tendencies in Transjordan" prevented the PEC's development of a Transjordanian power plant or delivery of electricity to more than Abdullah's residence and a few select locations.[17] Arab as well as Jewish workers built the plant, but only Jewish workers lived on site.[18] Emir Abdullah attended the opening ceremony in 1932.

The Israeli historian Zvi Ilan has written of Rutenberg's frustration with the British officers who stood in the way of significant land sales by Transjordanian Bedouin leaders to Jews for residential and industrial development.[19] Such land sales would have provided the basis for the binational Jewish-Arab communities that Rutenberg later envisioned. Ilan points out that Rutenberg joined the binationalist group Brit Shalom, where he came under the influence of Martin Buber and Judah Magnes. As an engineer among philosophers, Rutenberg drew plans in the binationalist spirit for the Zarqa River Valley in Transjordan. Rutenberg imagined a hydroelectric plant serving as an industrial center where Jewish and Arab management and labor would be entirely integrated. Managers and workers would live along the banks of the Zarqa with their families, although Jews and Arabs would live on opposite banks of the river in order to preserve their religious and cultural customs. Without rehearsing the whole contrary history, I will say that the population of the contemporary city of Zarqa is almost entirely made up of Palestinian exiles and their descendants and that the Zarqa River is so polluted that its water is not potable.

---

16. Sara Reguer, "Rutenberg and the Jordan River: A revolution in Hydroelectricity," *Middle Eastern Studies* 31, no. 4 (1995): 718.

17. Renate Dietrich, "Electrical Current and Nationalist Trends in Transjordan: Pinhas Rutenberg and the Electrification of Amman," *Die Welt des Islams* 43, no. 1 (2003): 90. See also Reguer, "Rutenberg and the Jordan River," 716. Swedenburg notes that during the 1936–39 Palestinian rebellion, "commanders of the armed peasantry instructed urban residents to stop using the electric power produced by the Anglo-Jewish Palestine Electric Company." Swedenburg, *Memories of Revolt*, 33.

18. Dietrich, "Electrical Current," 92.

19. From 1921 through 1932 Transjordanian leaders and Emir Abudallah voiced and even published their readiness to sell and jointly develop land with the Jews. An October 23, 1921, article in the newspaper *Al-Jazeera* expressed the impatient wish of many Arab landowners on the east bank to sell their lands to Jews. Zvi Ilan, *Attempts at Jewish Settlement in Trans-Jordan 1871–1947* (Jerusalem: Yad Ben Zvi, 1984), 363 (Hebrew).

Should the peace park materialize with a visitors' center at the former Naharayim power plant, the memorialization of Pinchas Rutenberg will require reexamination. Currently, Rutenberg is celebrated as a Zionist hero who had the technological ability to illuminate the Jewish national dream. One hears this narrative on tours to the cabin he shared with other founding fathers on Kibbutz Kinneret and in a film shown at the tourist site of Old Gesher. Old Gesher, the former grounds of Kibbutz Gesher, stands as the closest Israeli territory to the former power plant. Among its features is "the Naharayim Experience at Gesher," a screening room built in the former dining hall of the kibbutz complete with a miniature train to signal Rutenberg's electrification of the Jaffa-Jerusalem train line and simulated steam. The film tells a Zionist-inflected story of Rutenberg and Naharayim that ends with the bitter loss of Jewish homes in 1948. With no sense of the ramifications for Palestinian claims, the film ends with a peace treaty–enabled return to the ruins of homes lost in 1948 by a group of Israeli adults who lived in them as children.

Can Palestinian and Israeli narratives stand side by side at a peace park museum? Can a compelling mythology grow around a transborder site? Can the ruins of the Naharayim power plant house the memories of pioneers, refugees, and freedom fighters while pointing toward a binational or postnational future? These concerns seem to be shared by FOEME, since one of the power-point presentations available on their website lists "conflict of interpretation"—including Naharayim's status as a Jewish national symbol—and "political views in presentation of historical events" as possible constraints. It is almost too bad that they cannot claim the peace park as the authentic site of Jesus's baptism, since I am not sure that environmental commitments can quite overshadow political realities in the same way that religious commitments can. Most clear to me is the fact that the Israeli-Palestinian conflict cannot be solved in historical terms. For this reason, I have directed my attention to spatial, rather than historical, configurations.

Another way to raise my same concern is by asking what would happen to the state and its borders at the peace park? In the responses that I have been able to locate thus far, the question is phrased in terms of national interests. On the Israeli side, this means the question of security—how can we be safe without definitive, patrolled borders?—and on the Jordanian side the question is, how can we engage in a joint project with the occupier of the West Bank and Jerusalem? In such questions one hears the reiteration of tropes that, in part, reify the extant borders, but on a different level they interrogate the very need for borders. For example, who is ultimately responsible for the bodies engaging in eco-tourism at the peace park? If we imagine shared Jordanian-Israeli jurisdiction, then how far across or past the border can soldiers or police go in order to protect or apprehend? Would Jordanians ultimately be responsible for Jordanian nationals in the park as well as for tourists who entered the park from Jordan? Would the same go for Israelis? Would, in fact, the military be in charge of policing the park? On what terms would the Israeli and the Jordanian militaries collaborate? Who would define these terms? Where would Palestinians stand in the peace park? Who would protect or represent their interests?

Questions further multiply: Who is sovereign in a transborder zone? Who protects the water? Who polices the state and industry? Who benefits from the tourist money? Who pays for the development? The water question seems the easiest to answer. Because FOEME wants to create the peace park in order to rehabilitate the Jordan, this transnational organization should oversee the restoration project. But we return to some of the same problems when we ask who would fund the project. By proposing that parks be established and rehabilitation efforts begin in earnest on both sides of the border before the transborder potential is explored, FOEME avoids getting caught in a trap like the one I have set with my questions. I ask the questions not because I want to thwart the efforts—in fact I hope to support them with my work on the transborder mythology of the Jordan—but because I wonder whether ecological investments have the power to realign notions of the state or to oppose the largely unhindered extraction of resources and dumping of waste by multinational corporations. Evidence from recent negotiations surrounding the Copenhagen Accord and from the "green" campaigns spearheaded by multinational corporations suggests that ecological concerns are easily coopted. But, maybe, the Jordan River really is the gateway to heaven.

SUGGESTED READINGS

Hebrew Bible

Alt, Albrecht. "Josua." In *Kleine Schriften zur Geschichte des Volkes Israel*, 176–92. Munich: C. H. Beck, 1953.

Alter, Robert. *Genesis: Translation and Commentary*. New York: W. W. Norton, 1996.

———. "How Convention Helps Us Read: The Case of the Bible's Annunciation Type-Scene." *Prooftexts* 3 (1983): 115–30.

———. "Sodom as Nexus: The Web of Design in Biblical Narrative." In *The Book and the Text: The Bible and Literary Theory*. Ed. Regina Schwartz, 146–60. Cambridge: Basil Blackwell, 1990.

Assis, Elie. "The Choice to Serve God and Assist His People: Rahab and Yael." *Biblica* 85, no. 1 (2004): 82–90.

———. *From Moses to Joshua and From the Miraculous to the Ordinary: A Literary Analysis of the Conquest Narrative in the Book of Joshua*. Jerusalem: Hebrew University Magnes Press, 2005 (Hebrew).

Assmann, Jan. *Moses the Egyptian: The Memory of Egypt in Western Monotheism*. Cambridge: Harvard University Press, 1997.

———. *Religion and Cultural Memory*. Trans. Rodney Livingstone. Stanford: Stanford University Press, 2006.

Auld, A. Graeme. "From King to Prophet in Samuel and Kings." In *The Elusive Prophet: The Prophet as a Historical Person, Literary Character, and Anonymous Artist*. Ed. Johannes C. de Moor. Leiden: Brill, 2001.

———. *Joshua Retold: Synoptic Perspectives*. Edinburgh: T & T Clark, 1998.

Bakon, Shimon. "Elisha the Prophet." *Jewish Bible Quarterly* 29 (2001): 242–48.

Bal, Mieke. *Death and Dissymmetry: The Politics of Coherence in the Book of Judges*. Chicago: University of Chicago Press, 1988.

———. *Loving Yusuf: Conceptual Travels from Present to Past*. Chicago: University of Chicago Press, 2008.

Baly, D. *The Geography of the Bible*. New York: Harper and Row, 1974.

Bartlett, John. "The Conquest of Sihon's Kingdom: A Literary Re-examination." *Journal of Biblical Literature* 97, no. 3 (1978): 347–51.

———. *Edom and the Edomites*. Sheffield: Sheffield Academic Press, 1989.

———. "The Land of Seir and the Brotherhood of Edom." *Journal of Theological Studies* 20, no. 1 (1969): 1–20.

Baud, M. "La representation de l'espace en Egypte ancienne—cartographie d'un intinéraire d'expédition." *Mappe-monde* 3 (1989): 9–12.

Berman, Joshua A. *Created Equal: How the Bible Broke with Ancient Political Thought.* New York: Oxford University Press, 2008.

Berquist, Jon L. *Controlling Corporeality: The Body and the Household in Ancient Israel.* New Brunswick: Rutgers University Press, 2002.

———. "Critical Spatiality and the Construction of the Ancient World." In *'Imagining' Biblical Worlds: Studies in Spatial, Social, and Historical Constructs in Honor of James W. Flanagan.* Ed. D. M. Gunn and P. M. Nutt, 14–29. London: Sheffield Academic Press, 2002.

Biebertein, Klaus. *Josua-Jordan-Jericho: Archäologie, Geschichte und Theologie der Landnahmeerzählungen Josua 1–6.* Göttingen: Vandenhoeck & Ruprecht, 1995.

Bienkowski, Piotr. "Transjordan and Assyria." In *The Archaeology of Jordan and Beyond: Essays in Honor of James A. Sauer.* Ed. Lawrence E. Stager, Joseph A. Greene, and Michael D. Coogan, 44–53. Winona Lake, IN: Eisenbrauns, 2000.

Boer, P. A. H. de. "Egypt in the Old Testament: Some Aspects of an Ambivalent Assessment." *Oudtestamentische Studiën* 27 (1991): 152–67.

Boer, Roland. *Political Myth: On the Use and Abuse of Biblical Themes.* Durham: Duke University Press, 2009.

Boling, Robert G. "Levitical History and the Role of Joshua." In *The Word of the Lord Shall Go Forth.* Ed. C. L. Meyers and M. O'Connor, 241–61. Winona Lake, IN: Eisenbrauns, 1983.

———. "Some Conflated Readings in Joshua-Judges (Josh. 3:11, 14, 17; 23:4, 7,12; Judg. 10:4; 20:10)." *Vetus Testamentum* 16 (1966): 293–98.

Brodsky, Harold. "Ezekiel's Map of Restoration." In *Land and Community: Geography in Jewish Studies.* Ed. Harold Brodsky. Bethesda: University Press of Maryland, 1997.

Bronner, Leah. *The Stories of Elijah and Elisha.* Leiden: E. J. Brill, 1968.

Callahan, Allen Dwight. *The Talking Book: African Americans and the Bible.* New Haven: Yale University Press, 2006.

Carroll, Robert P. "The Myth of the Empty Land." *Semeia* 59 (1992): 79–93.

Caspi, Mishael Maswari, and Rachel Havrelock. *Women on the Biblical Road: Ruth, Naomi, and the Female Journey.* Lanham: University Press of America, 1996.

Coats, George W. "The Ark of the Covenant in Joshua: A Probe into the History of a Tradition." *Hebrew Annual Review* 9 (1985): 137–57.

———. "Lot: A Foil in the Abraham Saga." In *Understanding the Word: Essays in Honor of Bernhard W. Anderson.* Ed. J. T. Butler, 119–32. Sheffield: JSOTS Press, 1985.

———. *Rebellion in the Wilderness.* Nashville: Abingdon, 1968.

Cogan, Mordechai. *I Kings: A New Translation with Introduction and Commentary.* New York: Doubleday, 2000.

———. *Imperialism and Religion: Assyria, Judah, and Israel in the Eighth and Seventh Centuries B.C.E.* Missoula: Scholars Press, 1974.

Cohn, Robert L. "Before Israel: The Canaanites as Other in Biblical Tradition." In *The Other in Jewish Thought and History: Constructions of Jewish Culture and Identity.* Ed. Laurence J. Silberstein and Robert L. Cohn, 74–90. New York: New York University Press, 1994.

———. "Form and Perspective." *Vetus Testamentum* 31, fasc. 2 (April, 1983): 73–84.

———. "Negotiating (with) the Natives: Ancestors and Identity in Genesis." *Harvard Theological Review* 96, no. 2 (2003): 147–66.

———. *The Shape of Sacred Space: Four Biblical Studies.* Chico: Scholars Press, 1981.

Collins, Terence. *The Mantle of Elijah.* Sheffield: JSOT Press, 1993.

Coote, Robert B, ed. *Elijah and Elisha in Socioliterary Perspective.* Atlanta: Scholars Press, 1992.

Cross, Frank M. "The Ammonite Oppression of the Tribes of Gad and Reuben: Missing Verses from 1 Samuel 11 Found in 4QSamuel[a]." In *History, Historiography, and Interpretation: Studies in Biblical and Cuneiform Literatures.* Ed. H. Tadmor and M. Weinfeld, 148–58. Jerusalem: Magnes Press, 1986.

Crüsemann, Frank. "Human Solidarity and Ethnic Identity: Israel's Self-Definition in the Genealogical System of Genesis." *Ethnicity and the Bible.* Ed. Mark G. Brett, 58–76. Leiden: Brill, 1996.

Csapo, Eric. *Theories of Mythology.* Malden: Blackwell Publishing, 2005.

Darr, Katheryn Pfisterer. "The Wall around Paradise: Ezekielian Ideas about the Future." *Vetus Testamentum* 37, fasc. 3 (July 1987): 271–79.

De Vaux, Roland. "Notes d'histoire et de topographie Transjordaniennes." *Vivre et Penser: recherches d'exégès et d'histoire* 1 (1941): 25–29.

Dever, William G. "Nelson Glueck and the Other Half of the Holy Land." In *The Archaeology of Jordan and Beyond: Essays in Honor of James A. Sauer.* Ed. Lawrence E. Stager, Joseph A. Greene, and Michael D. Coogan, 114–21. Winona Lake, IN: Eisenbrauns, 2000.

Douglas, Mary. *In the Wilderness: The Doctrine of Defilement in the Book of Numbers.* Oxford: Oxford University Press, 2001.

Dozeman, Thomas. "The Yam-Sûp in the Exodus and the Crossing of the Jordan River." *Catholic Biblical Quarterly* 15 (1996): 407–16.

Fields, Weston W. *Sodom and Gomorrah: History and Motif in Biblical Narrative.* Sheffield: Sheffield Academic Press, 1997.

Fishbane, Michael. "The Well of Living Water: A Biblical Motif and Its Ancient Transformations." In *Sha'arei Talmon: Studies in the Bible, Qumran, and the Ancient Near East Presented to Shemaryahu Talmon.* Ed. E. Tov and M. Fishbane. Winona Lake, IN: Eisenbrauns, 1992.

Flanagan, James W. "Ancient Perceptions of Space/Perceptions of Ancient Space." *Semeia* 87 (1999): 15–44.

Fokkleman, J. P. "Genesis." In *The Literary Guide to the Bible.* Ed. Robert Alter and Frank Kermode, 36–56. Cambridge: Harvard University Press, 1987.

———. *Reading Biblical Narratives: An Introductory Guide.* Trans. Ineke Smit. Leiderdorp: Deo Publishing, 1999.

Friedman, Richard Elliott. *The Exile and Biblical Narrative: The Formation of the Deuteronomistic and Priestly Works.* Chico: Scholars Press, 1981.

———. "Sacred History and Theology: The Redaction of Torah." In *The Creation of Sacred Literature: Composition and Redaction of the Biblical Text.* Ed. R. E. Friedman, 25–34. Berkeley: University of California Press, 1981.

———. *Who Wrote the Bible?* San Francisco: Harper San Francisco, 1997.

Frymer-Kensky, Tikva. "Pollution, Purification, and Purgation in Biblical Israel." In *The Word of the Lord Shall Go Forth: Essays in Honor of David Noel Freedman in Celebration of His Sixtieth Birthday.* Ed. Carol L. Meyers and M. O'Connor, 399–414. Winona Lake, IN: Eisenbrauns, 1983.

Ginsberg, H. K. "Judah and the Transjordan States from 734 to 582 B.C.E." In *Alexander Marx Jubilee Volume*, 347–68. New York: Jewish Theological Seminary, 1950.

Glueck, Nelson. *The Other Side of the Jordan.* Cambridge: American Schools of Oriental Research, 1940.

———. *The River Jordan.* New York: McGraw-Hill, 1968.

Goodblatt, David. *Elements of Ancient Jewish Nationalism.* Cambridge: Cambridge University Press, 2006.

Gorman, Frank H. *The Ideology of Ritual: Space, Time, and Status in the Priestly Theology.* Sheffield: Sheffield Academic Press, 1990.

Gottwald, Norman K. "Early Israel as an Anti-Imperial Community." In *In the Shadow of Empire: Reclaiming the Bible as a History of Faithful Resistance.* Ed. Richard A. Horsley, 9–24. Louisville: Westminster/John Knox Press, 2008.

Gros Louis, Kenneth R. R. "Elijah and Elisha." Vol. 1, *Literary Interpretations of Biblical Narratives*, 177–90. Nashville: Abingdon Press.

Grosmark, Tziona. *Jordan and Its Sites: The Early Geography and History of Jordan*. Ed. Gavriel Barkai and Eli Shiller. Jerusalem: Ariel Publishers, 1995. (Hebrew).

Gunkel, Hermann. *Die Geschichten von Elisa*. Berlin: Berlag Karl Curtius, 1927.

———. *Genesis übersetst und erklärt*. Göttingen: Vandenhoeck & Ruprecht, 1977.

Hawk, L. Daniel. *Joshua*. Collegeville, MN: Liturgical Press, 2000.

———. "Strange Houseguests: Rahab, Lot, and the Dynamics of Deliverance." In *Reading Between Texts: Intertexuality and the Hebrew Bible*. Ed. Danna Nolan Fewell, 89–98. Louisville: Westminster/John Knox Press, 1992.

Hendel, Ronald S. "Israel Among the Nations: Biblical Culture in the Ancient Near East." In *Cultures of the Jews: A New History*. Ed. David Biale, 43–75. New York: Schocken Books, 2002.

———. "The Poetics of Myth in Genesis." In *The Seductiveness of Jewish Myth*. Ed. S. D. Breslauer, 157–70. Albany: SUNY Press, 1997.

———. "Sacrifice as a Cultural System: The Ritual Symbolism of Exodus 24, 3–8," *Zeitschrift für die Alttestamentliche Wissenschaft* 101, no. 3 (1989): 366–90.

———. "When the Sons of God Cavorted with the Daughters of Men." In *Understanding the Dead Sea Scrolls*. Ed. Hershel Shanks, 167–77. New York: Random House, 1992.

Hoppe, Leslie. *Joshua, Judges with an Excursus on Charismatic Leadership in Israel*. Wilmington: Michael Glazier, 1982.

Horowitz, Aharon. *The Jordan Rift Valley*. Exton: AA Balkemna, 2001.

Horowitz, Wayne. "Moab and Edom in the Sargon Geography." *Israel Exploration Journal* 43 (1993): 151–56.

Hutton, Jeremy M. "The Left Bank of the Jordan and the Rites of Passage: An Anthropological Interpretation of 2 Samuel XIX." *Vetus Testamentum* 56, no. 4 (2006): 470–84.

Jobling, David. *I Samuel*. Collegeville, MN: Liturgical Press, 1998.

Josipovici, Gabriel. *The Book of God: A Response to the Bible*. New Haven: Yale University Press, 1988.

Kaiser, Otto. "Zwischen den Fronten: Palästina in den Auseinanadersetzungen zwischen dem Perserreich und Ägypten in der esten Hälfte des 4. Jahrhunderts." In *Wort, Lied und Gottesspruch: Beiträge zur Psalmen und Propheten*, II. Ed. Josef Schreiner, 197–206. Würzberg: Echter Verlag.

Kallai, Zechariah. *Historical Geography of the Bible: The Tribal Territories of Israel*. Jerusalem: Magnes Press, 1986.

———. "The Reality of the Land and the Bible." *Das Land Israel in biblischer Zeit*. Göttingen: Vandenhoeck & Ruprecht, 1981: 76–90.

Kaminsky, Joel. "Did Election Imply the Mistreatment of Non-Israelites?" *Harvard Theological Rreview* 96, no. 4 (2003): 397–425.

———. *Yet I Loved Jacob: Reclaiming the Biblical Concept of Election*. Nashville: Abingdon Press, 2007.

Knohl, Israel. *The Sanctuary of Silence: The Priestly Torah and the Holiness School*. Minneapolis: Fortress Press, 1995.

Kugel, James L. "The Holiness of Israel and the Land in Second Temple Times." In *Texts, Temples, and Traditions: A Tribute to Menahem Haran*. Ed. Michael V. Fox, Victor Avigdor Hurowitz, Avi Hurvitz, Michael L. Klein, Baruch J. Schwartz, and Nili Shupak, 21–32. Winona Lake, IN: Eisenbrauns, 1996.

———. *In Potiphar's House: The Interpretive Life of Biblical Texts*. Cambridge: Harvard University Press, 1994.

Lemche, Niels Peter. *The Canaanites and Their Land: The Tradition of the Canaanites*. Sheffield: JSOT Press, 1991.

Levine, Nachman. "Twice as Much of Your Spirit: Pattern, Parallel, and Paronomasia in the Miracles of Elijah and Elisha." *JSOT* 85 (1999): 25–46.

Liverani, Mario. *Myth and Politics in Ancient Near Eastern Historiography.* Ed. Zainab Bahrani and Marc Van De Mieroop. Ithaca: Cornell University Press, 2004.

Loewenstamm, Samuel E. "The Death of Moses." In *Studies on the Testament of Abraham.* Ed. George W. E. Nickelsburg, 185–217. Missoula, Montana: Scholars Press, 1976.

MacDonald, Burton. "Early Edom: The Relation between the Literary and Archaeological Evidence." In *Scripture and Other Artifacts: Essays on the Bible and Archaeology in Honor of Philip J. King.* Ed. Michael Coogan, J. Cheryl Exum, and Lawrence Stager, 230–46. Louisville: Westminster/John Knox Press, 1994.

Machinist, Peter. "On Self-Consciousness in Mesopotamia." In *The Origins and Diversity of Axial Age Civilizations.* Ed. S. N. Eisenstadt, 183–202. Albany: State University of New York Press, 1986.

———. "Outsiders or Insiders: The Biblical View of Emergent Israel and Its Contexts." In *The Other in Jewish Thought and History: Constructions of Jewish Culture and Identity.* Ed. Laurence J. Silberstein and Robert L. Cohn, 31–60. New York: New York University Press, 1994.

———. "The Question of Distinctiveness in Ancient Israel: An Essay." In *Ah, Assyira . . . : Studies in Assyrian History and Ancient Near Eastern Historiography Presented to Hayim Tadmor.* Ed. M. Cogan and I. Eph'al, 196–212. Jerusalem: Magnes Press, 1991.

———. "The Twilight of Judah: In the Egyptian-Babylonian Maelstrom." *SVT* 28 (1955): 123–45.

McCarthy, Dennis J. "An Installation Genre?" *Journal of Biblical Literature* 90, no. 1 (1971): 31–41.

McKenzie, Steven L., and M. Patrick Graham, eds. *The History of Israel's Traditions: The Heritage of Martin Noth.* Sheffield: Sheffield Academic Press, 1994.

Mendels, Doron. *The Rise and Fall of Jewish Nationalism.* Grand Rapids: Eerdmans, 1992.

Mitchell, Gordon. *Together in the Land: A Reading of the Book of Joshua.* Sheffield: Sheffield Academic Press, 1993.

Montgomery, James A. "The Elisha Cycle . . ." *Journal of Biblical Literature* (1966): 241–54.

Mullen, E. Theodore. *Ethnic Myths and Pentateuchal Foundations: A New Approach to the Formation of the Pentateuch.* Atlanta: Scholars Press, 1997.

———. *Narrative History and Ethnic Boundaries: The Deuteronomistic Historian and the Creation of Israelite National Identity.* Atlanta: Scholars Press, 1993.

Na'aman, Nadav. "The Brook of Egypt and Assyrian Policy on the Border of Egypt." *Tel Aviv* 6, no. 1–2 (1979): 68–90.

———. "The Kingdom of Judah under Josiah." *Tel Aviv* 18, no. 1 (1991): 3–71.

Nelson, Richard D. *Joshua: A Commentary.* Louisville: Westminster/John Knox Press, 1997.

Noth, Martin. *The Deuteronomistic History.* Sheffield: JSOT Press, 1981.

———. *A History of Pentateuchal Traditions.* Trans. Bernhard W. Anderson. Chico, CA: Scholars Press, 1981.

———. "Israelitische Stämme zwischen Ammon and Moab." *Zeitschrift für die Alttestamentliche Wissenschaft* 60, no. 1–4 (1944): 11–57.

Olyan, Saul M. "'And with a Male You Shall Not Lie the Lying Down of a Woman': On the Meaning and Significance of Leviticus 18:22 and 20:13." *Journal of the History of Sexuality* 5, no. 2 (Oct. 1994): 197–206.

Pardes, Ilana. "Imagining the Birth of Ancient Israel: National Metaphors in the Bible." In *Cultures of the Jews: A New History.* Ed. David Biale, 9–42. New York: Schocken Books, 2002.

Peckham, Brian, "The Significance of the Book of Joshua in Noth's Theory of the Deuteronomistic History." In *The History of Israel's Traditions: The Heritage of Martin Noth.* Ed.

Steven L. McKenzie and M. Patrick Grahma, 213–34. Sheffield: Sheffield Academic Press, 1994.

Porter, J. Roy. "Thresholds in the Old Testament." In *Boundaries and Thresholds: Papers from a Colloquium of the Katharine Briggs Club*. Ed. Hilda Ellis Davidson. Exeter: Short Run Press, 1993.

Propp, William H. C. *Water in the Wilderness: A Biblical Motif and Its Mythological Background*. Atlanta: Scholars Press, 1987.

Rofé, Alexander. "The Classification of Prophetical Stories." *Journal of Biblical Literature* 89, no. 4 (1970): 427–40.

———. "Moses' Blessing, the Sanctuary at Nebo, and the Origin of the Levites." In *Studies in Bible and the Ancient Near East: Presented to Samuel E. Lowenstamm on His Seventieth Birthday*. Ed. Yitschak Avishur and Joshua Blau, 409–24. Jerusalem: E. Rubinstein, 1978.

Römer, Thomas. *The So-Called Deuteronomistic History: A Sociological, Historical, and Literary Introduction*. London: T & T Clark International, 2007.

Rouwhorst, Gerard. "Leviticus 12–15 in Early Christianity." *Purity and Holiness: The Heritage of Leviticus*. Ed. M. J. H. M. Poorthuis and J. Schwartz, 181–93. Leiden: Brill, 2000.

Sanders, Seth L. *Writing, Ritual, and Apocalypse*. Ph.D. thesis, Johns Hopkins University, 1999.

Schafer-Lichtenberger, Christa. "'Josua' und 'Elischa'—eine biblische Argumentation zur Begründung der Autorität und Legitimität des Nachfolgers." *Zeitschrift für die Alttestamentliche Wissenschaft* 101, no. 2 (1989): 198–222.

Schoville, Keith N. "Tackling an Extended Passage (Josh. 4:1–7)." *Bible Review* 9 (1993): 16–27.

Schwartz, Regina M. *The Curse of Cain: The Violent Legacy of Monotheism*. Chicago: University of Chicago Press, 1997.

Seebass, Horst. "'Holy' Land in the Old Testament: Numbers and Joshua." *Vetus Testamentum* 56, no. 1 (2006): 92–104.

Smith, Jonathan Z. "Map Is Not Territory." In *Map Is Not Territory: Studies in the History of Religions*, 289–309. Leiden: Brill, 1978.

Smith, Mark S. *The Priestly Vision of Genesis 1*. Minneapolis: Fortress Press, 2010.

Sternberg, Meir. *The Poetics of Biblical Narrative: Ideological Literature and the Drama of Reading*. Bloomington: Indiana University Press, 1985.

Stone, Lawson G. "Eglon's Belly and Ehud's Blade: A Reconsideration." *Journal of Bibilcal Literature* 128, no. 4 (2009): 649–63.

Tadmor, Hayim. "Monarchy and the Elite in Assyria and Babylonia: The Question of Royal Accountability." In *The Origins and Diversity of Axial Age Civilizations*. Ed. S. N. Eisenstadt, 203–24. Albany: State University of New York Press, 1986.

Talmon, Shemaryahu. "The 'Desert Motif' in the Bible and in Qumran Literature." In *Biblical Motifs, Origins, and Transformation*. Ed. Alexander Altmann, 31–63. Cambridge: Harvard University Press, 1966.

———. "Har and Midbar: An Antithetical Pair of Biblical Motifs." In *Figurative Language in the Ancient Near East*. Ed. M. Mindlin, M. J. Geller, and J. E. Wansbrough 117–42. London: School of Oriental and African Studies, University of London, 1987.

———. *Literary Studies in the Hebrew Bible: Form and Content*. Jerusalem: Magnes Press, 1993.

Turner, Victor. *The Ritual Process: Structure and Anti-Structure*. Ithaca: Cornell University Press, 1969.

Van Zyl, A. H. *The Moabites*. Leiden: Brill, 1960.

Von Rad, Gerhard. "The Promised Land and Yahweh's Land in the Hexateuch." In *The Problem of the Hexateuch and Other Essays*, 79–93. New York: McGraw-Hill Book Company, 1966.

Walzer, Michael. *Exodus and Revolution*. New York: Basic Books, 1985.

Weinfeld, Moshe, and Gershon Galil, eds. *Studies in Historical Geography and Biblical Historiography Presented to Zechariah Kallai*. Boston: Brill, 2000.
Weinfeld, Moshe. *Deuteronomy and the Deuteronomic School*. Oxford: Clarendon Press, 1972.
Yamada, Frank M. "Shibboleth and the Ma(r)king of Culture: Judges 12 and the Monolingualism of the Other." In *Derrida's Bible (Reading a Page of Scripture with a Little Help from Derrida)*. Ed. Yvonne Sherwood, 119–34. New York: Palgrave Macmillan, 2004.
Zakovitch, Yair. "Humor and Theology or the Successful Failure of Israelite Intelligence: A Literary-Folkloric Approach to Joshua 2." In *Text and Tradition: The Hebrew Bible and Folklore*. Ed. Susan Niditch, 75–98. Atlanta: Scholars Press, 1990.

Christianity/Rabbinic Judaism

Abrams, Israel. "How Did the Jews Baptize?" *Journal of Theological Studies* 12, no. 4 (1911): 609–12.
Allison, Dale C. "Elijah Must Come First." *Journal of Biblical Literature* 103, no. 2 (1984): 256–58.
Alon, Gedaliah. *The Jews in Their Land in the Talmudic Age (70–640 C.E.)*. Ed. and trans. Gershon Levi. Jerusalem: Magnes Press, 1984.
Anderson, Gary A. *The Genesis of Perfection: Adam and Eve in Jewish and Christian Imagination*. Louisville: Westminster/John Knox Press, 2001.
Anderson, Gary A., and Michael E. Stone, eds. *A Synopsis of the Books of Adam and Eve*. Atlanta: Scholars Press, 1994.
Bakhos, Carol. *Ishmael on the Border: Rabbinic Portrayals of the First Arab*. Albany: State University of New York Press, 2006.
Ben-Amos, Dan, "Analytical Categories and Ethnic Genres." In *Folklore Genres*, 215–42. Austin: University of Texas Press, 1976.
———. "Generic Distinctions in the Aggadah." In *Studies in Jewish Folklore*. Ed. Frank Talmage, 45–71. Cambridge, Mass, 1980.
———. "Towards a Definition of Folklore in Context." *Journal of American Folklore* 84 (1971): 3–15.
Blowers, Paul M. "Origen, the Rabbis, and the Bible: Toward a Picture of Judaism and Christianity in Third-Century Caesarea." In *Origen of Alexandria: His World and His Legacy*. Ed. Charles Kannengiesser and William L. Petersen, 96–116. Notre Dame: University of Notre Dame Press, 1988.
Boyarin, Daniel. "'Are There Any Jews in 'The History of Sexuality'?" *Journal of the History of Sexuality* 5, no. 3 (January 1995): 333–55.
———. *Borderlines: The Partition of Judaeo-Christianity*. Philadelphia: University of Pennsylvania Press, 2004.
———. "The Gospel of the *Memra*: Jewish Binitarianism and the Prologue to John." *Harvard Theological Review* 94, no. 3 (2001): 243–84.
———. *Intertextuality and the Reading of Midrash*. Bloomington: Indiana University Press, 1990.
———. *A Radical Jew: Paul and the Politics of Identity*. Berkeley: University of California Press, 1994.
———. *Unheroic Conduct: The Rise of Heterosexuality and the Invention of the Jewish Man*. Berkeley: University of California Press, 1997.
Brandt, W. *Die jüdischen Baptismen oder das religöse Waschen und Baden im Judentum mit Einschluss des Judenchristentums*. Giessen: Töpelmann, 1910.
Brooks, Roger. "Straw Dogs and Scholarly Ecumenisum: The Appropriate Jewish Background for the Study of Origen." In *Origen of Alexandria: His World and His Legacy*. Ed. Charles

Kannengiesser and William L. Petersen, 63–95. Notre Dame: University of Nortre Dame Press, 1988.

Brown, Peter. *The Body and Society: Men, Women, and Sexual Renunciation in Early Christianity.* New York: Columbia University Press, 1988.

Brown, Raymond E. "Jesus and Elisha." *Perspective* 7 (1971): 85–103.

———. "John the Baptist in the Gospel of John." *Catholic Biblical Quarterly* 22 (1960): 292–98.

Chilton, Bruce. "John the Purifier: His Immersion and His Death." *Harvard Theological Review* 57, no. 1 (2001): 247–67.

Cohen, Shaye J. D. "Is 'Proselyte Baptism' Mentioned in the Mishnah? The Interpretation of m. Pesahim 8:8 (= m. Eduyot 5:2)." In *Pursuing the Text: Studies in Honor of Ben Zion Wacholder on the Occasion of His Seventieth Birthday.* Ed. J. C. Reeves and J. Kampen, 278–92. Sheffield: Sheffield Academic Press, 1994.

———. "The Rabbinic Conversion Ceremony." *Journal of Jewish Studies* 41, no. 2 (1990): 177–203.

Crossan, John Dominic. *The Historical Jesus: The Life of a Mediterranean Jewish Peasant.* San Francisco: HarperSanFrancisco, 1991.

Dahl, N. A. "The Origin of Baptism." In *Interpretationes ad Vetus Testamentum pertinentes Sigmundo Mowinckel Septuagenario Missae.* Ed. Nils Alstrup and Arvid S. Kapelrud, 36–52. Oslo: Fabritus and Sonner, 1955.

Dapaah, Daniel S. *The Relationship between John the Baptist and Jesus of Nazareth: A Critical Study.* Lanham: University Press of America, 2005.

Davies, W. D. *The Territorial Dimension of Judaism.* Berkeley: University of California Press, 1982.

De Lange, Nicholas. *Origen and the Jews: Studies in Jewish-Christian Relations in Third-Century Palestine.* Cambridge: Cambridge University Press, 1976.

Flusser, David. "The Baptism of John and the Dead Sea Sect." In *Essays on the Dea Sea Scrolls: In Memory of E. L. Sukenik.* Ed. C. Rabin and Y. Yadin, 209–39. Jerusalem: Hekhal Ha-Sefer, 1961 (Hebrew).

Gafni, Isaiah M. *Land, Center, and Diaspora: Jewish Constructs in Late Antiquity.* Sheffield: Sheffield Academic Press, 1997.

Hasan-Rokem, Galit. "Myth." In *Contemporary Jewish Religious Thought: Original Essays on Critical Concepts, Movements and Beliefs.* Ed. Arthur A. Cohen and Paul Mendes-Flohr, 657–61. New York: Free Press, 1987.

———. "Narratives in Dialogue: A Folk Literary Perspective on Interreligious Contacts in the Holy Land in Rabbinic Literature of Late Antiquity First–Fifteenth Centuries CE." In *Sharing the Sacred: Religious Contacts and Conflicts in the Holy Land.* Ed. Arieh Kofsky and Guy G. Strouma, 109–29. Jerusalem: Yad Ben Zvi, 1998.

———. *Tales of the Neighborhood: Jewish Narrative Dialogues in Late Antiquity.* Berkeley: University of California Press, 2003.

———. *Web of Life: Folklore and Midrash in Rabbinic Literature.* Trans. Batya Stein. Stanford: Stanford University Press, 2000.

———. "Within Limits and Beyond: History and Body in Midrashic Texts." *International Folklore Review* 9 (1993): 5–12.

Hayes, Christine E. "Do Converts to Judaism Require Purification? M. Pes 8:8: An Interpretive Crux Solved." *Jewish Studies Quarterly* 9 (2002): 327–52.

Himmelfarb, Martha. *A Kingdom of Priests: Ancestry and Merit in Ancient Judaism.* Philadelphia: University of Pennsylvania Press, 2006.

Neusner, Jacob. *Genesis Rabbah: The Judaic Commentary on Genesis: A New American Translation.* Atlanta: Scholars Press for Brown Judaic Studies, 1985.

———. *The Idea of Purity in Ancient Judaism.* Leiden: E. J. Brill, 1973.

———. *The Judaic Law of Baptism: Tractate Miqva'ot in the Mishnah and the Tosephta: A Form-Analytical Translation and Commentary, and a Legal and Religious History.* Atlanta: Scholars Press, 1995.

Origen. *Homilies on Genesis and Exodus.* Trans. Ronald E. Heine. Washington, D.C.: Catholic University of America Press, 1982.

———. *Homilies on Leviticus1–16.* Trans. Gary Wayne Barkley. Washington, D.C.: Catholic University Press of America, 1982.

Robinson, John A. T., ed. "The Baptism of John and the Qumran Community." In *Twelve New Testament Studies*, 11–27. London, SCM, 1962.

———. "Elijah, John, and Jesus: An Essay in Detection." *New Testament Studies* 4 (1957–58): 263–81.

———. "The One Baptism." In *Twelve New Testament Studies*, 158–75. London: SCM Press, 1962.

Scobie, Charles H. H. *John the Baptist.* London: SCM, 1964.

Stein, Dina. "Believing Is Seeing: A Reading of Baba Bathra 73a–75b." *Jerusalem Studies in Hebrew Literature* 17 (1999): 9–32 (Hebrew).

———. "A King, a Queen, and the Riddle Between: Riddles and Interpretation in a Late Midrashic Text." In *Untying the Knot: On Riddles and Other Enigmatic Modes.* Ed. Galit Hasan-Rokem and David Shulman, 125–46. New York: Oxford University Press, 1996.

Thomas, Joseph. *Le Mouvement Baptiste en Palestine et Syrie (150 av. J.C.–300 ap. J.C.).* Gembloux: Duculot, 1935.

Wilken, Robert L. *The Land Called Holy: Palestine in Christian History and Thought.* New Haven: Yale University Press, 1992.

Yassif, Eli. *The Hebrew Folktale: History, Genre, Meaning.* Trans. Jacqueline S. Teitelbaum. Bloomington: Indiana University Press, 1999.

Middle East

Abu El-Haj, Nadia. *Facts on the Ground: Archaeological Practice and Territorial Self-Fashioning in Israeli Society.* Chicago: University of Chicago Press, 2001.

Alonso, Ana. "The Politics of Space, Time, and Substance: State Formation, Nationalism, and Ethnicity." *Annual Review of Anthropology* 23 (1994): 379–405.

Anderson, Benedict. *Imagined Communities: Reflections on the Origin and Spread of Nationalism.* London: Verso, 1983.

Anidjar, Gil. *The Jew, the Arab: A History of the Enemy.* Stanford: Stanford University Press, 2003.

Appadurai, Arjun. "Sovereignty without Territoriality: Notes for a Postnational Geography." *The Geography of Identity.* Ed. Patricia Yaeger, 40–58. Ann Arbor: University of Michigan Press, 1999.

Ateek, Naim S. "A Palestinian Perspective." In *Voices from the Margin: Interpreting the Bible in the Third World.* Ed. R. S. Sigirtharajah, 227–34. Maryknoll: Orbis Books, 2006.

Augustinovic, A. o.f.m. *"El-Khadr" and the Prophet Elijah.* Jerusalem: Franciscan Printing Press, 1972.

Baratz, Josef. *A Village by the Jordan.* New York: Sharon Books, 1957.

Bar-Itzhak, Haya. "'The Unknown Variable Hidden Underground' and the Zionist Idea: Rhetoric of Place in an Israeli Kibbutz and Cultural Interpretations." *Journal of American Folklore* 112, no. 446 (1999): 497–514.

Biger, Gideon. "The Boundaries of Israel-Palestine, Past, Present, and Future: A Critical Geographical View." *Israel Studies* 13, no. 1 (2008): 68–93.

———. "The Names and Boundaries of Eretz-Israel (Palestine) as Reflections of Stages in Its History." In *The Land That Became Israel: Studies in Historical Geography*. Ed. Ruth Kark, 1–22. New Haven: Yale University Press, 1990.

Boyarin, Jonathan. "Hegel's Zionism?" In *Remapping Memory: The Politics of TimeSpace*. Ed. Jonathan Boyarin, 137–60. Minneapolis: University of Minnesota Press, 1994.

Brown, Wendy. *Walled States, Waning Sovereignty*. New York: Zone Books, 2010.

Dawn, C. Ernest. "The Origins of Arab Nationalism." In *The Origins of Arab Nationalism*. Ed. Rashid Khalidi, Lisa Anderson, Muhammad Muslih, and Reeva S. Simon, 3–30. New York: Columbia University Press, 1991.

Feige, Michael. *Settling in the Hearts: Jewish Fundamentalism in the Occupied Territories*. Detroit: Wayne State University Press, 2009.

Fromkin, David. *A Peace to End All Peace: Creating the Modern Middle East 1914–1922*. New York: Henry Holt and Company, 1989.

Habibi, Amirah. *The Second Exodus: Critical Field Studies on the 1967 Exodus*. Beirut: Munazzamat al-Tahrir al-Filastiniyah, Markaz al-Abhath, 1970 (Arabic).

Halaby, Laila. *West of the Jordan*. Boston: Beacon Press, 2003.

Havrelock, Rachel. "Pioneers and Refugees: Arabs and Jews in the Jordan River Valley." In *Understanding Life in the Borderlands: Boundaries in Depth and in Motion*. Ed. I. William Zartman, 189–16. Athens: University of Georgia Press, 2010.

Karsh, Efraim, and Inari Karsh. *Empires of the Sand: The Struggle for Mastery in the Middle East 1789–1923*. Cambridge: Harvard University Press, 1999.

Katriel, Tamar. "Sites of Memory: Discourses of the Past in Israeli Pioneering Settlement Museums." In *Cultural Memory and the Construction of Identity*. Ed. Dan Ben-Amos and Liliane Weissberg, 99–135. Detroit: Wayne State University Press, 1999.

Khalidi, Rashid. *British Policy toward Syria and Palestine 1906–1914: A Study of the Antecedents of the Hussein-McMahon Correspondence, the Sykes-Picot Agreement, and the Balfour Declaration*. London: Ithaca Press, 1980.

———. "Contrasting Narratives of Palestinian Identity." In *The Geography of Identity*. Ed. Patricia Yaeger, 187–222. Ann Arbor: University of Michigan Press, 1999.

Khalili, Leila. "Commemorating Contested Lands." In *Exile and Return: Predicaments of Palestinians and Jews*. Ed. A. Lesch and I. Lustick, 19–40. Philadelphia: University of Pennsylvania Press, 2005.

Morris, Benny. *Israel's Border Wars 1949–1956*. Oxford: Clarendon Press, 1993.

Muhawi, Ibrahim, and Sharif Kanaana. *Speak Bird, Speak Again: Palestinian Arab Folktales*. Berkeley: University of California Press, 1989.

Muslih, Muhammad. "The Rise of Local Nationalism in the Arab East." In *The Origins of Arab Nationalism*. Ed. R. Khalidi, L. Anderson, M. Muslih, and R. Simon, 167–85. New York: Columbia University Press, 1991.

Newman, David. "Introduction—Geographic Discourses: The Changing Spatial and Territorial Dimensions of Israeli Politics and Society." *Israel Studies* 13, no. 1 (2008): 1–19.

———. "Metaphysical and Concrete Landscapes: The Geopiety of Homeland Socialization in the 'Land of Israel.'" In *Land and Community: Geography in Jewish Studies*. Ed. Harold Brodsky, 153–82. Bethesda: University Press of Maryland, 1997.

Qadurah, Yusuf Jurays. *Tarikh Madinat Ramallah* [The History of Ramallah]. New York: Hadah Publishing, 1954 (Arabic).

Rabin, Hayim, Yehuda Elitsur, Hayim Gevaryahu, and Ben-Tzion Luria, eds. *Inquiries into the Tanakh by the Study Group at the Home of David Ben Gurion: The Book of Joshua*. Jerusalem: Kiryat Sepher, 1971.

Rabinowitz, Dan. "In and Out of Territory." In *Grasping Land: Space and Place in Contempo-*

rary Israeli Discourse and Experience. Ed. E. Ben-Ari and Y. Bilu, 177–201. Albany: State University of New York Press, 1997.

———. "National Identity on the Frontier." In *Border Identities: Nation and State at International Frontiers*. Ed. Thomas M. Wilson and Hastings Donnan, 142–61. Cambridge: Cambridge University Press, 1998.

Renan, Ernst. "What Is a Nation?" In *Nation and Narration*. Ed. Homi Bhabha, 8–22. London: Routledge, 1990.

Schiff, Benjamin. *Refugees unto the Third Generation: UN Aid to the Palestinians*. Syracuse: Syracuse University Press, 1995.

Segev, Samuel. *Crossing the Jordan: Israel's Hard Road to Peace*. New York: St. Martin's Press, 1998.

Segev, Tom. "A Bitter Prize." *Foreign Affairs* 85, no. 3 (2006), http://www.foreignaffairs.com/articles/61726/tom-segev/a-bitter-prize.

———. *One Palestine, Complete: Jews and Arabs Under the British Mandate*. Trans. Haim Watzman. New York: Henry Holt, 2000.

Shapira, Anita. "The Bible and Israeli Identity." *AJS Review* 28, no. 1 (2004): 11–41.

———. "From the Palmach Generation to the Candle Children: Changing Patterns in Israeli Identity." *Partisan Review* 67, no. 4 (2000).

Sinai, Anne, and Allen Pollack, eds. *The Hashemite Kingdom of Jordan and the West Bank*. New York: American Academic Association for Peace in the Middle East, 1977.

Smith, Anthony D. *The Ethnic Origins of Nations*. Oxford: Basil Blackwell, 1986.

Troen, Selwyn Ilan. *Imagining Zion: Dreams, Designs, and Realities in a Century of Jewish Settlement*. New Haven: Yale University Press, 2003.

Wilson, Mary C. "The Hashemites, the Arab Revolt, and Arab Nationalism." In *The Origins of Arab Nationalism*. Ed. Rashid Khalidi, Lisa Anderson, Muhammad Muslih, and Reeva S. Simon, 204–24. New York: Columbia University Press, 1991.

INDEX

Italicized page numbers indicate maps.

Aaron and sons, 91n12, 144, 146, 150
Abraham: east-west binary and, 42n4, 50–51; God-person-land triangulation with, 43; God's covenant with, 41n2, 43, 49; Hagar as surrogate wife of, 41n2, 43, 51; Hebrews/*Ivrim*, use of term, 41; hereditary line of, 41, 43, 47, 64, 67, 170; homeland-diaspora environment and, 41, 68–69; Isaac and Ishmael mythic bond between rivals, 41; Ishmael and, 41, 43, 67; as legend, 50; Lot's mythic bond and rivalry with, 42–47, *48*, 49–51, 64; Sarah as wife of, 41, 43, 51, 58
Allenby/King Hussein Bridge border crossing, 268–73, 275
Alter, Robert, 12, 47n13, 142
American Palestine Exploration Society (APES), 223
Ammon/Ammonites: Creation narrative and, 42n4; Deuteronomistic tradition and, 35, 54–55; frontier justice and, 125–27; Gad as integrated into cults of Moab and, 121n26, 124n38; Israelites' relations with, 51, 55; Jabbok River as border between Gilead and, 80–81; Lot's inheritance and, 42n4; maps, *48*; as nation of perversion and deviance, 10, 40, 44, 47n13; Tobiad family's insider-outsider position and, 129–31
Amor/Amorites, 118, 120–23, 125–26
Anderson, Benedict, 4–5
animal tithing, 208, 209–12, *210*. See also *halakhah* (Jewish law)

APES (American Palestine Exploration Society), 223
apocalyptic associations with Jordan: Babylonian Talmud and, 13, 212–17; Jesus and John the Baptist and, 22, 23, 184n21, 186n24, 212; prophetic succession and, 137, 142; redemption for Jews and, 13, 140
apocalyptic visions, 22, 215
apprenticeship status, and prophetic succession, 138, 139, 141, 144, 144n9, 160
Arab nationalist territorial proposals, 223, *225*, 227, 231–32
Arafat, Yasir, 267–68, 276, 280–81
Aram/Arameans: Israelite/Aramean duality and, 64, 72, 74–75, 176; Israelites' conflicts/treaties with, 65, 71–73, *73*, 74–75, 78; Laban as from, 71; memorial site at Gilead treaty scene and, 71–72, 74–75; Northern Kingdom and, 65; Samaria siege by, 72n15. See also Syria
archaeological excavations, 16, 51, 221, 276, 279–80, 283
Ark of the Covenant, and crossing the Jordan, 90–92, 93. See also Temple
Assis, Elie, 90n10, 93n16, 109n5, 113n13
Assmann, Jan, 38–39
Assyrian Empire: Babylon, and Egypt's alliance with, 37; Deuteronomistic tradition influences by, 9n11, 35n58; hegemonic discourse and, 32–33, 38; Jordan as border and, 128–29; legal traditions and, 9n11; memorial sites and, 93n17; Northern Kingdom's battles against, 6, 11, 55, 77, 128; Transjordan and, 128–29

Babylon: Assyrian/Egyptian alliance against, 37; Egypt's power struggles with, 33, 36–39, 37–38; Israel as crossroads between Egypt and, 36–39; Israel's border in terms of, 22–23, 26n27, 31, 33, 34, 35–36; *Mappa Mundi* and, 18, 26–27, 35; memorial sites and, 93n17; mythic geography and, 18; Near East as conceived by, 26–27, 35; Nebuchadnezzar as ruler of, 23n17, 27–28, 35, 37, 124n39, 129n49; Southern Kingdom's conflicts with, 33–37, 124n39, 129

Babylonian Talmud: animal tithing and, 208, 209–12, 210; apocalyptic associations with Jordan and, 13, 212–17; Jews' reclamation of Jordan as baptismal signifier and, 195–96, 199–200, 201, 202n60, 203n61, 204; on the land of Israel, 210, 216–17; on purity and self-healing, 204–5; tall tales and, 208–9, 212–16. *See also* rabbinic narratives

Bal, Mieke, 44, 56–57, 123

Balaam, 51–52, 54, 60. *See also* Moab/Moabites

Balfour Declaration of 1917, 226–27, 233, 234

Bammel, Ernst, 164n60

baptism: dip/dipping and, 178n7, 184, 201; exclusion from World to Come for Jews and, 195–96, 200–201; heaven-earth mediation on vertical axis and, 13, 175–76; illnesses/healing and, 177–86, 197–98, 206; Jews' reclamation of Jordan's association with, 195–96, 199–200, 201, 202n60, 203n61, 204; Naaman association with, 180, 182, 184, 186–87, 192–93, 196, 197–98, 200–201, 204; rebirth after, 12, 13, 169, 174, 180, 186–87, 192–93; redemption for Christians through, 12, 13, 139, 161–69, 172–74, 176; repentance of sins for Christians through, 164–65, 184–86, 206; symbolic border of Jordan and, 13, 187, 190–92, 193; transformation for Jesus through, 168–70, 172. *See also* baptismal sites; Jesus and John the Baptist; John the Baptist; Naaman; purity of the body

baptismal sites: archaeological excavations in support of, 279–80, 283; Jordan state support through, 277, 279, 280, 281–82; Al-Maghtas site, 276, 278, 279–83, 285, *285*; materiality and authenticity of, 279–80, 282–83; multinational control and, 245, 276–82; Palestinian Christians and, 245; Qasr al-Yahud site, 276–78, 280–81, 283, 285, *285*; Yardenit site, 276–79, 283, 285, *285*

Barth, Frederik, 10

Barthes, Roland, 1, 80, 81, 82

Beit Midrash (House of Study), 198–99, 204

Ben-Amos, Dan, 94n19, 214n11

Ben-Gurion, David, 14, 246–47, 255, 256, 259n108

Ben Sira, 139–40

Bergen, Wesley, 160

Bible. *See* Biblical national myths; Hebrew Bible; New Testament

biblical canon (Tanakh), 6–7, 141, 282–83. *See also* Hebrew Bible; Torah

Biblical national myths, 5–7, 79. *See also* Deuteronomistic tradition; Elijah; Elisha as successor of Elijah; Hebrew Bible; Jesus and John the Baptist; John the Baptist; Joshua; Joshua as successor of Moses; Moab/Moabites; New Testament; Priestly tradition; sea-to-river maps; Transjordan

Blenkinsopp, Joseph, 153n28

body, the, 13, 206–7; Christian/Jewish binary and, 206; Deuteronomistic tradition and, 181; heaven-earth mediation and, 13, 175–76; Moabite women, and politics of, 44, 46, 49, 51, 52, 57; of Moses, 147; national identity imprinted on, 181, 182, 193–94; as physically ephemeral and politically immortal, 137, 147, 181; Priestly tradition and, 13, 176; sexuality of women and, 44, 45–46, 49, 60, 62–63. *See also* baptism; Naaman; purity of the body

Boling, Robert G., 89n9, 94n21, 111n7

border drawing, 7, 17, 64–65, 82–84, 125n40. *See also* baptismal sites; Europe/European hegemony; territory border(s); *and specific mandates*

borderland zones, 43n6, 52n22, 82, 273, 275. *See also* Gilead; territory border(s)

borders, territory. *See* territory border(s)

boundaries/maps, 10–11; ethno-nationalism and, 10; Gilead and, 126–27, 146; Israelites and, 21, 108, 109, *110*, 146; Jacob and, 10–11, 83–84; Joshua and, 11, 25n25,

87–88, 101–5; Moabite women, and gender dimension of, 10, 50–51, 52–54, 56–57, 63; Northern Kingdom and, 10–11; Persian period and, 129; Priestly tradition and, 8, 9, 13, *20*, 29–30, 78, 230; prophetic succession, and ritual, 135–36; Samaria and, *73*, *132*; Southern Kingdom and, 21, 146; Zionists and, 1–2, *228*, 232–33. *See also* territory border(s)
Boyarin, Daniel, 188–89, 193, 194, 200, 203, 248
Boyarin, Jonathan, 179n10
British mandate maps, 220–21, 223–27, *224*, *225*, *228*, 229–33, 269, 277, 286–87
Brodie, Thomas, 161n52

Canaan/Canaanites, 9, 20, 30n42, 99n27; Deuteronomistic tradition and, 11, 24n23; as homeland-diaspora environment, 71, 73, 75, 80–82; Israel's border in terms of Egypt and, 30–35; Jacob and, 71, 73, 75, 80, 81; maps, *48*; miraculous dimension of Joshua's crossing into, 91, 92; Moabite-Israelite relations versus, 55–58; Moses and, 19; Priestly tradition and, 9. *See also* Israel/Israelite tribes
Carrol, Robert, 138n5, 159
Christian Bible. *See* New Testament
Christianity/Christians: archaeological excavations and, 221; collective purification and, 12–13, 184–85, 188–89, 190–95, 206–7; heaven-earth mediation on vertical axis for, 12–13, 140, 157, 162–63, 170, 171n73, 172–73, 175–76; hegemonic discourse over Near East and, 277, 280–81; the holy land and, 206–7, 283; Jewish-Christian binary divisions in purity for, 12–13, 14, 187, 206–7; Jordan state's relations with, 277, 279–82; memorial sites and, 207n70; Muslims' relations with, 242–44, 277, 281; Palestinian, 241–45; papacy and, 46, 277, 280–81, 282; Priestly tradition and, 12–13, 14, 176; redemption through baptism for, 12, 13, 139, 161–69, 172–74, 176; repentance of sins through baptism for, 164–65, 184–86, 206; symbolic border between Jews and baptized, 190–92, 193; universalism and, 175. *See also* baptism; baptismal sites; Gentile(s); New Testament

Cisjordan and Transjordan division. *See* east-west binary; Transjordan; West Bank
cloak of Elijah, 155, 157–58, 161
Coats, George W., 86n2
Cohn, Robert L., 153n30, 155n35, 179n8, 182n16
collective purification of Christians, 12–13, 184–85, 188–89, 190–95, 206–7. *See also* collective purification of Jews; purity of the body
collective purification of Jews: Israel's border and, 22–23, 34–35; Joshua and, 90, 99; Priestly tradition and, 13, 34–35, 176; purity of the body, 12, 13, 176, 188–89, 195–201, 206. *See also* collective purification of Christians
Collins, Adela Yarbro, 165n62, 184n21
co-national status of Israelite tribes, 21, 53, 78–79, 95, 101–6. *See also* east-west binary; Israel/Israelite tribes; *and specific tribes*
Creation/Eden narrative, 22–24, 28, 42n4
Cross, Frank Moore, 88–89, 111n8
crossing the Jordan: Ark of the Covenant and, 90–92, 93; Elisha and, 155–56, 158–59; Israelite tribes' structure during, 92–95; maps as national criteria, 1–2; militaristic dimension to, 95–96, 99–100, 104–5, 246; miraculous dimension of, 89–93, 155–56, 158–60; national territory/memory/futurism, synthesis of, 86, 93–94, 96–101, 104–5; new beginning aspect of, 86; Palestinians' use bridge for, 268–73, 275; revisionist aspect of, 98, 99–100; Ruth and, 59, 154n31; spies' story read with, 98, 102–3, 105; transformation from, 138, 143–44, 150–52. *See also* exodus narratives
Crüsemann, Frank, 84

daughter of Jephthah, 56–57
daughters of Lot, 44, 46–47, 47n13, 49, 51
daughters of Zelophehad, 52–54, 57, 58, 61
David, King, 30, 50, 62–63
Davies, W. D., 175
Dawson, John, 193–95
death of the master type-scene, 141, 146, 149–50, 153, 156–57. *See also* impending death of the master type-scene; type-scenes

Deborah, Song of, 119
de Certeau, Michel, 68, 80n31
Degania kibbutz pioneers, 247–52, 254–56, 265, 274. *See also* Zionists
DeMaris, Richard, 163n59
Deuteronomistic tradition, 5, 6; Ammonites and, 35, 54–55; Amorites and, 118, 122–23; Assyrian Empire influences on, 9n11, 35n58; the body politic and, 181; Canaanites and, 11, 24n23; Edomites and, 35; Euphrates-to-sea maps and, 29–30; Ezekiel and, 34n56; hereditary line and, 137; Israel national myth and, 12, 13–14, 34, 219; Jerusalem and, 9; Jordan as border and, 13–14, 219; Joshua and, 85, 87; Levites and, 91n12; maps as national criteria and, 9; Moabites and, 35; Moabite women and, 35, 50, 54, 58; Palestine and, 219; prophetic succession and, 137–38; sea-to-Jordan maps and, 19–20, *20*; Transjordan and, 12, 30, 107n2. *See also* Torah
Deutsch, Nathaniel, 52n24
diaspora-homeland environment. *See* homeland-diaspora environment
dip/dipping, 178n7, 184, 201. *See also* baptism
disorder versus peace, 176–78, 183, 206
double portion of spirit, 139–40, 147, 156, 157, 158n44, 159. *See also* Elijah
Douglas, Mary, 177–78, 181
Druze, and half-tribe status, 117
Dundes, Alan, 4

E (Elohist) source, 5n7, 6, 69n9, 77n26, 77–78. *See also* Northern Kingdom/ tradition
east-west binary: Abraham and, 42n4, 50–51; co-national status of Israelites, 21, 53, 78–79, 95, 101–6; Joshua and, 101–5; Lot and, 42n4, 50–51, 101n31; Moabite women and, 49–51; Zionists and, 248–49
ecology of Jordan, 237, 276, 279, 283–89, *285*
Eden/Creation narrative, 22–24, 28, 42n4
Edom/Edomites: border drawing between Israel and, 64–65, 82–84; Deuteronomistic tradition and, 35; Esau and, 11, 64–67, 75n21, 82–84; Gilead/Israel and Esau/ Edom duality and, 64, 82–84; Israelites' competition with, 65–66, 83–84; Israelites' material cultural continuity with, 51, 82, 83; Jabbok River as border and, 80–81; Jacob and, 64–66, 82–84; Judea versus, 83n40; maps, *48*; Moab against Israel/ Judah alliance with, 56; as Other, 83–84; Solomon and, 63; Southern Kingdom and, 11, 66–67, 75n21, 82–83. *See also* Esau
Egeria, 283
Egypt: Assyrian alliance with, 37; Babylon's power struggles with, 33, 36–39, 37–38; Gaza Strip and, 237, 239, *240*; Israel as crossroads between Babylon and, 36–39; Israel's border in terms of, 30–35; river of, 20n6, 23–24, 26, 56; Syria/Israel's border in terms of, 31, 33
Eleazar the Priest, 149
Elijah: cloak of, 155, 157–58, 161; death of the master type-scene and, 141, 156–57; double portion of spirit of, 139–40, 147, 156, 157, 158n44, 159; Gilead and, 140, 153; heaven-earth mediation on vertical axis and, 12, 140–41, 156–57, 168, 183; immortality conferred by memory/futurism synthesis and, 12, 139–42, 144, 156; impending death of the master type-scene and, 154; Jesus's association with, 192; life and narrative of, 152–53, 160; miraculous events and, 153–54, 167; monarchical initiation connection with, 143, 152, 155, 160; visions and, 155–56, 159, 160. *See also* Elisha as successor of Elijah
Elisha as successor of Elijah, 152–54; apprenticeship status and, 138, 141, 144n9, 160; cloak of Elijah and, 157–58; crossing the Jordan and, 155–56, 158–59; death of the master type-scene and, 141, 156–57; double portion of spirit and, 139–40, 147, 156, 157, 158n44, 159; etymology of name, 139; exodus narratives parallels with, 155; Gehazi as corrupt insider/disciple of, 183–84, 196–204; God's divine aid and, 158; healing Naaman and, 178–82; immortality conferred by memory and futurism and, 12, 139–42, 144; impending death of the master type-scene and, 154; Jesus and John the Baptist comparisons with, 161, 162–63, 165–66, 167, 172; Jesus's correlation with, 171, 188, 192, 196–97; Joshua as successor of Moses compared with, 157n39; life and narrative of, 152–53; master/disciple relations

and, 139–40, 153, 159–60, 164; miracle of crossing on dry land and, 155–56, 158–60; miraculous events and, 153–54, 156n37, 158n44, 159–60, 167; monarchical initiation connection with, 153, 156, 157n38, 158, 160; Naaman/baptism association with, 196, 197–98, 200–201, 204; nonhereditary line and, 12, 157, 170–71; redemption for Jews and, 139, 162; self-healing by, 204–6; spirit transfer to disciple and, 144, 155–56, 157–60, 168; vision implementation by, 144, 153, 160–61; witnesses as presence of prophets, 154–57, 159

Elohist (E) source, 5n7, 6, 69n9, 77n26, 77–78. *See also* Northern Kingdom/tradition

Ephraim/Ephraimites, 21, 108, *110*, 117, 126–28, 146

Esau: border drawing between brothers and, 64–65, 82–84; Edom and, 11, 64–67, 75n21, 82–84; etiology of, 66; God's covenant with Abraham, 67; Isaac's blessing of hereditary line and, 66–67; mythic bond between rivals and, 41; Northern Kingdom links between Jacob and, 64; Rebecca as trickster, 66–67, 71; revenge against Jacob, 75–77, 79, 80n32, 81, 82–83; Southern Kingdom and, 11, 66–67, 75n21, 82–83. *See also* Abraham; Jacob

Esther, book of, 58

ethno-nationalism, 5, 10; boundaries/maps and, 10; Esther story and, 58; Ezra-Nehemiah myth and, 5, 6, 10, 59–60; Moabites and, 5–6, 10, 54, 58–63; Moabite women and, 10, 54, 63; Ruth and, 5–6, 58–63; Transjordan and, 106–7, 118, 120–23, 125–26, 130n57; universalism versus, 175

Euphrates-to-sea maps, 24–26, 28–30, 31, 35, 129

Europe/European hegemony: Balfour Declaration of 1917 and, 226–27, 233, 234; British mandate maps and, 220–21, 223–27, *224, 225, 228,* 229–33, 269, 277, 286–87; French mandate maps and, 221, 223–24, *225, 228,* 232; Jews as refugees from, 251–52; Jordan state and, 219, 223, *225,* 227, 231, 257, 261n114, 269; Lawrence map and, 223–24, *225;* Ottoman administrative districts map and, 220–21, *222,* 234, 250; Paris and San Remo peace conferences' territorial proposal, 229–32; Peel maps and, 235, *236,* 237; PEF maps and, 220–23, *224,* 232; Sykes-Picot map, 223–24, *225, 226,* 232; via territory borders in Near East, 223–24, *226,* 229; United Nations partition and, 237, *237, 239, 240. See also* hegemonic discourse

exodus narratives: crossing parallels with, 86, 89n7, 89n8, 92, 94n18, 95, 97–100, 109n5; Elisha's crossing parallels with, 155; Israel national myth and, 252n80; Joshua's crossing and, 99–100; Moses and, 100, 108, 109, 143, 146; Palestinian refugees and, 252–53, 257, 261–64

Ezekiel, book of: apocalyptic visions and, 22, 215; Creation narrative and, 24, 28; Deuteronomistic tradition and, 34n56; heaven-earth mediation on horizontal axis and, 28; imperial geography and, 29, 32, 33n52, 34; Israelite tribes' locations and, 21; Moabites and, 45; paradisical visions and, 21–23, 34n53; Priestly tradition and, 23–24, 34; sea-to-Jordan maps and, 20–22; Sodom and, 45

Ezra-Nehemiah myth, 5, 6, 10, 59–60, 129–30. *See also* Nehemiah

Fatah, 265–68. *See also* Palestine/Palestinians

feeding the people miracles, 166–67

fetishism, 237, 259n108

fida'iyin movement, 264–68. *See also* Palestine/Palestinians

FOEME (Friends of the Earth Middle East), 284–86, *285,* 288, 289

folklore traditions: Palestine and, 239, 241; Zionist pioneers versus, 241, 248

folklorist traditions: Biblical national myth classifications and, 6, 79; Jacob and, 79; Joshua and, 85–87, 88–89, 96–97, 100–101; purity of the body and, 177–78

foreigners/foreignness: Ezra-Nehemiah myth, and expulsion of women as, 5, 6, 10, 59–60; impurity/purity of the body and, 177n3; Jordan as porous border against biblical, 176, 181–82, 206; of Moabite women as asset, 5–6, 58–63; as profane, 177n3

308 • Index

founding mothers of Israel, 47n13, 49, 50–51, 53, 57–58, 76n22
freedom fighters, 219–20, 257, 264–68. *See also* Palestine/Palestinians
French mandate maps, 221, 223–24, 225, *228*, 232
Friends of the Earth Middle East (FOEME), 284–86, *285*, 288, 289
frontier/frontier justice: Ammonites and, 125–27; Ephraimites and, 126–28; Gilead and, 64, 125–27; Jephthah and, 125–27; in Transjordan, 123–28; West Bank as, 244–45, 260–61; Zionists' stories of, 249–50, 261
future privileges past, in Jesus and John the Baptist story, 140, 143, 162, 164–65. *See also* master/disciple relations
futurism/memory synthesis. *See* memory/futurism synthesis

Gad: co-national status of, 78–79, 95, 101–5, 106; insider-outsider position of, 101, 114, 117–18, 129n53; as integrated into cults of Moabites and Ammonites, 121n26, 124n38; memorial site built by, 111–17; Transjordan as location of, 21, 78, *110*, 111–17
Galilee, *132*, 134n66
Gaza Strip, 237, 239, *240*
Gehazi, 183–84, 196–204, 201–3. *See also* Elisha as successor of Elijah
Gelilot altar, 111–13
gender dimension/gender roles. *See* men; Moabite women; women
Gentile(s): Naaman as God-fearing, 180–81, 192, 198; purity of the body and, 177n3, 180–81, 187–88; sin and, 186. *See also* Islam/Muslims; Jews
geography, mythic. *See* mythic geography
Gilead: boundaries/maps and, *48*, 64, 73, 126–27, 146; contestation sites and, 71–73; Elijah and, 140, 153; Esau/Edom and Gilead/Israel duality and, 64, 82–84; frontiers/frontier justice and, 64, 125–27; half-tribe of Manasseh, and incorporation of, 118–20; Israelite/Aramean disputes and, 71–73, *73*, 74–75, 78; Israel national myth and, 56, 77–79, 118–20; Israel's border in terms of Egypt, 32n49; Jabbok River as border and, 80–81, 125; Jacob and, 82–83; location of, 74; Machir as Israelite-Aramean founding father and, 53, 78n29; memorial site at, 71–72, 74–75, 116; naming places and, 73, 74; Northern Kingdom and, 77–78
Gilgal, 88–89, 99, 100, 112–16, 154, 158n40
God: covenant between Abraham and, 41n2, 43, 49, 67; Elisha, and divine aid from, 158; God-person-land triangulation, 43; Israel's disorder in relations with, 176–77; Jesus's divine initiation as Son of, 163, 167, 169–71, 173; Joshua, and divine aid from, 95, 158; as king and monarchical initiations, 169; memorial sites as sacred sites of, 71n12; miraculous dimension of Joshua's crossing and, 90–93; sacred sites of, 71n12
Gospels. *See* New Testament
Great Britain, and mandate maps, 220–21, 223–27, *224*, *225*, *228*, 229–33, 269, 277, 286–87
Greater Syria map, 223, *225*, 231. *See also* Syria
Greeks, 130–31
Grosmark, Ziona, 129n51, 130n54
Gunkel, Hermann, 159

H (Holiness) Code source, 5n5, 6, 23–24, 185. *See also* Priestly tradition
Hagar, 41n2, 43, 51. *See also* Abraham
halakhah (Jewish law): animal tithing, 208, 209–12, *210*; self-healing and purity, 197–99, 204–6
Halbwachs, Maurice, 104n36, 112
half-tribe of Manasseh: boundaries/maps and, 146; co-national status and, 21, 53, 78–79, 95, 104, 106; Ephraimites and, 117; Gilead as part of Israel nation via, 118–20; insider-outsider position of, 101, 114, 117–18, 129n53; memorial site built in Transjordan by, 111–17; as recapitulation of bipartition of Israel, 117–18, 233–34; Transjordan as location of, 21, 78, *110*, 111–20
Halpern, Baruch, 24n23, 30n40
Halutzim (Zionist pioneers), 241, 246–50
Hammurabi laws, 186n24
Harley, J. B., 18
Hartman, Geoffrey, 77
Hashemite Kingdom of Jordan. *See* Jordan state

Hasmoneans, 131–32, *132*
Hayes, Christine, 177nn3–4
Hays, Peter, 80n31
healing. *See* illnesses/healing
heaven-earth mediation: baptism, and vertical axis for, 13, 175–76; the body and, 13, 175–76; cosmic boundary of the Jordan and, 26–28; Elijah, and vertical axis for, 12, 140–41, 156–57, 168, 183; Ezekiel, and horizontal axis for, 28; Jacob's wounds, and vertical axis for, 79, 80; Jesus, and vertical axis for, 168, 170, 171n73, 172; Jesus and John the Baptist story on vertical axis for, 12, 140, 162, 163, 173; Joshua, and horizontal axis for, 12, 140–41; New Testament, and vertical axis for, 157, 162
Hebrew Bible: Creation narrative and, 22–24, 28, 42n4; Gilead as contestation site and, 72; moral purity and, 185; New Testament dependence on, 136, 173; as past versus New Testament as future, 162, 172–73; prophetic succession and, 136, 137–42; redemption for Jews as return and, 12, 174; on religious body politic, 181; ritual purity and, 185, 189; Transjordan's ambivalent status and, 49; type-scene and, 142–43. *See also* Biblical national myths; Tanakh (biblical canon); Torah
Hebrews (*Ivrim*), use of term, 41
hegemonic discourse: Israel national myth and, 32–33, 38, 277, 280–81; over Near East by Christians, 277, 280–81; Palestine and, 277, 280–81; Zionists and, 14. *See also* Europe/European hegemony
Hendel, Ronald, 42n2, 71n11
hereditary line: Aaron's sons and, 91n12, 150; David and, 62–63; Deuteronomistic tradition and, 137; Esau and, 11, 66–67, 75n21, 82–83; "holy seed" concept and, 10, 59, 130; Isaac's blessing for, 66–67; Israelites' territory borders versus, 108, 125n40; Jacob and, 66–67, 68–69, 71; Jephthah and, 124–25, 129n53; land versus, 11, 64–69, 71, 75n21, 82–84; Lot, and territory borders versus, 42n4, 43, 47, 49, 64; monarchical initiation foundation in, 136–38, 145, 150; territory borders and, 108, 125n40, 130. *See also* nonhereditary line
Herod's kingdom, 131–34, *133*. *See also* Roman Empire

Holiness (H) Code source, 5n5, 6, 23–24, 185. *See also* Priestly tradition
"holy seed," 10, 59, 130. *See also* Ezra-Nehemiah myth; hereditary line
homeland-diaspora environment: Abraham and, 41, 68–69; borderland zones and, 82; Canaan as, 71, 73, 75, 80–82; Euphrates-to-sea maps and, 26, 29; Israel national myth and, 55; Jacob and, 66–68, 71, 73, 75, 77, 80, 81; Joshua's homeland ideology and, 85, 87, 88, 92n15, 101–5; Petra trek and, 258–60, 269n133; sea-to-Jordan maps and, 19; Syria and, 72n17; Transjordan and, 107–9, *110*, 111–17; wanderings connections with self and space and, 10–11, 68–71, 75–76, 77
homosexuality, and Sodom/Sodomites, 44–46
House of Study (Beit Midrash), 198–99, 204
Hutton, Jeremy, 107n2, 108n4
hydroelectric plant project, 286–88. *See also* ecology of Jordan
Hyrcanus, 130–31

Ilan, Zvi, 287
illnesses/healing: baptism and, 177–86, 178–82, 184–86, 197–98, 206; disorder in Israel and, 176–77; Jordan as border and, 177–83, 185, 206; Naaman and, 177–83, 185, 197–98, 206; purity of the body and, 177–83, 185, 206; self-, 197–99, 204–6; sin correlation with, 185, 186
immortality: the body as ephemeral physical and, 137, 147, 181; of Jesus, 12; memory/futurism synthesis and, 12, 139–44, 147, 156
impending death of the master type-scene, 144–48, 154, 163–67. *See also* death of the master type-scene; type-scenes
imperial geography, 28–31, 32–33; Babylon's conception of Near East and, 26–27, 35; David and, 30; Ezekiel and, 29, 32, 33n52, 34; Israel as crossroads and, 36–39; Israel national myth and, 33, 38–39; Israel's border in terms of Babylon and, 26n27, 33, 34, 35–36; Israel's border in terms of Egypt and, 30–35; maps as national criteria and, 5, 29–30; Solomon and, 30, 33n51
impurity/purity. *See* purity/impurity; purity of the body

310 • Index

insider-outsider position, 7; Gehazi as corrupt, 183–84, 196–204; of Israelites in Jerusalem, 16, 100–101; of Israelites in Transjordan, 101, 114, 117–18, 129n53; of Jacob, 77; of Joshua, 95, 101; of Naaman, 176–77, 182, 187; Southern Kingdom and, 10; Tobiad family's, 129–31

Intifada stories, 244, 245, 270n134. *See also* Palestine/Palestinians

Isaac, 41, 66–67, 71, 82. *See also* Abraham; Esau; Jacob

Ishmael/Ishmaelites, 41, 43, *48*, 67

Islam/Muslims: Christians' relations with, 242–44, 277, 281; Jerusalem and, 234n34; in Palestine, 14, 219–20, 224, 231, 234n34, 243–45, 277, 279; prophetic succession and, 14, 147n18. *See also* Gentile(s)

Israel/Israelite tribes: Ammonites' relations with, 51, 55; Aramean conflicts/treaties with, 65, 71–73, *73*, 74–75, 78; Aramean/Israelite duality and, 64, 72, 74–75, 176; Aramean relations with, 71–72; Biblical national myths and, 7; border in terms of Egypt for, 30–35; boundaries/maps and, 21, 108, 109, *110*, 146; child sacrifice and, 56–57; collective character of, 85, 86, 87, 90, 93, 98; co-national status of tribes and, 21, 53, 78–79, 95, 101–6; as crossroads between Egypt and Babylon, 36–39; disorder and, 176–77; Ezekiel and, 21; founding mothers of, 47n13, 49, 50–51, 53, 57–58, 76n22; hereditary line and, 108, 125n40, 130; as holy people, 130n55; insider-outsider position of tribes in Transjordan, 101, 114, 117–18, 129n53; "Israelite" as honorific, 99n27; Moabite-Israelite intermarriages and, 55, 62–63; as name of person and place, 79, 80n31, 81–82; in terms of Babylon, 22–23, 26n27, 31, 33, 34, 35–36; tribal structure of crossing, 92–95; women's immodest behavior and, 49. *See also* Israel nation/national myth; Zionists; *and specific biblical characters and tribes*

Israel nation/national myth, 14–16; archaeological excavations and, 279; the body imprinted with national identity, 181, 182, 193–94; British mandate maps and, 220–21, 223–27, *224*, *225*, *228*, 229–33, 269, 277, 286–87; Christian relations via baptismal sites, 277, 279; Deuteronomistic tradition and, 12, 13–14, 34, 219; exodus narratives and, 252n80; freedom fighters and, 219–20, 257–61; French mandate maps and, 221, 223–24, 225, *228*, 232; Gaza Strip and, 237, 239, *240*; Gilead and, 56, 77–79, 118–20; half-tribe of Manasseh as recapitulation of bipartition of, 117–18, 233–34; hegemonic discourse and, 32–33, 38, 277, 280–81; homeland-diaspora environment and, 55; hydroelectric plant project and, 286–88; imperial geography and, 33, 38–39; Israel's border in terms of Babylon, 31; Jordan as border and, 237, 239, *240*, 241, 275; Al-Maghtas baptismal site and, 276, 278, 279–83, 285; maps of Israel, *236*, 237, *238*; multiple traditions and, 4; myth(s) defined, 3–4; Northern Kingdom and, 65; peace park plan and, 284; Persian period and, 38–39, 58, 130; Petra trek and, 258–60, 269n133; pioneers and, 260–61; popes and, 277, 280–81, 282; Priestly tradition and, 34–35; Roman Empire and, 133–34; settlements as political strategy and, 134, 234, 235, 237, 248; tourism and, 5, 247, 268–69, 288; water rights and, 283–84; West Bank under control of, 219, 237, 260–61, 277–78; Yardenit baptismal site, 276–79, 283, 285, *285*. *See also* Biblical national myths; Europe/European hegemony; Israel/Israelite tribes; Jews; West Bank; Zionists

Ivrim (Hebrews), use of term, 41

Jabbok River, 80–81, 125
Jabotinsky, Ze'ev, 14n15, 234
Jacob, 5, 64–65, 84; boundaries/maps and, 10–11, 83–84; Edom and, 64–66, 82–84; Esau's revenge against, 75–77, 79, 80n32, 81, 82–83; folklorist traditions and, 79; as founding father, 68, 71, 79–80; Gilead and, 82–83; God's covenant with Abraham, 67; homeland-diaspora environment and, 66–68, 71, 73, 75, 77, 80, 81; insider-outsider position of, 77; Israel as place and personal name for, 79, 80n31, 81–82; Israelite/Aramean duality and, 72, 74–75, 176; Jabbok River and, 80–81;

Laban and, 64, 71–72, 74–76; land and hereditary lineage for, 64–69, 71; Leah as wife of, 71, 76; memorial sites and, 68–70, 71nn11–12, 73, 74, 77, 79–80; mythic bond between rivals and, 41; naming places and, 73, 74, 77, 79–80; Northern Kingdom links with, 5n7, 10–11, 64, 69n9, 78; porous borders and, 10–11, 78; Priestly tradition and, 83n40; Rachel as wife of, 71, 75–76; Rebecca as trickster and, 66–67, 71; Southern Kingdom and, 10–11, 64, 65; as trickster, 66–67, 71, 80n32; two camps of, 75–79, 85; visions/encounters with angels and, 70, 72, 76–77, 79–80, 81n34, 82, 154n33; wanderings' connections with self and space for, 10–11, 68–71, 75–76, 77; wounds as heaven-earth mediation, 79, 80. *See also* Esau; Gilead; Isaac; Laban; Northern Kingdom/tradition

Jacobs, Andrew, 192n36, 196n43

Jephthah, 56–57, 124–27, 129n53

Jerusalem: as center of Israel national myth, 94n21, 181; destruction of, 37; Deuteronomistic tradition and, 9; insider-outsider position of Israelites in, 16, 100–101; international model and, 224, *240*; Islam and, 234n34; maps, *73*, *110*, *228*, *240*; mythic geography and river under, 22–23; Northern versus Southern traditions and, 77n26, 134n66; Palestinian claims to East, 275, 282; Palestinian resident status in, 263–64; Palestinians flight from, 263–64; as shared space of Others, 16, 100–101; Southern Kingdom's conflicts with Babylon over rule of, 33–37, 124n39, 129; Temple in, 9, 11, 21, 22–23, 24n20, 59–60, 131, 207n70

Jesus: Elijah's association with, 192; Elisha's correlation with, 171, 188, 192, 196–97; Gehazi's correlation with, 201–3; God's divine initiation as Son of God, 163, 167, 169–71, 173; heaven-earth mediation on vertical axis, 168, 170, 171n73, 172; as heretical apostate in rabbinic narratives, 201–4; Holy Spirit to disciple and, 167–71, 173; illnesses/sin correlation and, 186; immortality of, 12; as Jordan River, 189–90; monarchical initiation connection with, 169–71, 173; Moses, and status of, 171; nonhereditary line and, 12, 162, 163, 167, 169–71. *See also* Jesus and John the Baptist

Jesus and John the Baptist, 161–63, 173–74; apocalyptic associations with Jordan and, 22, 23, 184n21, 186n24; "death" of the master type-scene and, 140, 165–66, 172–73; Elisha as successor of Elijah comparisons with, 161, 162–63, 165–66, 167, 172; etymology of name, 139; future privileges past and, 140, 143, 162, 164–65; heaven-earth mediation on vertical axis and, 12, 140, 162, 163, 173; Hebrew Bible/New Testament as past/future and, 162, 172–73; Herod and, 166; impending death of the master type-scene and, 163–67; Joshua as successor of Moses comparisons with, 162–63, 166n63, 167, 172; master/disciple inverted relations between, 139, 140, 163–64, 167, 173; miracle of feeding the people, 166–67, 204n67; monarchical initiation connections and, 169–71, 173; prophetic succession comparisons and, 135, 136, 167, 173; redemption through baptism and, 139, 161–69, 172–73; repentance of sin through baptism and, 164–65, 167, 172–73, 185, 190; symbolic and national boundary status of Jordan and, 134; transformation for Jesus through baptism and, 139, 140, 168–70, 172; vision/vision implementation and, 162–63, 165, 171; witnesses' presence at baptism and, 171–72. *See also* Jesus; John the Baptist

Jewish law (*halakhah*). See *halakhah* (Jewish law)

Jews: baptism/Jordan association, and reclamation by, 195–96, 199–200, 201, 202n60, 203n61, 204; Christian-Jewish binary divisions in purity for, 12–13, 14, 187, 206–7; conversion ceremony for, 198; exclusion from World to Come for, 195–96, 200–201; heaven-earth mediation on vertical axis for, 12, 79, 80, 140–41, 156–57, 168, 183; horizontal boundary of Jordan for, 13; the land correlation with purity of the body for, 179, 180–82, 183n18, 184, 207n70; purity, and self-healing for, 197–99, 204–6; as refugees from Europe, 251–52; sym-

Jews (*continued*)
  bolic border of Jordan between baptized Christians and, 190–92, 193. *See also* Gentile(s); Israel/Israelite tribes; redemption for Jews; Zionists
Jobling, David: on east-west binary, 11, 42n4, 49–51, 53, 57; on Jephthah and Moab, 126; on Manasseh tribe, 53, 117–18; on Moabite women, 53, 57; on spies' story read with Joshua's crossing, 102–3
John Hyrcanus, 130–31
John the Baptist: authority through baptism of Jesus and, 139, 171–72, 173; cloak of, 155, 157–58, 161; "death" of the master type-scene and, 140, 165–66, 172–73; Elijah association with, 166, 171n74, 173, 192; festival at Jordan for, 89, 245; future privileges past and, 140, 143, 162, 164–65; on Gentile sinners, 186; impending death of the master type-scene and, 163–67; master/disciple inverted relations between Jesus and, 139, 140, 167, 173; on Pharisees, 164n60, 170, 186; as prophet of old era, 140, 143, 161–62; realignment of Christian and Jewish binaries of self and Other, 187; on Sadducees, 170, 186; subordination of, 161–62, 164–65, 167; symbolic and national boundary status of Jordan, 134; visions and, 162. *See also* Jesus; Jesus and John the Baptist
Jordan, Mark, 46
Jordan as border, 19–20, *20*, 31, 218–20, 273–75. *See also* Biblical national myths; Deuteronomistic tradition; Elisha as successor of Elijah; Israel nation/national myth; John the Baptist; Joshua; Joshua as successor of Moses; Palestine/Palestinians; Priestly tradition; prophetic succession; Transjordan; Zionists
Jordan state: baptismal sites and, 277, 279–83; British mandate maps and, 269; Christians' relations via baptismal sites and, 277, 279–82; European hegemony and, 219, 223, *225*, 227, 231, 257, 261n114, 269; Greater Syria map and, 223, *225*, 231; hydroelectric plant project and, 286–88; Jordan as border and, 219, 223, *225*, 227, 231, 232–33, 257, 261n114, 269; Palestinian citizenship and, 261n114; Palestinian refugees and, 253, 260–61, 261n114; Palestinians' crossing on bridge and, 275; peace park plan and, 284, 286; water rights and, 284; West Bank and, 239, *240*, 257
Josephus, 28n37, 45, 130–31, 133, 139n7, 166n63, 216n15, 282–83
Joshua, book of, 85–87, 89, 105; boundaries/maps and, 11, 25n25, 87–88, 101–5; collective character of Israel and, 85, 86, 87, 90, 93, 98; collective purification of Jews and, 90, 99; Deuteronomistic tradition and, 85, 87; east-west binary and, 101–5; exodus narratives' parallels with, 86, 89n7, 89n8, 92, 94n18, 95, 97–100, 109n5; folklorist traditions in, 85–87, 88–89, 96–97, 100–101; Gelilot and, 111–13; Gilgal and, 88–89, 99, 100, 112–16, 154, 158n40; homeland ideology and, 85, 87, 88, 92n15, 101–5; insider-outsider position and, 95, 101; Jebusite Jerusalem and, 16, 100–101; Jordan as border and, 87–88, 101–4; the land of Israel and, 103–4; national-colonial projects and, 88; pioneering ancestors' stories and, 97–98; Shiloh memorial site/altar and, 112; variant versions versus plot of, 85–87, 100–101
Joshua as successor of Moses, 143–48; apprenticeship status and, 139, 141, 144; death of the master effects on status of, 149–50; Elisha as successor of Elijah compared with, 157n39; etymology of name, 139; God's divine aid to, 95, 158; heaven-earth mediation on horizontal axis and, 12, 140–41; immortality conferred by memory/futurism synthesis and, 12, 140–44; impending death of the master type-scene and, 144–48, 154; master/disciple relations and, 137, 138–39, 153, 164; militaristic dimension to crossing for, 95–96, 99–100, 104–5, 246; miracle of feeding the people and, 167; miraculous dimension of crossing for, 89–93; monarchical initiation connection with, 149, 150, 151n27; national territory/memory/futurism synthesis of crossing and, 86, 93–94, 96–101, 104–5; new beginning aspect of crossing for, 86; nonhereditary line and, 12, 138, 144, 145–46, 170–71, 174; power transfer and, 91,

92–94, 95–96, 145n13, 149, 158; prophet successor appointments and, 138; redemption for Jews and, 11, 139; revisionist aspect of crossing and, 98, 99–100; spies' story read with crossing by, 98, 102–3, 105; spirit transfer to disciple and, 144, 148–49, 168; Torah/prophets as memory/future and, 141, 145, 147–48, 151, 162; Torah's tenets promoted by, 151; transformation from crossing and, 138, 143–44, 150–52; tribal structure of crossing and, 92–95; vision implementation and, 144, 148, 152, 153, 160–61; witnesses' presence and, 149, 150–52, 159

Jubilees text, 130

Judah Maccabee, 131

Judea (former Southern Kingdom of Judah): Edom versus, 83n40; Levites versus, 24n20; Moabite conflicts with, 56; Persian Empire sponsorship of return of exiles to, 5, 10, 34, 50. *See also* Southern Kingdom of Judah (Judea)

judges, era of, 153, 160

Kantorowicz, Ernst, 12, 137

Karama commandos, 266–68, 274. *See also* Palestine/Palestinians

Katz, Kimberly, 281

King Hussein/Allenby Bridge border crossing, 268–73, 275

Kinneret kibbutz, 276–79, 283, 285

Klawans, Jonathan, 185

Knight, Douglas, 111n6

Kohanim (Sadducees): hereditary line and, 169; illnesses of the body and, 193; Jesus's conflicts with, 169; John the Baptist's conflicts with, 170, 186; Joshua's crossing and, 90–92, 94–96; monarchical initiation and, 169; power transfer from Moses to Joshua and, 145n13, 149, 158; prophetic succession and, 158. *See also* Levites

Kraeling, Carl, 161n51, 162n54, 164n60

Kugel, James, 130

Laban, 64, 71–72, 74–76. *See also* Gilead; Northern Kingdom/tradition

land/land of Israel, 17, 21; in Babylonian Talmud, 210, 216–17; borderland zones and, 43n6, 52n22, 82, 273, 275; God-person-land triangulation and, 43; hereditary lineage versus, 11, 64–69, 71, 75n21, 82–84; Joshua and, 103–4; naming places and, 17, 73, 74, 77, 79–80; pioneering ancestors' stories and, 17; purity of the body correlation with, 179, 180–82, 183n18, 184, 207n70; sacred nature of, 104, 130n55, 206–7, 283; west of Jordan, 21. *See also* border drawing; boundaries/maps; homeland-diaspora environment; Israel/Israelite tribes; Israel nation/national myth; maps as national criteria; memorial site(s); Northern Kingdom/tradition; Palestine/Palestinians; sea-to-river maps; Southern Kingdom of Judah (Judea); Transjordan; *and specific tribes*

land ownership, and Moabite women, 52–54, 56–57, 58, 61

late capitalism. *See* multinational control

Law, the (Al-Shariah; the Watering Place), 245

Lawrence, T. E., 223, 225

Lawrence map, 223–24, 225

Laws of Hammurabi, 186n24

Leah, 71, 76

legends/myths, 2–4. *See also* Biblical national myths; Deuteronomistic tradition; Israel nation/national myth; mythic geography; Palestine/Palestinians; Priestly tradition

Lévi-Strauss, Claude, 3, 80n31

Levites, 21, 24n20, 90–91, 136–38, 143–45, 145n13, 150, 169. *See also* Kohanim (Sadducees); Priestly tradition

Lot, 42–47, 48, 49–51, 64, 101n31

Machir, 53, 78n29, 118–20. *See also* Aram/Arameans

Maghtas baptismal site, Al-, 276, 278, 279–83, 285, 285

Manasseh. *See* half-tribe of Manasseh

Mappa Mundi, 18, 26–27, 35

maps as national criteria, 1–2, 5, 8, 9, 29–30. *See also* Biblical national myths; boundaries/maps; imperial geography; legends/myths; mythic geography

martyr commandos, 268. *See also* Palestine/Palestinians

master/disciple relations: future privileges past in, 140, 143, 162, 164–65; inversion of, 139, 140, 167, 173; past privileges future in, 136–40, 143, 153, 159–60, 164

Mazar, Benjamin, 31, 34
memorial site(s), 17, 96; Assyrian Empire and, 93n17; Babylon and, 93n17; Christianity and, 207n70; Gelilot altar as, 111–13; at Gilead treaty scene, 71–72, 74–75, 116; Jacob's construction of, 68–70, 71nn11–12, 73, 74, 77, 79–80; Laban's construction of, 71–72, 74–75; miraculous dimension of Joshua's crossing and, 93–94, 96, 97–98; as sacred sites of God, 71n12; Shiloh altar as, 29, 73, 78, 112, 112n9; in Transjordan, 111–17
memory/futurism synthesis: Hebrew Bible as past versus New Testament as future and, 162, 172–73; immortality conferred by, 12, 139–44, 147, 156; national territory and, 86, 93–94, 96–101, 104–5, 112–16; Torah as memory and, 141, 145, 147–48, 151, 162
men: as feminine and passive, 40–41, 54, 55–56; festival as subversion against, 56–57; Moabite women's actions versus agenda of, 52–54, 56–57. *See also* women
Mesha Inscription, 51n21, 79n30, 123–24. *See also* Moab/Moabites
Mesopotamian Empire. *See* Babylon
midrashic texts, 45, 92, 160, 189, 208–9, 213, 215n13
militaristic dimension to crossing, 95–96, 99–100, 104–5, 246
miraculous events: crossing on dry land by Elisha and, 155–56, 158–60; Elijah and, 153–54, 167; Elisha and, 153–54, 156n37, 158n44, 159–60, 167; feeding the people and, 166–67, 204n67; Joshua's crossing and, 89–98; purification of the body and, 176, 180, 182
mishnah texts, 135, 198n51, 208, 211n7
Moabite women: the body politic and, 44, 46, 49, 51, 52, 57; boundaries/maps and, 10, 50–51, 52–54, 56–57, 63; daughter of Jephthah and, 56–57; Deuteronomistic tradition and, 35, 50, 54, 58; east-west binary and, 49–51; ethno-nationalism and, 10, 54, 63; festival as subversion against men and, 56–57; foreignness as asset for, 5–6, 58–63; as founding mothers, 47n13, 49, 50–51, 53, 57–58; immodest behavior of, 49, 51–52; Israelite-Moabite intermarriages and, 55, 63; as landowners, 52–54, 56–57, 58, 61; men's agenda versus actions of, 52–54, 56–57; as Other, 49, 50, 56, 58; overview and source of ideological topos, 5–6, 40–41, 50, 54, 58, 63; Persian period and, 58–63; Priestly tradition and, 53, 54. *See also* Moab/Moabites; women; *and specific Moabite women*
Moab/Moabites, 40–41; Balaam as prophet of, 51–52, 54, 60; Deuteronomistic tradition and, 35; ethno-nationalism, 5–6, 10, 54, 58–63; etymology, 47n12; Ezekiel and, 45; feminine and passive men from, 40–41, 54, 55–56; Gad as integrated into cults of Ammonites and, 121n26, 124n38; Israelites' relations with, 55–58, 63; *Ivrim*/Hebrews, use of term, 41; material cultural continuity between Israel and, 51; Mesha Inscription and, 51n21, 79n30, 123–24; national identity and, 124; as nation of perversion and deviance, 10, 40, 44, 47n13; Sodom and, 42–47, 50; territory border and, 51, 52n22, 55, 56, 58, 59, 60, 121, 126. *See also* Abraham; Lot; Moabite women; Ruth
monarchs/monarchical initiations: Elijah's connection with, 143, 152, 155, 160; Elisha's connection with, 153, 156, 157n38, 158, 160; God as king and, 169; hereditary lineage foundation for, 136–38, 145, 150; Jesus's connection with, 169–71, 173; physically ephemeral and politically immortal body of, 137, 147; priesthood's connection with, 136–38, 143, 145, 169; prophetic succession connection with, 136–37, 149, 150, 151n27
moral purity/impurity, 177n3, 185. *See also* purity of the body; ritual purity/purification
Moses: the body/bodies of, 147; Canaan and, 19; death of the master type-scene and, 146, 153; gravesite of, 146, 147; immortality conferred by memory/futurism synthesis and, 12, 140–44, 147, 156n37; impending death of the master type-scene and, 144–48, 154; miracle of feeding the people by, 167; nonhereditary line and, 12, 138, 144, 145, 170–71, 174; redemption for Jews and, 162; sea-to-river maps and, 19, 41–46, 146–47; spies' story and, 98, 102–3; visions and, 22, 144, 146,

148, 160. *See also* Joshua as successor of Moses
mothers of Israel, founding, 47n13, 49, 50–51, 53, 57–58, 76n22
multinational control, 275–76, 288–89; archaeological excavations and, 276; baptismal sites and, 245, 276–82; peace park plan and, 284–86, *285*, 288–89; water rights and, 283–84; West Bank and, 288
Muslims/Islam. *See* Islam/Muslims
myth(s) defined, 3–4. *See also* legends/myths
mythic geography, 18–19; Babylon and, 18; cosmic boundary of the Jordan and, 26–28; Creation narrative and, 22–24, 28; Euphrates-to-sea maps and, 24–26, 28–30; river of Egypt and, 20n6, 23–24, 26; sea-to-Jordan maps and, 19–22, *20*, 211–12

Naaman: as anti-conversion/conversion narrative, 182, 184; baptism association with, 180, 182, 184, 186–87, 192–93, 196, 197–98, 200–201, 204; disorder and, 177; as God-fearing Gentile, 180–81, 192, 198; illness/healing and, 177–83, 185, 197–98, 206; insider-outsider position of, 176–77, 182, 187; Israelite superiority versus arrogance of, 179, 180–82; purity of the body and, 197–98; rebirth after immersion by, 180, 186–87, 192–93; servants of, 177–78, 179; sin and, 185
Nakbah of 1948, 252–54, 262. *See also* Palestinian refugees
Naksa, al- (the Setback), 261. *See also* Palestinian refugees
naming places/natural features, 17, 73, 74, 77, 79–80
Naomi, 59–62, 154n31
Naphtali, 146
national homelands. *See* homeland-diaspora environment; Jordan as border
national myths. *See* Israel nation/national myth; Palestine/Palestinians
national territory/memory/futurism synthesis, 86, 93–94, 96–101, 104–5, 112–16
Nebuchadnezzar, 23n17, 27–28, 35, 37, 124n39, 129n49. *See also* Babylon
Nehemiah, book of, 5, 6, 10, 59–60, 129–30, 249–50
Nelson, Richard D., 151n27

New Testament: as future versus Hebrew Bible as past, 162, 172–73; heaven-earth mediation on vertical axis and, 157, 162; Hebrew Bible dependence on, 136, 173; rebirth after baptism and, 12, 169, 174, 186–87; Sodom and, 46; type-scene and, 142. *See also* Jesus and John the Baptist; John the Baptist
nonhereditary line: Elisha and, 12, 157, 170–71; Jesus and John the Baptist and, 12, 162, 163, 167; Jesus as Son of God and, 169–71; Joshua and, 12, 138, 144, 145–46, 170–71, 174; Priestly tradition and, 16; prophetic succession and, 12, 136, 137–38, 144, 157n38, 174. *See also* hereditary line; prophetic succession
Noort, Ed, 29n39
Northern Kingdom/tradition, 5, 77n26; Assyrian Empire's battles against, 6, 11, 55, 77, 128; authorship and, 5n7, 6, 69n9; boundaries/maps and, 10–11; Elohist (E) source and, 5n7, 6, 69n9, 77n26, 77–78; Esau and Jacob links with, 64; Gilead as part of, 77–78; imperial geography and, 33, 38; Israelite/Aramean conflicts/treaties and, 65; national identity and, 65; porous borders and, 10–11, 78; Rachel as founding mother and, 76n22; territory borders and, 11; Transjordan as part of, 77–78. *See also* Aram/Arameans; Gilead
Noth, Martin: on Esau/Edom and Gilead/Israel duality, 83n40; on folklorist tradition in Joshua, 88–89; on Gad as integrated into Moabite and Ammonite cults, 121n26, 124n38; on Gelilot altar, 111n7; on Gilead, 74n18; on Jacob's two camps, 76n22; on Joshua as successor of Moses, 148n19; on Machir tradition, 118–19; on national and cannon formation, 7, 79n30
Nuzuh of 1967 ("Second Exodus"), 252–53, 257, 261–64. *See also* Palestinian refugees

Origen, 189–95
Other/Others: Christian and Jewish binaries' realignment of self and, 187; Edomites as, 83–84; Jebusite Jerusalem as shared space of, 16, 100–101; Moabite women as, 49, 50, 56, 58; Transjordan as site of, 11–12, 60, 65, 119–20, 128–34, *132*, *133*; women in Transjordan as, 49, 50, 56, 58, 119–20

Ottoman administrative districts map, 220–21, 222, 234, 250
Ottosson, Magnus, 74n19, 75n21, 83n38, 118n19, 121n27, 122n31, 154n33

P (Priestly) source, 5n5, 24, 112n9. *See also* Priestly tradition
Palestine Exploration Fund (PEF), 220–23, *224*, 232
Palestine Liberation Organization (PLO), 235n35, 263, 265–66, 267, 268n29
Palestine/Palestinians, 14–15; archaeological excavations and, 279; bridge border crossing for, 268–73, 275; British mandate maps and, 220–21, 223–27, *224*, *225*, *228*, 229–33, 269, 277, 286–87; Christian, 241–45; Deuteronomistic tradition and, 219; Fatah and, 265–68; *fida'iyin* movement and, 264–68; folklore traditions and, 239, 241; freedom fighters and, 219–20, 257, 264–68; French mandate maps and, 221, 223–24, *225*, *228*, 232; hegemonic discourse and, 277, 280–81; Intifada stories and, 244, 245, 270n134; Jordan as border and, 237, 239, 275–76; Karama commandos and, 266–68, 274; maps of Palestine, 2, *240*; martyr commandos and, 268; mothers/adult women and, 270n134; Muslims and, 14, 219–20, 224, 231, 234n34, 243–45, 277, 279; Ottoman administrative districts map and, 220–21, *222*, 234, 250; peace park plan and, 284; popes and, 277, 280–81, 282; redemption for, 273; returnees across Jordan and, 253, 271–73; riots as political strategy and, 235, 237, 239, 254–55; statelessness and, 219, 264–65; territorial proposals and, 231–32
Palestinian Christians, 245
Palestinian National Council, 267, 268n29
Palestinian pioneers: Bir Zeit's founding and, 244–45; Christian and Muslim relations and, 242–44; Haddadin migration and, 241–45, 263n121; Ramallah's founding by, 241–44; returnees crossing bridge over Jordan as, 271–73
Palestinian refugees: citizenship for, 261n114; Jordan state and, 253, 260–61, 261n114; *Nakbah* of 1948 and, 252–54, 262; *al-Naksa* and, 261; *Nuzuh* of 1967 and, 252–53, 257, 261–64; resettlement plans for, 262, 264n122; Samakh battle with Zionists and, 254–55; West Bank and, 256, 260–61
paradisical visions, 21–23, 34n53. *See also* visions
Paris and San Remo peace conferences' territorial proposal, 229–32
Passover narrative of Exodus. *See* exodus narratives
peace park plan, 284–86, *285*, 288–89
Peel Commission Report maps, 235, *236*, 237
Peel maps, 235, *236*, 237
PEF (Palestine Exploration Fund), 220–23, *224*, 232
Persian period: boundaries/maps and, 129; Esther story and, 58; Ezra-Nehemiah myth and, 5, 6, 10, 59–60, 129–30; imperial geography and, 38–39; Israel national myth and, 38–39, 58, 130; Moabite women and, 58–63; return of exiles to Judea during, 5, 10, 34, 50
Petra *tiyul* (trek), 258–60, 269n133
Pharisees, 164n60, 170, 186
Phineas the son of Eleazar the Priest, 112, 114–16, 124
pioneering ancestors' stories, 17, 97–98
PLO (Palestine Liberation Organization), 235n35, 263, 265–66, 267, 268n29
political strategies: riots, 235, 237, 239, 254–55; settlements, 134, 234, 235, 237, 248
Polzin, Robert, 86n3, 97n25, 101
popes/papacy, 46, 277, 280–81, 282
Porter, J. Roy, 150n25, 151n27
priesthood. *See* Kohanim (Sadducees); Levites; Priestly tradition
Priestly (P) source, 5n5, 24, 112n9. *See also* Priestly tradition
Priestly tradition, 5, 12–13, 24; the body and, 13, 176; boundaries/maps and, 8, 9, 13, *20*, 29–30, 78, 230; Canaanites and, 9; Christian-Jewish binary divisions in purity and, 12, 13; Christians and, 12–13, 14, 176; collective purification of Jews and, 13, 34–35, 176; Creation narrative and, 23nn18–19, 23–24; Ezekiel and, 23–24, 34; Holiness (H) Code source for, 5n5, 6, 23–24; Israel national myths and, 34–35; Israel's border in terms of Egypt

and, 31–33, 34–35; Jacob's reburial and, 83n40; Kohanim and, 90–91; the land as holy and, 130n55; Levites and, 90–91; Moabite women and, 53, 54; nonhereditary line and, 16; Priestly (P) source for, 5n5, 24, 112n9; ritual purity and, 176, 177, 178nn6–7, 184, 184n21, 193; Shiloh altar and, 29, 73, 78, 112n9. *See also* Deuteronomistic tradition; Hebrew Bible; Kohanim (Sadducees); Levites; Torah

prophetic succession, 12, 135–37, 173–74; apocalyptic associations with Jordan and, 137, 142; boundaries/maps of ritual and, 135–36; Deuteronomistic tradition and, 137–38; Hebrew Bible and, 136, 137–42; Islam and, 14, 147n18; master/disciple relations and, 136, 137, 138, 143, 153; monarchical initiation connection with, 136–37, 150, 151n27; monarchs' connection with, 136–37, 149, 150, 151n27; New Testament and Hebrew Bible dependence and, 136, 173; nonhereditary line and, 12, 136, 137–38, 144, 157n38, 174; redemption for Jews and, 139, 142, 162, 174; transformation from crossing river and, 135–36; vision implementation and, 138, 153. *See also* Elisha as successor of Elijah; Jesus and John the Baptist; Joshua as successor of Moses; type-scenes

Propp, William H., 99n28

Pseudo-Philo, 157n39

purity/impurity: moral, 177n3, 185; ritual, 176, 177, 178nn6–7, 184, 185, 193. *See also* purity of the body

purity of the body: baptism and, 12, 13, 180, 184–85, 186–87, 192–93; Christian-Jewish binary divisions in purity and, 12–13, 187, 206–7; collective purification of Christians and, 12–13, 184–85, 188–89, 190–95, 206–7; collective purification of Jews and, 12, 13, 176, 188–89, 195–201, 206; dip/dipping and, 178n7, 184, 201; disorder versus peace and, 176–78, 183, 206; folklorist traditions and, 177–78; foreigners and, 177n3; Gehazi as corrupt insider/disciple and, 183–84, 196–204; Gentiles and, 177n3, 180–81, 187–88; healing after baptism and, 177–86, 197–98, 206; illnesses and, 177–83, 185, 206; insider-outsider position and, 176–77, 182, 187; Israelites as clean/pure and, 179, 180–82; Jordan as border and symbol of, 13, 176; the land correlation for Jews with, 179, 180–82, 183n18, 184, 207n70; the land correlation with, 179, 180–82, 183n18, 184, 207n70; miraculous dimension of Jordan and, 176, 180, 182; moral impurity for Jews and, 185; Naaman/baptism association and, 180, 182, 184, 186–87, 192–93, 196, 197–98, 200–201, 204; ritual purity and, 176, 177, 178nn6–7, 184, 185, 193; self-healing for Jews and, 197–99, 204–6; servants of Naaman and, 177–78, 179. *See also* baptism; body, the; Naaman; purity/impurity

Qasr al-Yahud baptismal site, 276–78, 280–81, 283, 285, *285*

rabbinic narratives: on baptism and exclusion from World to Come, 195–96, 200–201; on baptism and purity of the body, 204; Christian-Jewish binary divisions in purity and, 12, 13; on Elisha's association with Naaman and baptism, 196, 197–98, 200–201, 204; on Elisha's correlation with Jesus, 196–97; on Gehazi as corrupt insider/disciple, 196–201; on Jesus as heretical, 201–4; on Naaman's healing and lack of purification, 197–98. *See also* Babylonian Talmud

Rachel, 71, 75–76

Rahab, 52n22, 139n7. *See also* Moab/Moabites

Ramallah, 241–44. *See also* Palestine/Palestinians; West Bank

Rashi, 49n14, 211n7

Rebecca, 66–67, 71, 82

rebirth, after baptism, 12, 13, 169, 174, 180, 186–87, 192–93

redemption: for Christians through baptism, 12, 13, 139, 161–69, 172–74, 176; for Palestinians, 273. *See also* redemption for Jews

redemption for Jews, 12–14, 273; apocalyptic associations with Jordan and, 13, 140; Joshua as successor of Moses and, 11, 139; prophetic succession and, 139, 142, 162, 174; return discourse and, 12, 174; tall tales of, 208–9, 212–16; Zionists and, 14. *See also* Israel/Israelite tribes; Jews

refugees, and Jews from Europe, 251–52. *See also* Palestinian refugees
repentance for sins, 164–65, 167, 172–73, 184–85, 190–91, 206. *See also* sin(s)
resettlement plans, for Palestinian refugees, 262, 264n122
return discourse: exiles' return to Judea and, 5, 10, 34, 50; Palestinians and, 253, 271–73; redemption for Jews and, 12, 174
Reuben: co-national status of, 78–79, 95, 101–5, 106; insider-outsider position of, 101, 114, 117–18, 129n53; memorial site built by, 111–17; Transjordan location of, 21, 78, *110*, 111–17
revisionist aspect of crossing, 98, 99–100
riots as political strategy, 235, 237, 239, 254–55
ritual purity/purification, 176, 177, 178nn6–7, 184, 185, 193. *See also* purity of the body
river of Egypt, 20n6, 23–24, 26, 56
Rofé, Alexander, 159, 181, 182n16
Roman Empire, 131–34, *133*
Routledge, Bruce, 47n12, 50n17, 124
Rutenberg, Pinchas, 279n4, 286–88
Ruth, book of, 5–6, 10, 44, 46, 58–63, 154n31

Sadducees (Kohanim). *See* Kohanim (Sadducees); Levites
Said, Edward, 239, 275–76
Samakh battles, 254–56
Samaria, 72n15, *73*, 129, 131, *132*, 132n64, 134, 177, 255n90
Sanders, E. P., 165n60
Sanders, Seth, 51n21, 93n17
San Remo and Paris peace conferences' territorial proposal, 229–32
Sarah, 41, 43, 51, 58. *See also* Abraham
Saydon, Paul P., 89n9, 94n20
Schäfer, Peter, 202n60
sea-to-river maps, 17–18, 38–39; Euphrates and, 24–26, 28–30, 31, 35, 129; Jordan maps and, 19–20, *20*, 31; Moses and, 146–47
"Second Exodus" (*Nuzuh* of 1967), 252–53, 257, 261–64. *See also* Palestinian refugees
Seleucids, 131
self-healing, 197–99, 204–6. See also *halakhah* (Jewish law); illnesses/healing
Setback, the (*al-Naksa*), 261. *See also* Palestinian refugees

settlements as political strategy, 134, 234, 235, 237, 248
Shariah, Al- (the Watering Place; the Law), 245
Sherwood, Aaron, 94n18
Shiloh, 29, *73*, 78, 112, 112n9
Sidon/Sidonites, 10, 63, 65, *73*, *110*, 191, *228*
sin(s), 164–65, 167, 172–73, 184–85, 190, 206
Smith, Jonathan Z., 13, 21–22, 207
Smith, Mark, 151n27
Smith-Christopher, Daniel, 59n35, 60n36
Sodom/Sodomites, 42–47, 49, 50
Solomon, King, 30, 33n51, 63
Song of Deborah, 119
Sontag, Susan, 206
Southern Kingdom of Judah (Judea): Assyrian Empire influences on, 9n11; Babylon's conflicts with, 33–37, 124n39, 129; boundaries/maps of, 21, 146; Esau's association with, 11, 66–67, 75n21, 82–83; Herod's kingdom and, 132n64, *133*; imperial geography and, 33, 38; insider-outsider position and, 10; Jacob and, 10–11, 65; Leah as founding mother of, 76n22; Moabite-Israelite intermarriages and, 62–63; Northern refugees and, 6; Ruth and, 58, 62–63; territory borders and, 10–11; traditions, 77n26. *See also* Edom/Edomites; Esau; Jacob; Jerusalem; Judea (former Southern Kingdom of Judah)
spies' story read with crossing, 98, 102–3, 105
spirit transfer to disciple: Elisha and, 144, 155–56, 157–60, 168; Jesus's receipt of Holy Spirit and, 167–71, 173; Joshua and, 144, 148–49, 168
Stein, Dina, 123n35, 209n2
Sweeney, Marvin, 151n27
Sykes-Picot map, 223–24, 225, *226*, 232
Syria, 31, 33, 72n17, 223, *225*, 231, 257, 284. *See also* Aram/Arameans; Gilead

tall tales, 208–9, 212–16. *See also* Babylonian Talmud
talmudic texts. *See* Babylonian Talmud; rabbinic narratives
Tanakh (biblical canon), 6–7, 141, 282–83. *See also* Hebrew Bible; Torah
Temple, in Jerusalem, 9, 21, 22–23, 24n20, 59–60, 131, 207n70. *See also* Ark of the Covenant

territory border(s): Abraham's hereditary line versus Lot's claims through, 43, 47, 49, 64; "everyday life" negotiations and, 68; hereditary line and, 108, 125n40, 130; Moabites and, 51, 52n22, 55, 56, 58, 59, 60, 121; as porous, 10–11, 78, 176, 181–82, 206; river of Egypt and, 56; women's participation due to relaxed ethnic, 58. *See also* boundaries/maps

texts, 3–4, 157. See also *specific texts*

tithing, animal, 208, 209–12, 210

tiyul (trek) to Petra, 258–60, 269n133

Tobiad family, 129–31

Torah: as gift to Israelites, 90; Levites, and written, 91n12, 145, 150; as memory, 141, 145, 147–48, 151, 162. *See also* Deuteronomistic tradition; Hebrew Bible; Priestly tradition

tourism, 5, 247, 268–69, 288

transformation at Jordan, 135–36, 138, 139, 140, 143–44, 150–52, 168–70, 172

Transjordan, 11–12, 32, 106–7; ambivalent status of, 32, 49; Assyrian Empire and, 128–29; co-national status of Israelites and, 101–6; Deuteronomistic tradition and, 12, 30, 107n2; ethno-nationalism and, 106–7, 118, 120–23, 125–26, 130n57; frontier justice in, 123–28; Gad tribe in, 21, 78, *110*, 116; gender coded as female and, 50–51; half-tribe of Manasseh in, 21, 78, *110*, 111–20; homeland-diaspora environment and, 107–9, *110*, 111–17; insider-outsider position of Israelites in, 101, 114, 117–18, 129n53; Ishmaelites and, 41, 43, *48*, 67; Joshua's battles to conquer land in, 12, 107, 108–9, 111; Judea state in context of, 131–34, *132*, *133*; memorial sites in, 111–17; Northern Kingdom and, 77–78; as Other site, 11–12, 60, 65, 119–20, 128–34, *132*, *133*; Reuben tribe in, 21, 78, *110*, 111–17; sacred/profane nature of, 77–78, 119, 122n29; settlements for Jews and, 134, 234; territory borders and, 232–33; women as Other in, 49, 50, 56, 58, 119–20. *See also* east-west binary

trek (*tiyul*) to Petra, 258–60, 269n133

Tuell, Steven, 24n21

Turner, Victor, 135–36, 152

Twelve Tribes of Israel. *See* Israel/Israelite tribes; *and specific tribes*

type-scenes: death of the master, 141, 146, 149–50, 153, 156–57; Hebrew Bible, 142–43; impending death of the master, 144–48, 154, 163–67; New Testament, 142; witnesses' presence, 149, 150–52, 154–57, 159, 171–72. *See also* master/disciple relations; spirit transfer to disciple

United Nations partition, 117–18, 233–34, 237, *237*, 239, *240*

universalism, 175. *See also* Christianity/Christians

Van Der Toorn, Karen, 72n15, 90n11, 99n27

Van Gennep, Arnold, 136

vision implementation: actions during era of judges versus, 153, 160; by Elisha, 144, 153, 160–61; by Jesus, 162, 165; by Joshua, 144, 148, 152, 153, 160–61; prophetic succession and, 138, 153. *See also* visions

visions: apocalyptic, 22, 215; Elijah's, 155–56, 159, 160; Ezekiel's, 21–23, 22, 34n53, 215; Jacob's, 70, 72, 76–77, 79–80, 81n34, 82, 154n33; John the Baptist's, 162; Moses's, 22, 144, 146, 148, 160; paradisical, 21–23, 34n53. *See also* vision implementation

Wagenaar, Jan, 94n18

Waheed, Mohammed, 279, 283

wanderings' connections with self and space, 10–11, 68–71, 75–76, 77. *See also* homeland-diaspora environment

Watering Place, the (Al-Shariah; the Law), 245

water rights, 283–84. *See also* ecology of Jordan

Wazana, Nili, 28n36, 30n41, 34–35

Weinfeld, Moshe, 29–30, 31, 34, 38, 78

Weitzman, Steven, 125n40

West Bank: Fatah and, 266; as frontier, 244–45, 260–61; Israeli control of, 219, 237, 260–61, 277–78; Jordan state and, 239, *240*, 257; Al-Maghtas baptismal site and, 276, 278, 279–83, 285; multinational control and, 288; Palestinian refugees and, 256, 260–61; Ramallah as center of, 241–44. *See also* east-west binary

White Paper of 1922, 233–35

Wilson, Robert, 138, 159n46

witnesses as presence, 149, 150–52, 154–57, 159, 171–72

women: ethnic territory borders as relaxed, and role of, 58; Ezra-Nehemiah myth, and expulsion of foreign, 5, 6, 10, 59–60; as founding mothers of Israel, 47n13, 49, 50–51, 53, 57–58, 76n22; immodest behavior of, 49; "Moabite" men with feminine and passive attributes of, 40–41, 54, 55–56; as Other in Transjordan, 49, 50–51, 56, 58, 119–20; Palestinian, 270n134; sexuality of, 44, 45–46, 49, 60, 62–63; Transjordan coded as female and, 50–51. *See also* men; Moabite women

World to Come discourse, 195–96

Wright, G. Ernest, 89n9, 94n21, 111n7

Yerushalmi, Yosef Hayim, 97

Zakovitch, Yair, 156n37, 178n7, 183n18

Zelophehad's daughters, 52–54, 57, 58, 61

Zerubavel, Eviatar, 157n38

Zionists: boundaries/maps and, 1–2, *228*, 232–33; Degania kibbutz and, 247–52, 254–56, 265, 274; east-west binary and, 248–49; folklore traditions versus pioneer, 241, 248; frontier stories and, 249–50, 261; on half-tribe of Manasseh as recapitulation of bipartition, 233–34; hegemonic discourse and, 14; hydroelectric plant project and, 286–88; Jordan as border and, 233–34; Kinneret kibbutz and, 277; pioneer, 241, 246–52, 254–56, 265, 274; redemption for Jews, 14; Samakh battles between Palestinians and, 254–56; settlements as political strategy and, 235, 237, 248; territorial proposal by, 14, 227, *228*, 229–32; United Nations partition and, 233–34, 237, 239, *240*; White Paper of 1922 and, 233–35, 234. *See also* Israel nation/national myth; Palestine/Palestinians